Math Toolkit for Real-Time Programming

Jack W. Crenshaw

CRC Press
Taylor & Francis Group
6000 Broken Sound Parkway NW, Suite 300
Boca Raton, FL 33487-2742

First issued in hardback 2018

© 2000 by CMP Media, Inc.,
CRC Press is an imprint of Taylor & Francis Group, an Informa business

No claim to original U.S. Government works

ISBN 13: 978-1-138-41247-7 (hbk)
ISBN 13: 978-1-929629-09-1 (pbk)

Visit the Taylor & Francis Web site at
http://www.taylorandfrancis.com

and the CRC Press Web site at
http://www.crcpress.com

Designations used by companies to distinguish their products are often claimed as trade- marks. In all instances where CMP Books is aware of a trademark claim, the product name appears in initial capital letters, in all capital letters, or in accordance with the ven- dor's capitalization preference. Readers should contact the appropriate companies for more complete information on trademarks and trademark registrations. All trademarks and registered trademarks in this book are the property of their respective holders.

The programs in this book are presented for instructional value. The programs have been carefully tested, but are not guaranteed for any particular purpose. The publisher does not offer any warranties and does not guarantee the accuracy, adequacy, or completeness of any information herein and is not responsible for any errors or omissions. The pub- lisher assumes no liability for damages resulting from the use of the information in this book or for any infringement of the intellectual property rights of third parties that would result from the use of this information.

Cover art created by Janet Phares.

Table of Contents

Supplementary Resources Disclaimer

Additional resources were previously made available for this title on CD. However, as CD has become a less accessible format, all resources have been moved to a more convenient online download option.

You can find these resources available here: www.routledge.com/9781138412477

Please note: Where this title mentions the associated disc, please use the downloadable resources instead.

Preface

Who This Book Is For

If you bought this book to learn about the latest methods for writing Java applets or to learn how to write your own VBx controls, take it back and get a refund. This book is not about Windows programming, at least not yet. Of all the programmers in the world, the great majority seem to be writing programs that do need Java applets and VBx controls. This book, however, is for the rest of us: that tiny percentage who write software for real-time, embedded systems. This is the software that keeps the world working, the airplanes flying, the cars motoring, and all the other machinery and electronic gadgets doing what they must do out in the real world.

Computers are for Computing

These days, computers seem to be everywhere and used for everything, from running the air conditioner in your car to passing money around the world in stock and commodities markets and bank transactions, to helping lonely hearts lover wannabes meet in Internet chat rooms, to entertaining Junior (or Mom and Dad) with ever more accurate and breathtaking graphics — some rated XXX. At one time, though, there was no doubt about what computers were for.

Computers were for computing. When Blaise Pascal and others designed their first mechanical calculators, it was to help them do arithmetic. When Charles Babbage designed the original computer in 1823, using mechanisms like cams, cogwheels, and shuttles, he didn't call it his Difference Engine for nothing. In Babbage's day, the only way to perform numerical computations was by hand, and the worst part about computing things by hand is that people, being human, tend to make errors. An error partway through a long calculation ruins the rest and requires one to go back to the point of the error and start over. If the calculation takes on the order of a lifetime, you can see how the discovery of an error way back toward the beginning could ruin your whole day. It is hardly a surprise, then, that Babbage and his royal backers sought a faster and more reliable way of performing calculations. They didn't get it — the Babbage engine was never completed, but the dream remained alive.

The dream became more of an imperative during World War II. As usual, the team with the best weapons had the best chance of winning. Artillery firing tables took high priority, as did bombsights for airplanes. The Norden bombsight was perhaps the best kept secret of the war, next to The Bomb itself. Its miracles were accomplished through mechanical means, much like the difference engine conceived by Babbage, except that the Norden bombsight was an analog, not digital, computer. (The minor detail that it fit into an airplane also helped a lot.)

Researchers at Harvard, Yale, MIT, and other institutions recognized the value of computing things digitally and automatically. From their efforts came the original electric (i.e., relay) and electronic computers that showed up just at the end of the war. In the 1950s and 1960s, digital computers exploded on the scene, changing the world forever. Still, there was no doubt what those first computers were for. They were used to compute the trajectories of missiles, airplanes, and spacecraft. This was, in fact, the point where I entered the scene, calculating trajectories for NASA's Apollo missions, including the abort trajectories used for Apollo 13. Other people were using similar computers for more mundane, but still essential, functions, such as actuarial tables used by insurance companies. Computers were being used for computing, and this seemed quite proper at the time.

Shortly afterwards, however, a funny thing happened. In the process of computing difference equations and other tabular data, programmers found the need to do logic as well as arithmetic. If you're calculating something in a loop, you at least need to know when to quit. Often, this meant comparing two numbers and doing something different, depending on their values. The branch was born, which begat the IF test, which begat the DO, FOR, and WHILE loops.

Somewhere along the line, a bright person figured out, "Hey, if this thing can make decisions, maybe I can get it to make mine!" Someone surely must have envisioned a computer that could take all the world's data, be it financial, political, or scientific, and reduce it to the ultimate degree: a simple output that said, "Buy Microsoft," or "Push the red button," or "Run!" The concept of non-numerical programming was born.

The concept gained a lot more momentum with the advent of systems programming tools like assemblers, compilers, and editors. You won't find much mathematics going on inside a C++ compiler or Microsoft Word, beyond perhaps counting line and column numbers. You certainly won't find any computing in Internet browsers. Today, the great majority of software running on computers is of the non-numeric type. Math seems to have almost been forgotten in the shuffle.

This tendency is reflected in the computer science curricula taught in universities. As is proper, most of these curricula stress the non-numeric uses of computers. Most do still pay at least lip service to numeric computing (today, it is called numerical analysis). However, it's possible at some universities to get a degree in Computer Science without taking one single course in math.

The problem is, despite the success of non-numeric computing, we still need the other kind. Airplanes still need to navigate to their destinations; those trig tables still need to be computed; spacecraft still need to be guided; and, sadly enough, we still need to know where the bombs are going to fall.

To this need for numeric computing has been added another extremely important one: control systems. Only a few decades ago, the very term, "control system" implied something mechanical, like the damper on a furnace, the governor on a steam engine, or the autopilot on an airplane. Mechanical sensors converted some quantity like temperature or direction into a movement, and this movement was used to steer the system back on track.

After World War II, mechanical cams, wheels, integrators, and the like were replaced by electronic analogs — vacuum tubes, resistors, and capacitors. The whole discipline of feedback theory led to a gadget called the operational amplifier that's still with us today and probably always will be. Until the 1970s or so, most control systems still relied on electronic analog parts. But during the late 1960s, aerospace companies, backed by the Defense Department, developed ruggedized minicomputers capable of withstanding the rigors of space and quietly inserted them in military aircraft and missiles.

Today, you see digital computers in places that were once the domain of analog components. Instead of using analog methods to effect the control, designers tend to measure the analog signal, convert it to digital, process it digitally, and convert it back again. The epitome of this conversion may well lie hidden inside your CD disk player. The age of digital control is upon is. If you doubt it, look under the hood of your car.

As a result of the history of computing, we now have two distinct disciplines: the non-numeric computing, which represents by far the great majority of all computer applications, and the numeric computing, used in embedded systems. Most programmers do the first kind of programming, but the need is great, and getting greater, for people who can do the second kind. It's this second kind of programming that this book is all about. I'll be talking about embedded systems and the software that makes them go.

Most embedded systems require at least a little math. Some require it in great gobs of digital signal processing and numerical calculus. The thrust of this book, then, is twofold: it's about the software that goes into embedded systems and the math that lies behind the software.

About This Book

As many readers know, for the last five years or so I've been writing articles and a column for computer magazines, mostly for Miller Freeman's *Embedded Systems Programming* (ESP). My column, "Programmer's Toolbox," first appeared in the February 1992 issue of ESP (Vol. 5, #2) and has been appearing pretty much nonstop ever since. Although the title gives no hint, my emphasis has been on computer math, particularly the application of advanced methods to solve practical problems like simulation, analysis, and the control of processes via embedded systems.

Many of my articles are available on the CD-ROMs from ESP and *Software Development*. However, they're necessarily scattered across the disk, and it's not always easy to find the particular article one needs. Also, as I've drifted from subject to subject, sometimes in midcolumn, the seams between subjects have been more jangling than I might have liked.

For the last few years, I've been getting ever more insistent pleas from readers who would like to see all of my articles gathered together into a book, with the loose ends tied up and the joining seams smoothed over a bit. This book is the result.

Some of my articles have been almost exclusively of a programming nature; others, almost pure math with only the slightest hint of software. The first decision I had to make as plans for the book went forward was how, and whether, to separate the software from the math. Should I have two sections — one on software and one on math — or perhaps two volumes or even two separate books? Or should I try somehow to merge the two subjects? In the end, the latter decision won out. Although some topics are much more math-intensive than others, they are all aimed at the same ultimate goal: to provide software tools to solve practical problems, mostly in real-time applications. Because some algorithms need more math than others, it may seem to you that math dominates some chapters, the software, others. But the general layout of each chapter is the same: First, present the problem, then the math that leads to a solution, then the software solution. You'll need at least a rudimentary knowledge of math and some simple arithmetic. But I'll be explaining the other concepts as I go.

Some readers may find the juxtaposition of theory and software confusing. Some may prefer to have the software collected all together into nice, useful black boxes. For those folks, the software is available alone on CD-ROM. However, I will caution you that you will be missing a lot, and perhaps even misusing the software, if you don't read the explanations accompanying the software in each chapter. I'm a firm believer in the notion that everything of any complexity needs a good explanation — even a simple-minded, overly detailed one. Nobody ever insulted me by explaining the obvious, but many have confused me by leaving things out. I find that most folks appreciate my attempts to explain even the seemingly obvious. Like me, they'd rather be told things they already know than not be told things they don't know.

On each topic of this book, I've worked hard to explain not only what the algorithms do, but where they came from and why they're written the way they are. An old saying goes something like, give a man a fish, and tomorrow he'll be back for another; give him a fishing pole, teach him how to use it, and he will be able to catch his own. I've tried hard to teach my readers how to fish. You can get canned software from dozens, if not thousands, of shrink-wrap software houses. You'll get precious little explanation as to how it works. By contrast, if my approach works and you understand my explanations, you'll not only get software you can use, but you'll also know how to modify it for new applications or to write your own tools for things I've left out.

About Programming Style

Since we're going to be talking about math and algorithms that will ultimately be used in computer software, it's inevitable that I show you programming examples. In those examples, you will not only see how the algorithms unfold, you will also get to see examples of my programming style. This is not a Bad Thing; a secondary purpose of this book is to teach programming style as it applies to embedded systems.

Needless to say, the style I'm most likely to teach is the one I use. Over the years, I've learned a lot of things about how and how not to write software. Some of it came from textbooks, but most came from trial and error. After some 40 years of writing software, I think I'm finally beginning to get the hang of it, and I think I have some things to teach.

In their seminal book, *Software Tools*, (Addison-Wesley, 1976) Brian Kernighan and P.J. Plauger (K & P) suggest that one of the reasons that some programmers — and experienced ones, at that — write bad programs is because no one's ever showed them examples of good ones. They said, "We don't think that it is possible to learn to program well by reading platitudes about good programming." Instead, their approach was to show by example, presenting useful software tools developed using good programming practices. They also presented quite a number of examples, taken from programming textbooks, on how *not* to write software. They showed what was wrong with these examples, and how to fix them. By doing so, they killed two birds with one stone: they added tools to everyone's toolboxes and also taught good programming practices.

This book is offered in the same spirit. The subject matter is different — K & P concentrated mostly on UNIX-like text filters and text processing, whereas we'll be talking about math algorithms and embedded software — but the principle is the same.

For those of you too young to remember, it may surprise you (or even shock you) to hear that there was a time when calling a subroutine was considered poor programming practice — it wasted clock cycles. I can remember being taken to task very strongly, by a professional FORTRAN programmer who took great offense at my programming style. He saw my heavy use of modularity as extravagantly wasteful. He'd intone, "180 microseconds per subroutine call." My response, "I can wait," did not amuse.

I was lucky; the fellow who taught me FORTRAN taught me how to write subroutines before he taught me how to write main programs. I learned modularity early on, because I couldn't do anything else.

In those days, I soon developed a library of functions for doing such things as vector and matrix math, function root-solving, etc. This was back when punched cards were the storage medium of choice, and the library went into a convenient desk drawer as card decks, marked with names like "Vectors" and "Rotation." In a way, I guess these decks constituted the original version of the Ada "package."

That library followed me wherever I went. When I changed jobs, the software I'd developed for a given employer naturally stayed behind, but rest assured, a copy went along with me. I considered that library to be my toolbox, and I also considered it my duty to supply my own tools, in the same spirit that an automechanic or plumber is expected to bring his or her tools to the job. Many of those routines continue to follow me around, even today. The storage media are different, of course, and the languages have changed, but the idea hasn't changed. Good ideas never grow old.

A few years ago, I was asked to teach a class in FORTRAN as an adjunct professor. One look at the textbook that the school faculty had already chosen for the class told me that it was *not* a book I wanted to use. The examples were textbook examples right out of K & P's other seminal book, *Elements of Programming Style, 2nd edition* (McGraw-Hill, 1978), of how *not* to write FORTRAN. As is the case with so many textbooks, each program was just that: a stand alone program (and not a well-written one at that) for solving a particular problem. The issues of maintainability, reusability, modularity, etc., were not addressed at all. How could they be? Functions and subroutines weren't introduced until Chapter 9.

I decided to take a different tack, and taught the students how to write functions and subroutines on day one. Then I assigned projects involving small subroutines, beginning with trivial ones like abs, min, max, etc., and working up to more complex ones. Each student was supposed to keep a notebook containing the best solutions.

Each day, I'd ask someone to present their solution. Then we'd discuss the issues until we arrived at what we considered to be the best (I had more than one vote, of course), and that solution, not the students' original ones, was the one entered into everyone's notebook. The end result was, as in K & P's case, twofold. Not only did the students learn good programming practices like modularity and information hiding, they had a ready-made library of top-quality, useful routines to take to their first jobs. I blatantly admit that my goal was to create as many Crenshaw clones as possible. I don't know how well the school appreciated my efforts — I know they didn't ask me back — but the effect on the students had to be similar to the one the judge gets when they instruct the jury to forget they heard a particularly damaging piece of testimony. Try as they might, I suspect those students are still thinking in terms of small, reusable subroutines and functions. A little bit of Crenshaw now goes with each of them.

Do I hope to do the same with you? You bet I do. In giving you examples of math algorithms, it's inevitable that I show you a bit of the style I like to use. That's fine with me, and for good reason: the methods and styles have been thoroughly proven over many decades of practical software applications. You may already have a style of your own, and don't want to change it. That's fine with me also. Nobody wants to turn you *all* into Crenshaw clones. If you take a look at my style, and decide yours is better, good for you. We all have to adopt styles that work for us, and I don't really expect the cloning to fully take. Nevertheless if, at the end of this book, I will have given you some new ideas or new perspectives on programming style, I will have done my job.

On Readability

One aspect of my style will, I hope, stand out early on: I tend to go for simplicity and readability as opposed to tricky code. I'm a strong believer in the KISS principle (Keep It Simple, Simon). When faced with the choice of using a tricky but efficient method, or a straightforward but slower method, I will almost always opt for simplicity over efficiency. After all, in these days of 800 MHz processors, I can always find a few more clock cycles, but programming time (and time spent fixing bugs in the tricky code) is harder to come by. Many people talk about efficiency in terms of CPU clock cycles, but unless I'm coding for an application that's going to be extremely time-critical, there's another thing I find far more important: the time it takes to bring a program from concept to reality. To quote K & P again, "First make it run, *then* make it run faster."

A corollary to this concept requires that programs be written in small, easily understandable and easily maintainable chunks. The importance of the cost of program maintenance is only now becoming truly realized. Unless you want to be maintaining your next program for the rest of your life, picking through obscure branches and case statements, and wondering why a certain construct is in there at all, you'll take my advice and write your programs, even embedded programs, in highly modular form.

About the Programs

When I first began the Toolbox column, my intent was to make it multilingual, presenting examples in Pascal, C, C++, Ada, and even more exotic languages. This turned out to be a totally impractical dream. All I succeeded in doing was confusing everyone. Worse yet, I ended up with tools that couldn't play together, because they were written in different languages.

In this book, I'll be sticking almost exclusively to C++. Don't try to read into this decision any great love for the language. There are many nice features of C++, and many others I absolutely detest. Come to think of it, though, the same is true of other languages. I have no great, favorite language, though Pascal probably comes the closest. I find that the most rabid partisans for a certain language are those who have never (or rarely) programmed in any other. Once you've programmed a certain number of languages, writing in one or the other is no big deal, and the debates as to the pros and cons of each can become a little less heated. I'll be using C++ for my examples for quite a practical reason: It's the language that's most popular at the moment.

At the same time, I must admit that this choice is not without its downside. At this writing, few real-time embedded systems are being written in C++, and for good reason: the language is still in a state of flux, and some of its more exotic features present significant risk of performance problems. Over the last few years, C and C++ language expert P.J. Plauger has been engaged in an effort to define a subset of C++ for embedded systems. At this writing, that effort is complete; however, no major software vendor yet has a compiler available for that standard.

About the Author

I can't really say that I began with the computer age at its very beginning. I never worked with Admiral (then Captain) Grace Hopper, John von Neumann, Eckert and Mauchly, or the other greats who started the whole computer thing off. But I did manage to catch the very next wave. I wrote my first computer program in 1956 for an IBM 650 computer (we didn't actually have one, the professor had to desk-check our program). I studied lunar trajectories using an IBM 702, the forerunner of the 704/709x series. The programs were written in assembly language because that's all there was. No one had yet heard of FORTRAN, much less Pascal, C, or C++. I still remember, as though it was yesterday, seeing a fellow programmer writing something on a coding pad in a language that looked almost readable to an ordinary mortal. I asked him what he was doing. He said, "I'm writing a program in FORTRAN." That was my first awareness that compilers even existed.

When I graduated from college in 1959, I interviewed with, among others, a government agency called NACA. A few weeks later, I received an offer letter, on which someone had struck out the "C" with a ball-point pen and written an "S" above it. It's fair to say that I was at NASA from the beginning. I lived just down the road from Gus Grissom, who used to practice his scuba skills in the community swimming pool. Wally Schirra lived there too. Because of the astronaut's strange hours, we often passed each other as I was driving to and from work, and we'd give each other a fellow-sports-car honk and wave.

I never really intended to be a computer programmer, much less a software engineer or computer scientist, terms that hadn't even been invented at the time. My formal training is in Physics, and I joined NASA to help put men on the Moon; as a sci-fi buff of long standing, I felt I could do nothing else. My early days at NASA involved nothing more complex than a Friden electric calculator and a slide rule. Few people who remember the early NASA projects — Mercury, Gemini, and Apollo — realize the extent to which it was all done without computers. True, we never could have guided the Eagle to the surface of the Moon without its flight computer and the software that went into it. In fact, many of the most significant developments in real-time software and digital guidance and control systems evolved out of that effort. But all the early work was done with slide rules, calculators, and hand-drawn graphs.

Over the years, I developed a collection of useful, general-purpose tools, which I carried around in a wooden card case, just as a pool shark carries his favorite cue stick. The box followed me from job to job, and I considered the contents the tools of my trade, just as real as those of a carpenter, mechanic, or plumber.

During my years in the space program, most of the work involved math-intensive computations. We were doing things like simulating the motion of spacecraft or their attitude control systems. We were learning as we went along, and making up algorithms along the way because we were going down roads nobody had trod before. Many of the numeric algorithms originally came from those developed by astronomers for hand computations. But we soon found that they were not always suitable for computers. For example, when an astronomer, computing by hand, came across special cases, such as division by a number near zero, he could see the problem coming and devise a work-around. Or when using numerical integration, a person computing by hand could choose a step size appropriate to the problem. The computer couldn't handle special considerations without being told how to recognize and deal with them. As a result, many of the algorithms came out remarkably different from their classical forms. Many new, and sometimes surprising, techniques also came out that had no classical counterparts. As I discovered and sometimes invented new algorithms, they found their way into that ever-growing toolbox of programs.

I was into microprocessors early on. When Fairchild came out with their first low-cost integrated circuits ($1.50 per flip-flop bit), I was dreaming of a home computer. Before Intel shook the world with its first microprocessor, the 4004, I was writing code for a less general-purpose chip, but one still capable of becoming a computer.

In late 1974, MITS shook the world again with the advent of the $349, 8080-based Altair, the computer that *really* started the PC revolution, no matter what Steve Jobs says. By that time, I had already written floating-point software for its parent, the 8008. I wanted to catch this wave in the worst way, and I did, sort of. For a time, I managed a company that had one of the very first computer kits after the Altair. The first assembly language programs I wrote for hire were pretty ambitious: a real-time controller for a cold forge, written for the 4040, and a real-time guidance system, using floating-point math and a Kalman filter, to allow an 8080 to steer a communications satellite dish. Both worked and outperformed their requirements.

In 1975, before Bill Gates and Paul Allen had written Altair (later Microsoft) BASIC, we were building and selling real-time systems based on microprocessors, as well as a hobby computer kit. So why ain't I at least a millionaire, if not a billionaire? Answer: some luck, and a lot of hard work. You have to work hard, with great concentration, to snatch defeat from the jaws of victory, and that's exactly what my partners and I did. I expect we weren't alone.

I am relating this ancient history to emphasize that I think I bring to the table a unique perspective, colored by the learning process as the industry has developed and changed. Although I don't advocate that all other programmers of embedded systems go through the torture of those early days, I do believe that doing so has given me certain insights into and understanding of what's required to build successful, reliable, and efficient embedded systems. I'll be sharing those insights and understandings with you as I go. This will not be your ordinary how-to book, where the author gives canned solutions without explanation — solutions that you can either use or not, but cannot easily bend to your needs. Rather, I'll share alternative solutions and the trade-offs between them so that you can use my code and have the understanding necessary to generate your own, with a thorough knowledge of why some things work well and some don't. That knowledge comes from the experience I've gained in my travels.

Foundations

Chapter 1

Getting the Constants Right

It may seem strange that I'd spend any time at all discussing the proper way to define a numeric constant. After all, any beginning text on any programming language will tell you what constitutes a legal definition of a constant. Decimal integer constants consist of one to N decimal digits (where N varies from language to language); hex or binary numbers begin with 0x or 0b, and floating-point constants have a decimal point and possible exponent field. Hardly the kind of stuff to devote many words to. Yet, you'd be surprised how many computer programs end up broken or behaving erratically (which is functionally equivalent to being broken) because someone got a constant wrong. Part of the purpose of this chapter is to illustrate, as graphically as I can, how even the simplest concepts of programming can get you into trouble if not handled with care.

I assume you're advanced enough in your knowledge of programming to know that embedding constants as literals inside the code is generally a Bad Idea. One of the main reasons is that you run the risk of getting it wrong, either via a typo or simple sloppy programming, with multiple instances of the same constant. I once worked on the maintenance of a

FORTRAN program that had no fewer than five values for π, ranging from 12 digits down to, incredibly, three (3.14). This is clearly a Bad Idea, because if you assign a variable the value π in one place and test it in another place for equality to π, the test will likely fail. Likewise, if you're solving for an arctangent, whose output values are given as offsets from some submultiple of π, you're going to see discontinuities in the output of the function.

Because of the problems inherent in such things, programmers have been taught, almost from day one, to assign names to literal constants and use the names, rather than the literals, where they are needed. It goes without saying that such constants should also be correct and exact to a tolerance within the resolution of the computer word length.

I must stop at this point to say that, in my opinion, one can go way overboard on this issue of assigning names to constants. One would never (well, hardly ever) write:

```
#define    ZERO    0
#define    ONE     1
```

However, sometimes people get so carried away with the notion of avoiding embedded literal constants that they will go to extreme lengths, and the only effect is obfuscation of the results of an operation. When I see the statement

```
for(i = START_VALUE; i <= END_VALUE; i++)
```

I can see that I'm going to be doing something in a counted loop, but how many times? To find out, I must go track down START_VALUE and END_VALUE, which may well be in files separate from the one I'm working on — most likely in a C header file or, because nature is perverse, even in separate header files. This does nothing toward helping me to understand the program, much less maintain it. In cases where the numbers are not likely to change, the far simpler, and more transparent

```
for(i = 0; i <3; i++)
```

is much easier to understand. The lesson to be learned here is, do things properly, but don't be a slave to a convention just because it's conventional. Know when to use a literal and when not to. In general, the numbers that deserve to have names are either the long ones like π, that are hard to write, or the ones that are likely to change in the future, like the size of data buffers. For obvious values like the dimensions of vectors in a 3-D universe, keep it simple and use the literal.

In any case, having established that at least some literal constants deserve to be named, the problem still remains how to best assign these names to them and, far more importantly, how to communicate their values to the places where they are needed. It may surprise you to learn that these questions have plagued programmers since the earliest days of FORTRAN. Back when all programs were monstrous, monolithic, in a single file, and in assembly language, there was no problem; in assembly, all labels are global, so duplicate names never occurred, and the value of the constant was universally accessible from anywhere in the program.

#define **Constants to Avoid Common Errors**

I said that one would "hardly ever" define a simple literal like 1. Is this always true? Sometimes it is useful to assign floating-point constants for common numbers such as 1, 2, or 3 by including the following statements in the header file.

```
#define One     1.0
#define Two     2.0
#define Three   3.0
```

It may seem frivolous to define a constant called One, and I suppose it is. The idea does have some merit, though. A common programming error made by beginners (and sometimes us old-timers, too) is to leave the decimal point out of a literal constant, resulting in statements such as

```
x = x + 1;
```

where x is a floating-point number. By doing this, you're inviting the compiler to generate a wasteful run-time conversion of the 1 from integer to float. Many compilers optimize such foolishness out, but some do not. I've found that most modern compilers let me use integer-like constants, like 0 and 1, and are smart enough to convert them to the right type. However, to take the chance is bad programming practice, so perhaps there's some value to setting up 0 and 1 as floating-point constants, just to avoid the mistake.

Sadly, some FORTRAN programmers continue to use monstrous, monolithic, single-file programs, which is hardly surprising when one considers that this is the way the language is taught to them. In FORTRAN, however, programmers had, almost for the first time, the introduction of modularity via subroutines, and the principles of information hiding (not yet voiced as such by David Parnas, but still vaguely grasped as a Good Idea) prevented subroutines from having access to global data unless that access was explicitly granted. The mechanism for granting access to variables was to put them in a COMMON area. So, for example, I might write

```
COMMON PI, HALFPI, TWOPI,…
```

There was, however, a serious flaw in this mechanism. Early FORTRAN compilers provided no mechanism for assigning literal values to the constants. These constants were, in fact, not really constants, they were variables, and someone had to take responsibility for giving them values. Typically, the main program would have an initialization section, as shown in Listing 1.1.

Listing 1.1 Initializing constants in FORTRAN.

```
PI = 3.141592654
HALFPI = 1.570796327
TWOPI = 6.283185307
DPR = 57.29577951        {conversion from radians to degrees}
RPD = 1.74532952e-2      {and vice versa}
```

And so on. With a proper grasp of the concepts of modular programming, the main program might even call an initialization routine to take care of the initialization.

```
CALL CINIT
```

Related Constants

You might have noticed that the constants in the example are all related to each other and, ultimately, to π. One advantage of a FORTRAN initialization section or subroutine is that the assignment statements were truly assignment statements. That is, they represented executable code that could include expressions. A better implementation of the initialization code is shown in Listing 1.2.

Listing 1.2   **Computed constants in FORTRAN.**

```
PI = 3.141592654
HALFPI = PI / 2.0
TWOPI = 2.0 * PI
DPR = 180.0 / PI
RPD = PI / 180.0
```

This code not only makes it quite clear what the variables are, but more importantly, it leaves me with only one constant to actually type, decreasing the risk of making an error. If I went to a machine with different precision, changing just the one number got me back in business. The only remaining problems with this approach were simply the bother of duplicating the COM-MON statement in every subroutine that used the constants and remembering to call the initialization subroutine early on.

Recognizing the need to initialize variables, the designers of FORTRAN IV introduced the concept of the Block Data module. This was the one module in which assignment to COMMON variables was allowed, and it was guaranteed to set the values of all of them before any of the program was executed. This relieved the programmer of the need to call an initializer. FORTRAN IV also provided named common blocks, so I could create a single block named CONST with only a handful of constant values. (Earlier versions had just a single COMMON area, so the statement had to include every single global variable in the program.)

In one sense, however, the Block Data module represented a significant step backward. This was because the "assignment" statements in Block Data only looked like assignments. They weren't, really, and did not represent executable code. Therefore, the code in Listing 1.2 would not work in a FORTRAN IV Block Data module.

Time Marches On

Except for hard-core scientific programmers, FORTRAN is a dead language. However, the need to use named constants, which involves not only assigning them values but also computing certain constants based upon the values of others and making them available to the modules that need them, lives on. You might think that the designers of modern languages like C and C++ would have recognized this need and provided ample support for it, but you'd be wrong. Most of us find ourselves settling on partial solutions that do not completely satisfy.

In C or C++, the obvious mechanism for defining a constant is the preprocessor #define command.

```
#define pi 3.141592654
```

You can put these definitions at the top of each C file that needs them, but that leaves you not much better off than before. Unless the project has only one such file, you must duplicate the lines in each file of your project. This still leaves you with multiple definitions of the constants; you've only moved them from the middle of each file to the top.

The obvious next step is to put all the definitions in a central place — namely, a header file, as in Listing 1.3. Note that I've put an #ifndef guard around the definitions. It is always good practice in C to avoid multiple definitions if one header file includes another. Always use this mechanism in your header files.

Note that I've reverted to separate literals for each constant. This is because the #define statement, like the FORTRAN Block Data assignments, doesn't really generate code. As I'm sure you know, the preprocessor merely substitutes the character strings inline before the compiler sees the code. If you put executable code into the #defines, you run the risk of generating new executable code everywhere the "constant" is referenced.

Listing 1.3   A header file for constants.

```
// file constant.h
// definitions for common constants

#ifndef CONSTANTS
    #define CONSTANTS

    #define PI                      3.141592654
    #define HALFPI                  1.570796327
    #define TWOPI                   6.283185307
    #define RADIANS_PER_DEGREE      1.74532952e-2
    #define DEGREES_PER_RADIAN      57.29577951
    #define ROOT_2                  1.414213562
    #define SIN_45                  0.707106781

#endif
```

For constants such as TWOPI, this is not such a big deal. If you write

```
#define HALFPI    (PI/2.0)
```

the expression will indeed be substituted into the code wherever HALFPI is used (note the use of the parentheses, which are essential when using the preprocessor in this way). However, most compilers are smart enough to recognize common, constant subexpressions, even with all optimizations turned off, so they will compute these constants at compile time, not run time, and execution speed will not be affected. Check the output of your own compiler before using this trick too heavily.

But consider the constant ROOT_2. I'd like to write the following statement.

```
#define ROOT_2 (sqrt(2.0))
```

However, this contains a call to a library subroutine. It is guaranteed to generate run-time code. What's more, you'd better include <math.h> in every file in which ROOT_2 is referenced, or you're going to get a compilation error.

Does Anyone Do It Right?

I'm dwelling on this business of defining constants simply to show you that, even for so simple a problem as defining π or its related constants, there are no simple, pat solutions, even in a language as advanced as C++. To my knowledge, only two languages give the mechanisms you really need to do it properly, and one of them is rather clumsy to use.

The Ada language allows you to separate code into packages. Each package can contain data global to the package, global to the world, or local. It can also include tasks and functions. Finally — and this is the secret that makes it so useful — a package can have a package body that is executed as the application is started. This provides a place to put the code to initialize global constants. The only problem here is that you must refer to the constant with the package prefix: constant.pi, for example.

Borland's Object Pascal, as used in Delphi, does it even better. Borland Pascal allows for the use of units, one of which can contain all the global constants. As in the Ada package, the Borland unit can have an executable part, which is guaranteed to run before the main program begins to execute. Listing 1.4 shows a total, permanent solution in Borland Pascal. As far as

I'm concerned, this is the ideal solution to the problem. The only catch is, you can only do it in Borland Pascal.

Listing 1.4   The Pascal solution.

```
Unit Constants;Interface
Const One   = 1.0;
Const Two   = 2.0;
Const Three = 3.0;
Const Four  = 4.0;
Const Five  = 5.0;
Const Ten   = 10.0;
Const Half   = One/Two;
Const Third  = One/Three;
Const Fourth = One/Four;
Const Fifth  = One/Five;
Const Tenth  = One/Ten;

{In Borland Pascal, Pi is predefined}
Const TwoPi               = Two * Pi;
Const Pi_Over_Two         = Pi/Two;
Const Pi_Over_Three       = Pi/Three;
Const Pi_Over_Four        = Pi/Four;
Const Degrees_Per_Radian = 180.0/Pi;
Const Radians_Per_Degree = Pi/180.0;

{These constants will be filled by startup code}
Const Root_Two:   real = 0.0;
Const Root_Three: real = 0.0;
Const Root_Five:  real = 0.0;
Const Golden:     real = 0.0;
Const Sin_45:     real = 0.0;
Const Sin_30      = Half;
Const Cos_30:     real = 0.0;

Implementation
```

Listing 1.4   The Pascal solution. (continued)

```
Begin
    Root_Two      := Sqrt(Two);
    Root_Three    := Sqrt(Three);
    Root_Five     := Sqrt(Five);
    Golden        := (Root_Five + One)/Two;
    Sin_45        := One/Root_Two;
    Cos_30        := Root_Three/Two;
End.
```

You could use the Unit Constants, for example, to create functions that convert degrees to radians. I can't remember how many lines of FORTRAN code I've seen devoted to the simple task of converting angles from degrees to radians and back.

```
PSI   = PSID/57.295780
THETA = THETAD/57.295780
PHI   = PHID/57.295708
```

Not only does it look messy, but you only need one typo (as I deliberately committed above) to mess things up.

The Unit Constants in Listing 1.4 allows you to create the following functions.

```
{ Convert angle from radians to degrees }
Function Degrees(A: Real): Real;
Begin
    Degrees := Degrees_Per_Radian * A;
End;

{ Convert angle from degrees to radians }
Function Radians(A: Real): Real;
Begin
    Radians := Radians_Per_Degree * A;
End;
```

To use them, you need only write the following.

```
Psi := Radians(PsiD);
```

If these two routines are in a module that uses constants, you can be sure that the conversion will always be correct to the appropriate degree of precision. Note that the function call has the look of a type cast, which is sort of what it is.

You might wince at the inefficiency of a single-line function call to do this job, but in this case it's perfectly justified. Remember, this kind of conversion is typically used only during input and output, so efficiency is not much of an issue.

A C++ Solution

Can I do anything with C++ objects to get the equivalent of the package and unit initializations? Yes and no. I can certainly define a class called Constants with one and only one instance of the class.

```
class Constants{

public
    double pi;
    double twopi;
    Constants( );
    ~Constants( );

} constant;
```

I can then put the code — any code — to initialize these variables into the class constructor. This solution is workable, but I am back to having to qualify each constant as I use it, just as in Ada: constant.pi. The syntax gets a bit awkward. Also, there must surely be some run-time overhead associated with having the constants inside a class object.

Fortunately, if you're using C++, you don't need to bother with either classes or preprocessor #defines. The reason is that, unlike C, C++ allows for executable code inside initializers. Listing 1.5 is my solution for making constants available in C++ programs. Notice that it's an executable file, a

.cpp file, not a header file. However, it has no executable code (except what's buried inside the initializers).

Listing 1.5   The C++ solution.

```
/* file const.cpp
 *
 * defines commonly used constants for use by other modules
 *
 */
double pi            = 3.141592654;
double halfpi        = pi / 2.0;
double twopi         = 2.0 * pi;
double radians_per_degree = pi/180.0;
double degrees_per_radian = 180.0/pi;
double root_2        = sqrt(2.0);
double sin_45        = root_2 / 2.0;
```

Don't forget that you must make the constants visible to the code that references it by including a header file (Listing 1.6).

Listing 1.6   The header file.

```
/* File constant.h
 *
 * header file for file constant.cpp
 */

#ifndef CONSTANT_H
    #define CONSTANT_H

extern double pi;
extern double halfpi;
extern double twopi;
extern double radians_per_degree;
extern double degrees_per_radian;
extern double root_2;
extern double sin_45;

#endif
```

This solution is as near to perfect as you're likely to get with C++ or any other language in the near future. It's my standard approach. There remains only one nagging problem: the issue of data precision. I've shown the "constants" as double-precision floating-point numbers, but you may prefer to define the variables as `float`s to save space (though I don't recommend it; in these days of multimegabyte memory, get reckless and use the extra 14 bytes). Some compilers also allow the type `long double`, which gives access to the 80-bit internal format of the Intel math coprocessor. This type naturally requires more storage, but again, who cares? More to the point, because `long double` is the natural format for the math coprocessor, moving data into and out of it is faster if no conversions are required, so you might actually get a faster program with the higher precision.

There is no way around it, though, mixing data types can be a real pain in the neck, because the compiler is going to flood you with warnings about implicit type conversions and potential loss of accuracy. My best advice is, pick one precision, use it for all the floating-point variables in your program, and write the `constant.cpp` file to match. If you do decide to go with `long double`, don't forget that you must call the math functions by different names — usually `sqrtl`, and so on. You also must put an `L` suffix on all literals.

Header File or Executable?

There's another approach to defining constants that avoids the need to write a separate header file and includes the `.cpp` file in the project make list. This is to put the statements in the header file, as shown in Listing 1.7.

Note that I now have executable code in the header file. This is usually a no-no, but in this case it seems reasonable. Depending on the compiler you use, you might get a warning message that the header file contains executable code, but the program will still work just fine.

Listing 1.7   Constants in header file.

```
/* File constant.h
 *
 * defines commonly used constants for use by other modules
 *
 */
```

Listing 1.7   Constants in header file. (continued)

```
#ifndef CONSTANT_H
#define CONSTANT_H

static double const pi                    = 3.141592654;
static double const halfpi                = pi / 2.0;
static double const twopi                 = 2.0 * pi;
static double const radians_per_degree    = pi/180.0;
static double const degrees_per_radian    = 180.0/pi;
static double const root_2                = sqrt(2.0);
static double const sin_45                = root_2 / 2.0;

#endif
```

Think a moment about the implications of using the header file shown above. Remember, this file will be included in every source file that references it. This means that the variables pi, halfpi, and so on will be defined as different, local variables in each and every file. In other words, each source file will have its own local copy of each constant. That's the reason for the keyword static: to keep multiple copies of the constants from driving the linker crazy.

It may seem at first that having different local copies of the constants is a terrible idea. The whole point is to avoid multiple definitions that might be different. But in this case, they can't be different because they're all generated by the same header file. You're wasting the storage space of those multiple copies, which in this example amounts to 56 bytes per reference. If this bothers you, don't use the method, but you must admit, that's a small amount of memory.

The sole advantage of putting the constants in the header file is to avoid having to link constant.obj into every program, and by inference, to include it in the project make file. In essence, you're trading off memory space against convenience. I guess in the end the choice all depends on how many programs you build using the file and how many separate source files are in each. For my purposes, I'm content to put the constants in the .cpp file instead of the header file, and that's the way it's used in the library accompanying this book. Feel free, however, to use the header file approach if you prefer.

Gilding the Lily

In Listing 1.5, I managed to get the number of constants I must explicitly specify down to one: π. In the listing, I've fixed it to 10 digits. That's a nice number, but those are too many digits for type float, and not enough for type double. The first type is accurate only to six or seven digits and the latter to 12 digits. Things really begin to get interesting if I use long double, where I must specify π to 22 digits. Then it's time for a quick run to the handbook.

To get around this problem, there's an alternative approach, which Intel recommends. Remember that the math coprocessor stores a value of π internally; it must in order to solve for arctangents. Instead of trying to remember π to 22 digits and key it in, why not ask the coprocessor what value it recommends?

```
double pi    = 4.0 * atan(1.0);
```

Clearly, this will work regardless of the precision you choose, the word length of the processor, or anything else. It also has the great advantage that you are absolutely guaranteed not to have problems with discontinuities as you cross quadrant boundaries. You may think it is wasteful to use an arctangent function to define a constant, but remember, this is only done once during initialization of the program.

The code shown in this chapter is included in Appendix A, as is all other general-purpose code. Bear in mind that I have shown two versions of the file constant.h. You should use only one of them, either the one in Listing 1.6 or Listing 1.7.

Chapter 2

A Few Easy Pieces

As long as I'm building a math library, I'll start with some basic functions. In this process, you'll be following in the footsteps of my old FORTRAN students — but fear not: this is not some useless textbook exercise (for that matter, neither was theirs). You'll be using the functions given here extensively in what follows.

About Function Calls

Lately I've been bumping into, and therefore wondering about, some peculiar aspects of the definitions of some very fundamental math functions. The peculiarity revolves around two facts:

Fact 1 As I'm sure you know, all C library functions using floating-point variables are set up to use `double`, rather than mere `float`, variables for both input arguments and return values. All `float` (or integer, of course) variables are promoted to `double` when the function is called and demoted again, if necessary, on the way out.

A few years ago, this used to be a good argument for staying with FOR-TRAN to do number crunching. If the precision of `float` is good enough for your task, you don't want the compiler to burn unnecessary clock cycles in type conversions at every single function call. Aside from the overhead of the type conversions, the higher precision required by double also makes the required series expansions longer than they need to be and the functions therefore significantly slower. Why did the authors of C decide to write the library using only double variables? I have no idea, but I have a theory:

- They didn't want to bother writing two versions of every function, one for float and one for double data types.

- They weren't primarily the kind of people who did a lot of number crunching, so they didn't care that they were burdening the language with unnecessarily slow performance.

Whatever the true reason, this peculiarity is one reason many C programs run at about half the speed of their FORTRAN counterparts. And that's one of the reasons why you still see scientists doing their serious programming in FORTRAN (the other is C's lack of conformant arrays; more on this later).

Some programming shops, committed to using floating point and C in real-time situations but needing all the speed they can get, have gone so far as to write their own math libraries, redefining all the math functions using float variables.

More recently, the pendulum has swung the other way. Many programmers are now using at least 486, if not Pentium, processors, even in real-time applications. Others are using Motorola processors like the 68xxx in similar applications. Both families include math coprocessors, either as options or built into the CPU chip. These math coprocessors typically have all the precision they need to do double-precision arithmetic with no noticeable degradation in performance.

The Intel math coprocessor uses an 80-bit format, which is usually supported (though not guaranteed to be) by the C type long double. In this situation, you'll get the most efficient code by defining all of your variables as long double. Suddenly, the C convention doesn't deliver too much precision — it doesn't deliver enough.

Fact 2   The C preprocessor supports macros with passed parameters. However, it doesn't know or care about variable types (which, I hasten to point out, means that you can abandon all hope of compile-time type checking using macros). The fundamental math functions min() and max() are defined as macros in the header file math.h.

```
#define max(a,b)    (((a) > (b)) ? (a) : (b))
#define min(a,b)    (((a) < (b)) ? (a) : (b))
```

Because of this definition, I can use these "functions" for any numeric type. I'll be using these two functions heavily.

I must tell you that I'm no fan of the C preprocessor. In fact, I'm President, Founder, and sole member of the Society to Eliminate Preprocessor Commands in C (SEPCC). I suppose the idea that the preprocessor could generate pseudofunctions seemed like a good one at the time. In fact, as in the case of min()/max(), there are times when this type blindness can be downright advantageous — sort of a poor man's function overloading.

For those not intimately familiar with C++, one of its nicer features is that it allows you to use the same operators for different types of arguments, including user-defined types. You're used to seeing this done automatically for built-in types. The code fragment

```
x = y * z
```

works as expected, whether x, y, and z are integers or floating-point numbers. C++ allows you to extend this same capability to other types. However, as noted in the text, this only works if the compiler can figure out which operator you intended to use, and that can only be done if the arguments are of different types.

Operator overloading is perhaps my favorite feature of C++. However, it is also one of the main reasons C++ programs can have significant hidden inefficiencies.

As noted, you lose any hope of having the compiler catch type conversion errors. However, the alternatives can be even more horrid, as the next function illustrates.

The simple function abs(), seemingly similar to min() and max() in complexity, functionality, and usefulness, is not a macro but a true function. This means that it is type sensitive. To deal with the different types, the C founding fathers provided at least three, if not four, flavors of the function: abs(), fabs(), labs(), and fabsl(). (This latter name, with qualifying characters both front and rear, earns my award as the ugliest function name in all C-dom, especially since the trailing letter is, in many fonts, indistinguishable from a one.)

Because abs() is such a simple function, it can easily be defined as a macro:

```
#define abs(x) (((x) < 0)? -(x): (x))
```

It could be used for all numeric types, but for some reason, the founding fathers elected not to do so. Could this be because zero really has an implied type? Should I use 0.0L, for long double? Inquiring minds want to know. In any case, I've found that most modern compilers let me use integer-like constants, like 0 and 1, and are smart enough to convert them to the right type. Perhaps the presence of a literal constant in abs() is the reason it was defined as a function instead of a macro. If so, the reasons are lost in the mists of time and are no longer applicable.

What's in a Name?

Although I detest preprocessor commands, I'm also no fan of the idea of multiple function names to perform the same function. As it happens, I'm also President, Founder, and sole member of the Society to Eliminate Function Name Prefixes and Suffixes (SEFNPS). The whole idea seems to me a coding mistake waiting to happen.

In a previous life writing FORTRAN programs, I learned that the type of a variable, and the return type of a function, was determined via a naming convention. Integer variables and functions had names beginning with characters I through N. All others were assumed to have floating-point values.

As a matter of fact, even before that, the very first version of FORTRAN (before FORTRAN II, there was no need to call it FORTRAN I) was so primitive that it required *every* function name to begin with an F; otherwise, it couldn't distinguish between a function call and a dimensioned variable. I suppose that this FORTRAN would have had FIABS(), FFABS(), and FFABSL().

When double-precision floats came along in FORTRAN, we had the same problem as we have now with C. While we could declare variables either real or double precision, we had to use unique names to distinguish the functions and subroutines. Although you could declare variables either real or double precision, you had to use unique names to distinguish the functions and subroutines. I can't tell you how many times I've written math libraries for FORTRAN programs, only to have to create another copy with names ending in D, or some similar mechanism, when the program migrated to double precision. I didn't like the idea then, and I don't like it still.

In C++ you can overload function names — a feature I consider one of the great advantages of the language — despite the fact that the compiler sometimes has trouble figuring out which function to call. This means, in particular, that a function such as abs() can have one name and work for all numeric types.

Unfortunately, this is not done in the standard C++ library. To assist in portability, C++ compilers use the same math libraries as C. You can always overload the function yourself, either by defining a new, overloaded function or by using a template (assuming you can find a compiler whose implementation of templates is stable).

Alternatively, you can always define your own macro for abs(), as in the example above. This approach works for both C and C++. As with other macros, you lose compile-time type checking, but you gain in simplicity and in portability from platform to platform. Or so I thought.

Recently, I was developing a C math library that included a routine to minimize the value of a function. Because of my innate distaste for multiple names and the potential for error that lurks there, I included in my library the macro definition for abs(). I thought I was making the process simpler and more rational. Unfortunately, things didn't work out that way. The function minimizer I was building, using Microsoft Visual C/C++, was failing to converge. It took a session with CodeView to reveal that the compiler was, for some reason, calling the integer library function for abs() despite my "overloaded" macro. The reason has never been fully explained; when I rebooted Windows 95 and tried again the next day, the darned thing worked as it should have in the first place. Chalk that one up to Microsoft strangeness. But the whole experience left me wondering what other pitfalls lurked in those library definitions.

Digging into things a bit more led me to the peculiar "broken symmetry" between min()/max() and abs(). The macros min and max are defined in <stdlib.h>. The abs() function, however, is defined in both <stdlib.h> and <math.h>. (To my knowledge, this is the only function defined in both header files. What happens if you include both of them? I think the first one prevails, because both files have #ifdef guards around the definitions.)

My confusion was compounded when I tried the same code on Borland's compiler and found that the two compilers behaved differently. If you use min() or max(), and fail to include <stdlib.h>, both the Microsoft and Borland compilers flag the error properly and refuse to link. However, if you try the same thing with abs(), both compilers will link, and each produces a bogus executable file. The Microsoft compiler at least gives an error message, but it's only a warning, and the .exe file is still generated.

The Borland compiler gives nary an error message and compiles cleanly, but the .exe file is still wrong; the "function," when called, returns a zero result, regardless of its input. To make matters even more interesting, try changing the file name from *.c to *.cpp. The Borland compiler will link in the library function for abs() (the integer version), even though you have not declared it anywhere. To my knowledge, this is the only case where you can call an undeclared function and get away with it.

The Borland help file gives a clue to what's going on with abs(): It says that you are free to write a macro version of it, but you must first use the statement

```
#undef abs
```

to turn off the existing definition. Ironically, the Borland compiler doesn't really seem to care if I do this or not. If I include the macro, it's used properly whether I use the #undef or not. On the other hand, the #undef does seem to help the Microsoft compiler, which appears to use the macro reliably; without it, it seems to do so only on odd days of the month or when the moon is full. Go figure.

If there's a lesson to be learned from this, I think it's simply that you should not take *anything* for granted, even such seemingly simple and familiar functions as min(), max(), and abs(). If you have the slightest doubt, read the header files and be sure you know what function you're using. Also, check the compiler's output. Don't trust it to do the right thing. In any case, the tools in this book use the macro abs. You'd better use the #undef too, just to be safe and sure.

What's Your Sign?

Sometimes it's useful to be able to assign the sign of one variable to the magnitude of another. I find this happens quite often in dealing with trig functions, as with the following function, taken right from the FORTRAN standard library:

```
#define sign(x, y) (((y)<0)? (-(abs(x))): (abs(x)))
```

Note the syntax: the value returned has the magnitude of x and the sign of y. Sometimes, you simply want to return a constant, such as 1, with the sign set according to the sign of some other variable. No problem; just call sign() with x = 1.

The Modulo Function

There's another simple function that I use a lot, and it also comes straight from FORTRAN. It's called the modulo function [mod() in FORTRAN]. C/C++ has a similar function in the operator %, as in

```
x = y % z;
```

In theory, this function is supposed to return the remainder of y, after it's been divided by z. This C/C++, however, only works for integer arguments, while the mod() function is useful for floats as well. A handy usage is to reduce an angle to the range 0° to 360° degrees before doing anything else with it.

This construct is supposed to limit the first argument to a range given by the second. Thus, for example,

```
mod (1, 4) = 1
mod (7, 2) = 1
mod (5, 3) = 2
mod (4, 2) = 0
```

and so on. In other words, the result of mod(x, y) must always be between 0 and y.

However, if you look in the appropriate language manuals, you'll see both mod() and % are defined by the following code.

```
int n = (int)(x/y);
x -= y * n;
```

Unfortunately, this code doesn't implement the rules we'd like to have. Consider, for example, the following expression.

```
-5 % 2
```

By the preceding definitions, the result should be between 0 and 2; it should, in fact, be 1. Try it, however, on any C compiler and you'll get -1. Thus, the result lies outside the range 0 to y, which to me seems a violation of the definition. The FORTRAN mod( ) works exactly the same way.

Is this a bug? That depends. If you use the definition given in the two lines of code above, which is the way the function is usually defined, it's not a bug. This is precisely the result one would expect from those two lines. If, however, you want the function to work like the mathematical definition of the modulo function, it's incorrect, since the latter requires that the result be bounded by 0 and whatever the second argument is. To me, that's a bug.

I use mod( ) for a lot of purposes, but one of them is to place angles in the proper quadrant. If I tell the computer to give me an angle between 0° and 360°, I dang well don't expect to see it coming back with -20°. I've struggled with this problem ever since my very first FORTRAN program, circa 1960. My solution has always been the same: write a replacement for the library function. Listing 2.1 is a mod( ) that works correctly for all arguments.

Listing 2.1   The mod( ) function.

```
// Return a number modulo the second argument
double mod(double x, double y){
    if(y == 0) return x;
    long i = (long)(x/y);
    if(x*y < 0)--i;
    x = x-((double)i)*y;
    if(x==y)x -= y;
    return x;
}
```

Chapter 3

Dealing with Errors

Every high-order language worthy of the name includes a library of arithmetic functions like abs(), sqrt(), and so on. Having those functions available, instead of having to write your own, is one of the very significant advantages of programming in a high-order language like C. In a moment, I'll discuss what you must do if such functions aren't available and you must roll your own. But for now, the good news is that any high-order language provides a ready-made library of fundamental functions. The bad news is, they can't be trusted.

The reason is simple enough: historically, these functions were designed to be fast at all costs. Error checking tends to be minimal to nonexistent. The programmer is expected to make sure that the arguments passed to the functions are such that the functions will work.

Consider the lowly square root function. The square of a real number is always positive, so the square root of a negative number is undefined. Try calling the square root function with a negative argument, and the program will abort with an error message like

```
Runtime error 207 at 0B74:00CA
```

That's hardly a helpful error message in the best of circumstances. About all it tells you is that a floating-point error occurred somewhere. To the programmer, the message is, "back to the debugger." To the average end user, it means, "call the vendor."

But consider the implications for, say, the manager of a complex and expensive chemical processing plant whose plant just shut down in midprocess leaving a large batch of half-cooked chemicals still reacting. In this case, the message means, "call your lawyer, then run like hell."

In Project Apollo, the landings of the Lunar Excursion Module (LEM) were computer assisted. Anyone who's ever played the old computer game Lunar Lander knows that the thrust schedule for descending and landing with zero velocity is tricky and best left to a computer. During that first historic landing of Apollo 11, astronauts Neil Armstrong and Buzz Aldrin depended on the computer to get them down. But when they were still rocketing toward the lunar surface, 500 feet up, still falling at great speed, and decelerating under a very precisely designed thrust program, they were suddenly greeted by the cheerful message

```
Alarm 1201
```

followed shortly by

```
Alarm 1202
```

Roughly translated, this means, "I've just gone out to lunch. You're on your own, kid." Having all the right stuff, Armstrong stoically took over and landed the LEM manually. The astronauts never voiced their reactions, but I know how I'd feel on learning that my transportation seemed to be failing, with the nearest service station 240,000 miles away. It's probably very much the same reaction that you'd get from that chemical plant manager. If I'd been them, the first person I'd want to visit, once I got home, would be the programmer who wrote that code.

The need for error checking in real-time embedded systems is very different than that for batch programming. The error mechanisms in most library functions have a legacy to the days of off-line batch programs, in which the system manager mentality prevailed; that is, "The programmer screwed up. Kick him off the machine and make room for the next one."

An embedded system, on the other hand, should keep running at all costs. That chemical plant or lunar lander must keep working, no matter what. It's not surprising, then, that the error-handling behavior of the standard library functions should often need help for embedded applications. Any good embedded systems programmer knows better than to trust the library functions blindly. Quite often, you'll find yourself intercepting the errors before they get to the library functions.

Appropriate Responses

If a global halt is an inappropriate response in an embedded system, what is the appropriate one? That depends on the application. Usually, it means flagging the error and continuing with some reasonable value for the function result. Surprisingly, it often means doing almost nothing at all.

To show you what I mean, I will begin with floating-point arithmetic. The important thing to remember is that floating-point numbers are, by their very nature, approximations. It is impossible to represent most numbers exactly in floating-point form. The results of floating-point computations aren't always exactly as you expect. Numbers that are supposed to be 1.0 end up as 0.9999999. Numbers that are supposed to be 0.0 end up as 1.0e-13.

Now, zero times any number is still zero. Zero times zero is still zero. But 1.0e-27 times 1.0e-27 is *underflow.*

Almost all computers and coprocessors detect and report (often via interrupt) the underflow condition. When using PC-based FORTRAN, my programs occasionally halted on such a condition. This is not acceptable in embedded systems programming.

So what can you do? Simple. For all practical purposes, anything too small to represent is zero in my book. So a reasonable response is simply to replace the number with 0 and keep right on going. No error message is even needed.

Many compilers take care of underflows automatically. Others don't. If yours doesn't, you need to arrange an interrupt handler to trap the error and fix it.

Tom Ochs, a friend and outspoken critic of floating-point arithmetic, would chide me for being so sloppy and glossing over potential errors. In some cases, perhaps he's right, but in most, it's the correct response. If this offends your sensibilities, as it does Tom's, it's OK to send an error message somewhere, but be aware that it could cause your customer to worry unnecessarily. If you have a special case where underflow really does indicate an error, your program needs special treatment. But for a library routine that's going to be used in a general-purpose fashion, stick with the simple solution.

Belt and Suspenders

As the underflow example indicates, the typical error-handling mechanism of the library routines is unacceptable for embedded systems. To compensate for that, you're going to end up handling the errors externally. In other words, you must trap the errors before they get to the library routines and deal with them in ways appropriate for real-time systems. That's the belt and suspenders approach.

Because the system must keep going in the face of errors, you'll usually replace the erroneous number with something reasonable. The only question, then, is whether the program should report the error.

There are two or three schools of thought on the way to handle such errors. Some programmers like to have each routine report its own error to the console. The Unix functions tend to work this way. Others feel that library routines should never produce output but, at the very most, should return an error code for the caller to handle.

For real-time embedded systems, my typical approach is the same as for the underflow handler: fix the number and don't even report the error. At first glance, this may seem almost criminally negligent, but it's often the best response, as the next example illustrates.

I'll return to the square root function. You may think a negative argument is a serious problem — one that implies a bug in the program, and therefore one that needs serious action. But suppose the number should have been 0, but merely came out as a very small negative number because of round-off error. In such a case, crying wolf is inappropriate. The program is working properly, and the function should simply return a result of zero.

C++ Features and Frailties

Listing 3.1 gives a safe square root routine that takes this action for any negative number, large or small. No error message is written. This code is part of my math library, jmath.cpp, which is included in Appendix A.

Listing 3.1   A safe square root.

```
// Safe square root
double square_root(double x){
  if(x <= 0) return 0;
  return sqrt(x);
}
```

On the Case

A word here about the name of the function in Listing 3.1 is appropriate. Every function in a C program needs a unique name. C++ function names can be overloaded, but only if their arguments have different types. In this case, they don't, so overloading doesn't work.

C and C++ are supposed to be case sensitive. Variable names and functions are considered different if they have different combinations of upper- and lowercase characters. The variables Foo, foo, FOO, and fOO are all different variables. The Microsoft Visual C++ environment, however, doesn't follow this rule. Visual C++ uses the standard Microsoft linker, which doesn't understand or distinguish case differences.

When I started writing safe functions, I wondered what I should call them. Names like my_sqrt() and my_atan() seemed trite. Sometimes I tacked on a leading j (for Jack), but that solution always seemed a bit egotistical.

The other day, I thought I had the perfect solution. I'd simply use the same name as the library function, but with an uppercase first letter: Sqrt() instead of sqrt(). This seemed like the perfect solution: familiar, easy to remember names, but also unmistakably unique and recognizably my own. With this approach, I not only could get more robust behavior, I could also eliminate the hated type-qualifying characters in the function names. So, originally, I called the function in Listing 3.1 Sqrt().

Imagine my surprise when I tried this on the Microsoft compiler and found that it insisted on calling the library function instead. A quick session with CodeView revealed the awful truth: Microsoft was calling _sqrt(), which refers to the library function. It was ignoring the case in my function name!

The problem is, I didn't realize the Microsoft linker wouldn't distinguish case differences. Thus, even though the names sqrt() and Sqrt() are supposed to be unique according to the C rules of engagement, the Microsoft linker misses this subtlety and sees them as the same.

I also dealt with yet another "feature" of the Microsoft development environment. In every language I've ever used since programming began, any function that's supplied locally by the user took precedence over library functions with the same name. Not so with Visual C++. That system gives the library function preference and ignores the user-supplied function. I learned this the hard way when trying to overload abs().

I've never figured a work around for this one. To avoid the problem and any future compiler strangeness, I took the next best option and chose a function name I was sure wouldn't confuse the linker. At the moment, I'm reduced to using unique but longer names, like square_root() and arcsine(), that the compiler has no choice but to recognize as uniquely mine.

I should say a word here about multiple floating-point formats. C and C++ provide for at least two formats (float and double), and possibly three (long double). We've already talked about the way the languages promote floats to doubles in function calls.

The Intel math coprocessor uses an internal 80-bit format, and some compilers assign this format to the type called long double. The distinction is important, because the library routines for long doubles have different names. Remember, the designers of the math library chose not to use overloading. So we have functions like sin, cos, sqrt, and atan, for double precision numbers, and their long double equivalents, sinl, cosl, sqrtl, and atanl. It's FORTRAN déjà vu all over again. The choice of names is, in my opinion, unfortunate because lowercase "L" is indistinguishable from the numeral "1" in most fonts.

If you choose to do your arithmetic in long double format, be aware that you *must* call the long double function. Otherwise, the compiler will truncate your carefully computed, high-precision long double number to double, call the double library function, and convert the (now incorrect, or at least inaccurate) value back again.

At one time, my own library file, jmath.cpp, had overloaded functions for all three precisions. For example, I had the three functions:

```
float square_root(float x){
   if(x <= 0) return 0;
   return sqrt(x);
}


double square_root(double x){
   if(x <= 0) return 0;
   return sqrt(x);
}


long double square_root(long double x){
   if(x <= 0) return 0;
   return sqrtl(x);
}
```

This approach, however, turned out to be something less that satisfactory. For starters, the first function is virtually useless. The float version still performs a type-conversion because sqrt is a double-precision function. I've accomplished no performance improvement by using the single precision. What's more, I've found that if you have an overloaded function name, with both float and double versions, C compilers tend to promote the float argument and call the `double` function, anyhow. In short, the compiler can't be trusted to call the right function.

The third function does at least make sure that the correct function is called, and the number is not truncated. If you have a compiler that really, truly uses `long doubles`, you may want to seriously consider including over-loadings for this type. However, my interest in `long-double` precision waned quickly when I learned that the compilers I use for 32-bit applications do not support long double anymore. The 16-bit compilers did, but the 32-bit ones did not. Why? Don't ask me; ask Borland and Microsoft.

I ran into trouble with type conversions; not in the code of `jmath.cpp`, but in the definitions of the constants. If I'm going to use `long doubles`, then I should also make the constants in `constant.h` `long double`. But whenever I did that, I kept getting compiler errors warning me that constants were being truncated.

In the end, `long double` arithmetic proved to be more trouble than it was worth, at least to me. So the code in the remainder of this book assumes dou-ble-precision arithmetic in all cases. This allowed me to simplify things quite a bit, albeit with a considerable loss of generality.

At this point, someone is bound to ask, "Why not use a template?" The answer is Keep It Simple, Simon (KISS). Although I have been known to use C++ templates from time to time, I hate to read them. They are about as readable as Sanskrit. I also have never been comfortable with the idea of sticking executable code in header files, which is where many compilers insist that templates be put. P.J. Plauger says his subset embedded C++ won't have templates, and that's no great loss, so I'll avoid them in this book. So much for the Standard Template Library (STL).

Is It Safe?

I must admit that the approach used in `square_root()` — simply ignoring negative arguments — can mask true program bugs. If the number com-puted really is a large negative number, it probably means that something is seriously wrong with the program. But it doesn't seem fair to ask the library functions to find program bugs. Instead of sprinkling error messages through a library of general-purpose routines, wouldn't it be better to debug the program? If the application program is truly robust and thoroughly tested, the condition should never occur.

If it really offends your sensibilities to let such errors go unreported, you might modify the function to allow a small "epsilon" area around zero and to report an error for negative numbers outside this region. Be warned, though, that now you have to define what the term "small" means, and that's usually specific to the problem.

Functions that can only deal with limited ranges of input parameters should be guarded by tests to assure the input is within bounds. Examples are the inverse sine and cosine. Their arguments should never exceed ±1.0, but in practice, when we're computing the arguments, we may stray just over the limit. We can't afford the fatal error that results, so we'll write safe versions that limit the arguments to the legal range. Similar problems can occur with functions that can produce infinite results, like the tangent or hyperbolic functions. The factorial and exponential functions don't produce true infinities but can still cause overflow for modest-sized inputs (87.0, for Exp( ), in single precision). So again, the input values should be tested to make sure they're within bounds. Safe versions of these functions are shown in Listing 3.2. Note that the argument limit for the argument of the exponential is extremely conservative. The value shown is the value that gives approximately BIG_FLOAT. In reality, the function is defined as returning a double, which can be considerably larger than BIG_FLOAT, so the argument could actually be quite a bit larger, around 710. However, I've stuck with the more conservative value based upon the philosophy that if you're taking e to the power 46, you must be doing something very, very peculiar, and deserve to get the wrong answer!

In the case of the four-quadrant arctangent, the library function atan2 gives reasonable and accurate answers except for the special case where both arguments are zero. In practice, this should never happen, but if it does, we should still return a reasonable value rather than crash. The function shown tests the value of x, and returns either plus or minus $\pi/2$. If $x = 0$, there is no need to call atan2 at all, so the safe function is actually faster for this special case.

The factorial function is not shown because you should never be using it. Proper coding can and should eliminate the need for it.

Listing 3.2   Simple and safe library functions.

```
// Safe inverse sine
double arcsin(double x)
{
    if(x>1)
        return halfpi;
    if(x<-1)
```

```
            return -halfpi;
        return asin(x);
}


// Safe inverse cosine
double arccos(double x)
{
    return(halfpi - arcsin(x));
}


// safe, four-quadrant arctan
double arctan2(double y, double x)
{
    if(y == 0)
        return(x>=0)? 0: -pi;
    return atan2(y, x);
}


// safe exponential function
double exponential(double x){
        if(x > 46)
                return BIG_FLOAT;
        return exp(x);
}
```

Taking Exception

A lot of the headaches from error tests could be eliminated if the programming language supported exceptions. I'm a big fan of exceptions, when they're implemented properly. They are wonderful way to deal with exceptional situations without cluttering up the mainstream code with a lot of tests. Unfortunately, I haven't found a single compiler in which exceptions have been implemented properly. Most implementations generate very, very inefficient code — far too inefficient to be useful in a real-time program.

Exception Exemption

For those not familiar with the term, an *exception* in a programming language is something that happens outside the linear, sequential processing normally associated with software. Perhaps the form most familiar is the BASIC "ON ERROR" mechanism. The general idea is that, in the normal course of processing, some error may occur that is abnormal; it could be something really serious, like a divide-by-zero error — usually cause for terminating the program. Or, it could be something that happens under normal processing, but only infrequently, like a checksum error in a data transfer.

In any case, the one thing we cannot allow is for the program to crash. The main reason compiler writers allow fatal errors to terminate programs is that it's difficult to define a general-purpose behavior that will be satisfactory in the particular context where the error occurs.

The general idea behind exception handling is that when and where an exception occurs, the person best able to make an informed decision as to how to handle it is the programmer, who wrote the code and understands the context. Therefore, languages that support exceptions allow the programmer to write an exception handler that will "do the right thing" for the particular situation. As far as I know, all such languages allow an exception to propagate back up the call tree; if the exception is not handled inside a given function, it's passed back up the line to the calling function, etc. Only if it reaches the top of the program, still unhandled, do we let the default action occur.

One point of debate is the question: what should the program do after the exception is handled? Some languages allow processing to resume right after the point where the exception occurred; others abort the current function. This is an issue of great importance, and has a profound impact upon the usability of exceptions in real-world programs.

The efficiency issue hinges on the question: do you consider an exception to be truly an *exception* — that is, an event that is not likely to happen often, but must be dealt with when it does — or do you consider it an *error*, calling for drastic action up to and including program abort. There are plenty of situations in embedded programs where the first definition makes sense; for example, protocol errors in a serial communication device. This is definitely a situation that must be dealt with, but it must be dealt with efficiently, and then the program must continue to operate normally.

Unfortunately, I fear that most compiler writers see the exception in light of the second viewpoint: a near- or truly fatal error. Because they see them this way, they see no reason to implement them efficiently. I mean, if the program is going down, why care how long it takes to do so?

Ironically, one language that does support exceptions, and supports them efficiently at that, is BASIC, with its ON condition mechanism. I say ironically because BASIC is hardly what one usually thinks of as a candidate for embedded systems programming.

Another language that supports exceptions is Ada. In fact, to my knowledge Ada was the first language to make exceptions an integral part of the language. (Trivia question: What was the second? Answer: Borland's short-lived and much-missed Turbo Modula 2, which was available on CP/M machines only and dropped by Borland because they never got a PC version working.)

Again, though, there's a certain irony, because even though exceptions were designed into the language, and I know in my heart that the language designers intended for them to be implemented efficiently (mainly because they told me so), that's not how they actually got implemented. At least the earlier Ada compilers suffered from the exception-as-fatal-error syndrome, and their writers implemented exceptions so inefficiently that they were impractical to use. In most Ada shops, in fact, the use of them was verboten. I suspect, but cannot prove, that this mistake was made because the Ada compiler developers were not real-time programmers, and didn't appreciate the intent of the language designers.

Unless you've been stranded on the Moon, you probably know that the current standard for C++ provides for exceptions, and most compilers now support them. The exception facility in C++ seems to be in quite a state of flux, with compiler vendors struggling to keep up with recent changes to the standards. However, from everything I've heard, they're at least as inefficient in C++ as the early Ada compilers. What's more, because of the way exceptions are defined in C++, there is little or no hope that they will ever be more efficient. Exceptions, as I've already noted, are eliminated from the upcoming embedded subset language.

Because C doesn't support exceptions, C++ implements them poorly, and the embedded subset won't support them at all, I'm afraid the great majority of embedded systems programmers will have to make do with if tests sprinkled throughout the code, as in the listings of this chapter.

The Functions that Time Forgot

The math library functions typically supplied in current programming languages include sine, cosine, arctangent, exponential, and log. But other related functions need to be supported, too. Many FORTRAN compilers, for example, do not support the functions tangent, inverse sine, inverse cosine, or hyperbolic functions. If you need those functions, and the language doesn't supply them, you have no choice but to write them yourself. The routines shown in Listing 3.3 fill the gaps. The safe-function tests are built-in.

The tangent function is defined as

[3.1] $\tan(x) = \sin(x)/\cos(x)$

This one is often left out of libraries for the very reason that it's unbounded; if $x = \pm 90°$, $\tan(x) = \pm\infty$. If you blindly implement Equation [3.1], you'll get a divide-by-zero error for that case. Again, the trick is to test for the condition and return a reasonable value — in this case, a "machine infinity" — that is, a very large number. It should *not* be the maximum floating-point number, for a very good reason: you might be doing further arithmetic with it, which could cause overflow if it's too big. To take care of this, I've added the definition

```
const long double BIG_FLOAT  = 1.0e20;
```

to the constant.cpp file described in Chapter 1. BIG_FLOAT is a big number, all right, but not so big that we're likely to encounter overflow exceptions. In Listing 3.3 for the tangent function, note the call to the jlib function sign().

C and C++ libraries usually do support the inverse sine and cosine functions, but they were unsafe. Because the output of the sine and cosine functions can't exceed the range –1 to 1, I can't allow values outside that range to be passed to the inverse functions. Normally, they wouldn't be, but in many cases floating-point round off can be enough to return a value just slightly out of range. This is the same kind of problem I encountered with the square root, and I can solve it the same way by limiting the argument to a safe range. Safe versions of the arcsine and arccosine are included in Listing 3.2.

Some languages, notably many versions of FORTRAN, don't support the inverse sine and cosine functions directly. If you're programming in assembly language, you'll be writing your own functions anyway. Those of you who don't have access to these functions can always generate your own based on the formulas that follow, beginning with the inverse sine.

[3.2] $\sin^{-1}(x) = \tan^{-1}\left(\dfrac{x}{\sqrt{1-x^2}}\right)$

Again, you can't allow arguments outside the range ±1.0, so the routine tests for that case before calling the square root. Note that calling the safe version, square_root(), described earlier in this chapter, doesn't help here: it would just lead to a divide-by-zero exception.

It's worth pointing out that the accuracy of Equation [3.2] goes to pot when x is near ±1.0. It's tempting to try to fix that, perhaps by some transformation in certain ranges. Forget it. The problem is not with the math, but with the geometry. The sine function has a maximum value of 1.0 at 90°, which means that the function is flat near 90°, which in turn means that the inverse is poorly defined there. There is no cure; it's a consequence of the shape of the sine curve.

A listing of the inverse sine, for those whose programming language doesn't support it, is shown in Listing 3.3. It is inherently safe, because we test the value of x before we do anything. Remember, you don't need this function if your programming language already supports it. You only need the safe version of Listing 3.2.

An equation similar to Equation [3.2] is also available for the inverse cosine.

$$[3.3] \qquad \cos^{-1}(x) = \tan^{-1}\left(\frac{\sqrt{1-x^2}}{x}\right).$$

However, Equation [3.3] is not the one to use. First of all, it returns a value in the range ±90°, which is wrong for the arccosine. Its principal value should be in the range 0° to 180°. Worse yet, the accuracy of the equation goes sour at $x = 0.0$, which is where it really should be the most accurate. A far better approach is also a simpler one: use the trig identity

$$\cos(x) = \sin(90-x),$$

which leads to

$$\cos^{-1}(x) = 90 - \sin^{-1}(x),$$

which is the algorithm implemented in Listing 3.3.

Doing it in Four Quadrants

I'll close this chapter with one of the most useful trig functions ever created: the four-quadrant arctangent. It's so useful precisely because neither the standard atan() nor its companions asin() and acos() return results in all four quadrants. They cannot, because they don't have enough information. Because

$$\sin(x) = \sin(\pi - x)$$

the value of the sine is not enough to determine which quadrant the result should be in. Accordingly, both atan() and asin() return results only in quadrants one and four (−90° to +90°), whereas acos() returns results in quadrants one and two (0° to 180°). Typically, programmers have to use other information [like the sign of the cosine if asin() is called] to decide externally if the returned value should be adjusted.

The four-quadrant arctangent function solves all these problems by taking two arguments, which represent or are at least proportional to, the cosine and the sine of the desired angle (in that order). From the two arguments, it can figure out in which quadrant the result should be placed with no further help from the programmer.

Why is the four-quadrant arctangent so valuable? Simply because there are so many problems in which the resulting angle can come out in any of the four quadrants. Often, in such cases, the math associated with the problem makes it easier to compute the sine and the cosine separately, rather than the tangent. Or, more precisely, it's easy to compute values which are *proportional* to these functions, but not necessarily normalized. We *could* then use an arcsine, say, and place the angle in the right quadrant by examining the sign of the cosine (what a tongue-twister!). But the four-quadrant arctangent function does it so much more cleanly.

This function is a staple of FORTRAN programs, where it's known as atan2(). The name doesn't hold much meaning, except to FORTRAN programmers [you'd think they'd at least call it ATAN4()], but it's so well known by that name that it also appears by the same name in the C library.

The atan2() function takes two arguments and is defined as

$$\texttt{atan2(s, c)} = \tan^{-1}(s/c)$$

with the understanding that the signs of both s and c are examined, not just the sign of the ratio, to make sure the angle comes out in the correct quadrant. (Actually, both atan2 and atan do a lot more than that. To avoid inaccuracies in the result when the argument is small, they use different algorithms in different situations.) Note carefully the order: the s (sine function) value then c. Note also that it is not necessary to normalize s and c to the range −1 to 1; as long as they are proportional, the function will take care of the rest.

The library function atan2() works as advertised if either argument is 0, but not both. To take care of that one exceptional case, you must test for the condition and return a reasonable answer.

What's a reasonable answer? Because the arguments provide no information at all about what the angle should be, you can return any value you choose. Zero seems as good a choice as any.

In Listing 3.3, I've shown two implementations of the function, which is here called arctan2() to avoid name conflicts. The first assumes that a library function atan2() is available. The second is for those unfortunate few who must roll their own. I might mention that I tried a lot of tricky ways to implement this function, making maximum use of the ±90° range of atan(). None of them were nearly as clean as the straightforward approach taken in the final version. There's a lesson to be learned here: KISS (Keep It Simple, Simon).

Listing 3.3 Functions that time forgot.

```
/* The Functions Time Forgot
 *
 * Standard C libraries now include the functions which follow,
 * so they are not needed.  Use them only if your system does not
 * support them, or use them as templates for programming the
 * functions in some other programming languaqe.  Note that the
 * names are the same as those of the C library routines.
 */


// tangent function
double tan(double x){
    double s = sin(x);
    double c = cos(x);
    if(c != 0)
        return s/c;
    return BIG_FLOAT * s;
}

// Inverse sine
double asin(double x){
    if(abs(x) >= 1.0)
```

```
            return sign(halfpi, x);
        return atan(x/sqrt(1.0 - x*x));
    }


// Inverse cosine
double acos(double x)
{
    return(halfpi - arcsin(x));
}
// Four-quadrant arctan (doesn't use atan2)
double arctan2(double y, double x){
    double retval;
    if(x == 0)
        retval = halfpi;
    else
        retval = atan(abs(y/x));
    if(x < 0)
        retval = pi - retval;
    if(y < 0)
        retval = -retval;
    return retval;
}
```

Fundamental Functions

Chapter 4

Square Root

A fundamental function is like an old, comfortable pair of slippers. You never pay much attention to it until it turns up missing. If you're doing some calculations and want a square root or sine function, you simply press the appropriate key on your calculator, and there's the answer. If you write programs in a high-order language like C, C++, Pascal, or Ada, you write sqrt(x) or sin(x), and that's the end of it. You get your answer, and you probably don't much care how you got it. I mean, even a three dollar calculator has a square root key, right? So how hard could it be?

Not hard at all, unless, like your slippers, you suddenly find yourself in a situation where the fundamental functions aren't available. Perhaps you're writing in assembly language for a small microcontroller or you're programming using integer arithmetic to gain speed. Suddenly, the problem begins to loom larger. Because the fundamental functions are ... well ... *fundamental*, they tend to get used quite a few times in a given program. Get them wrong, or write them inefficiently, and performance is going to suffer.

Most people don't make it their life's work to generate fundamental functions — we leave that to the writers of run-time libraries — so it's difficult to know where to turn for guidance when you need it. Most engineering or mathematical handbooks will describe the basic algorithms, but sometimes the path from algorithm to finished code can be a hazardous one.

If you implement the algorithm blindly, without understanding its capabilities and limitations, you could end up in big trouble. I once wrote a real-time program in assembly language for a 10MHz Z-8000. Among many other calculations, it had to compute a sine, cosine, square root, and arctangent, all within 500 microseconds. Believe me, getting those functions right was important!

The whole thrust of this book is the reduction of math theory to practical, useful software suitable for embedded applications. In this chapter, I will begin to look at the fundamental functions, beginning with the square root. I'll look at the fundamental basis for its computation, assuming that you're programming in C++ and using floating-point arithmetic. I'll examine the practical problems associated with the implementation, and see what you have to do to reduce the algorithm to practice in useful software. Then I'll show you how to optimize the algorithm for maximum performance. Finally, for those concerned with speed, I'll show you how to implement the same algorithm in integer arithmetic. Subsequent chapters will address other fundamental functions.

I'll begin with what surely must be the most fundamental of all the so-called fundamental functions: the square root. At least once a month, someone will post a message on CompuServe's Science and Math forum asking how to compute a square root. The answer is always the same: Newton's method. You'll find the formula in any handbook.

$$[4.1] \qquad x = \frac{1}{2}\left(\frac{a}{x} + x\right)$$

where a is the input argument for which you want the square root, and x is the desired result. The equation is to be applied repetitively, beginning with some initial guess, x_0, for the root. At each step, you use this value to calculate an improved estimate of x. At some point, you decide the solution is good enough, and the process terminates.

Note that the formula involves the average of two numbers: a guess, x, and the value that you get by dividing x into the argument. (Hint: don't start with a guess of zero!) If you have trouble remembering the formula, never fear; I'm going to show you how to derive it. You can take this lesson, one of the best lessons I learned from my training in physics, to the bank:

It's better to know how to derive the formula you need than to try to memorize it.

To derive Newton's method for square root, I will begin with an initial estimate, x_0. I know that this is not the true root, but I also know, or at least hope, that it's close. I can write the true root as the sum of the initial estimate plus an error term.

[4.2] $\qquad x = x_0 + d$

This is, you'll recall, the true root, so its square should equal our input, a.

$$a = x^2$$

From Equation [4.2], this is

$$a = (x_0 + d)^2$$

or

[4.3] $\qquad a = x_0^2 + 2x_0 d + d^2.$

Don't forget that the initial estimate is supposed to be close to the final root. This means that the error d is hopefully small. If this is true, its square is smaller yet and can be safely ignored. Equation [4.3] then becomes

$$a \approx x_0^2 + 2x_0 d,$$

which I can now solve for d, or more precisely, an estimate of it.

[4.4] $\qquad d \approx \dfrac{a - x_0^2}{2x_0}$

Now that I have d, I can plug it into Equation [4.4] to get an estimate of x.

$$x \approx x_0 + \dfrac{a - x_0^2}{2x_0}$$

Simplifying gives

$$x \approx \dfrac{2x_0^2 + a - x_0^2}{2x_0}$$

or

[4.5] $\qquad x \approx \dfrac{x_0^2 + a}{2x_0}.$

Dividing both numerator and denominator by x_0 gives

$$x \approx \frac{1}{2}\left(\frac{a}{x_0} + x_0\right).$$

This formula is equivalent to Equation [4.1]. The squiggly equals sign shows that this is an *approximation* to x. It is not exact, because I left out the d^2 term in the derivation. This, in turn, implies that I must keep applying the formula until the function *converges* on a satisfactory solution. This process is called *iteration*.

The previous statement seems like a simple one, but its implications can be vast and opens up lots of interesting questions, like

- How many times should I iterate?
- How do I know when to stop?
- How do I know the process will converge?
- How long before I reach the true root?

Answering the last question first, the theoretical answer is, forever. Each iteration gives us, at best, only an *approximation* to the root. Like the person in Zeno's Paradox, who moves half the distance to the goal at each step, I'll never actually quite get there. In practice, however, I can get arbitrarily close. In computer arithmetic, Newton's method will eventually get within the precision of the computer word, which means that no further improvement is possible. I will represent the root to its closest possible value, which is all I can ever hope for, and the process will converge on the true root, for all practical purposes.

To demonstrate Newton's method, I'll compute the square root of two, and I'll start with the rather horrible first guess of four. Table 4.1 gives the results.

Table 4.1   Newton's method: an example.

Trial	Guess	Square	New Guess	Error
1	4	16	2.250000000	0.835786438
2	2.250000000	5.062500000	1.569444444	0.155230882
3	1.569444444	2.463155863	1.421890364	0.007676802
4	1.421890363	2.021772207	1.414234286	0.000020724

Trial	Guess	Square	New Guess	Error
5	1.414234285	2.000058616	1.414213563	0.000000001
6	1.414213563	2.000000002	1.414213562	0.000000000
7	1.414213562	1.999999999	1.414213563	0.000000001
8	1.414213563	2.000000002	1.414213562	0.000000000
9	1.414213562	1.999999999	etc.	

This first example has some valuable lessons. First, you can see that convergence is rapid, even with a bad initial guess. In the vicinity of the root, convergence is quadratic, which means that you double the number of correct bits at each iteration (note how the number of zeros in the error doubles in steps 3 through 6). That seems to imply that even for a 32-bit result, you should need a maximum of six iterations to converge. Not bad. Also, note that you "sneak up" on the root from above — the guess is always a little larger than the actual root, at least until you get to step 7. I'll leave it to you to prove that when starting with a small initial guess the first iteration is high and you'd still approach the root asymptotically from above.

Perhaps most important, at least in this example, is that the process never converges. As you can see, an oscillation starts at step 7 because of round-off error in my calculator. When using Borland C++, it doesn't occur.

I've often seen the advice to use Equation [4.1] and iterate until the root stops changing. As you can see, this is bad advice. It doesn't always work. Try it with your system and use it if it works. But be aware that it's a dangerous practice.

I said that the convergence is quadratic as long as you're in the vicinity of the root. But what, exactly, does that mean, and what are the implications? To see that, I'll take a truly awful first guess (Table 4.2).

Table 4.2   Result of a poor initial guess.

Trial	Guess
1	1.000000 e30
2	5.000000 e29
3	2.500000 e29
4	1.250000 e29
5	6.250000 e28

Trial	Guess
6	3.125000 e28
7	1.562500 e28
8	etc.

I won't belabor the point by completing this table for all of 105 steps it takes to converge on the answer. I think you can see what's happening. Far from doubling the number of good bits at each step, all you're doing is halving the guess each time. That's exactly what you should expect, given Equation [4.5]. Because the value of a is completely negligible compared to the square of x, the formula in this case reduces to

$$[4.6] \qquad x \approx \frac{x_0}{2}.$$

At this rate you can expect to use roughly three iterations just to reduce the exponent by one decimal digit. It's only when the initial guess is close to the root that you see the quadratic convergence. A method that takes 105 iterations is not going to be very welcome in a real-time system.

You may think that this is an artificial example; that it's dumb to start with such a large initial guess. But what if you start with 1.0 and the *input* value is 1.0e30? From Equation [4.5], the next guess will be 5.0e29, and you're back in the soup.

It is amazing that such a seemingly innocent formula can lead to so many problems. From the examples thus far, you can see that

- you need a criterion for knowing when to stop the iteration; testing for two equal guesses is not guaranteed to work,
- to get the fast, quadratic convergence, you need a good initial guess, and
- the intermediate values (after the first step) are always higher than the actual root.

In the remainder of this chapter, I'll address these three points.

The Convergence Criterion

One way to decide when to quit is to look at the error. Instead of solving directly for x using Equation [4.5], you can solve for d using Equation [4.4] then compute x from Equation [4.3]. You can compute the relative error (RE),

$$[4.7] \qquad RE = \frac{d}{x},$$

and quit when it's less than some value. That's the safest way, but it's also rather extravagant in CPU time, and that can be an issue. In fact, the fastest and simplest, way of deciding when to stop is often simply to take a fixed number of iterations. This eliminates all the tests, and it's often faster to perform a few extra multiplies than to do a lot of testing and branching. A method that iterates a fixed number of times is also more acceptable in a real-time system.

The best way to limit the number of iterations is simply by starting with a good initial guess.

The Initial Guess

Arriving at an initial guess is tough because the range of floating-point numbers is so wide. You've already seen what happens if the guess is orders of magnitude off, but that example also provides a clue towards how to improve your guess. When the input number was 1.0e30, the answer was 1.0e15: the same number with half the exponent. This will be true whenever the input number is 1.0 times any even power of 10 (or two, in binary). Even if the mantissa isn't 1.0 or the exponent isn't even, halving the exponent still gets you a lot closer to the true root — close enough, in fact, for quadratic convergence to work.

Playing around with the exponent and mantissa separately means hacking the floating-point word, but that's the kind of thing you can do in assembly language. In fact, you can do it in most languages, although the algorithm will not be portable, being dependent on the internal floating-point format.

Floating-Point Number Formats

Before I go any further, I should explain some features of floating-point number formats. The purpose of using floating-point numbers is basically to give a wider dynamic range. A 16-bit integer, for example, can hold nonzero (positive) numbers from 1 through 32,767, a range of 2^{15}, or, if you prefer, 90dB. A 32-bit integer can hold numbers up to 2^{31}, giving a dynamic range of 186dB. On the other hand, the same 32 bits used to encode a floating-point number gives a dynamic range from roughly 10^{-38} to 10^{+38}, which is a factor of 10^{76} or a whopping 1,520dB.

Of course, there's no such thing as a free lunch, so to gain this enormous dynamic range, you must give something up, and that something is bits of precision in the number. The floating-point number gets its dynamic range by splitting the number into two parts: a fraction, called the *mantissa*, and an *exponent*, which is usually, but not always, a power of two. You can think of the exponent as defining the general range of the number (large or small), whereas the mantissa is the value within that range.

The integer part of a logarithm defines the size of a number as a power of 10. The fractional part defines the value of the number from $1/10$ to 1. The typical floating-point number retains the idea of storing the integral part of the log to define a power of two (not 10), but it stores the fractional part as a true fraction, not as its logarithm. This fractional part is called the *mantissa*. The mantissa always has an implied decimal point to its far left, so it's a true fraction.

Allocating the bits of the word to the various parts involves a trade-off. The more bits set aside for the exponent, the greater the dynamic range but the fewer bits remaining to store the mantissa and, therefore, the lower the precision. You can estimate the number of decimal digits of accuracy by counting the number of bits in the mantissa; because eight is roughly the same size as 10, figure three bits per digit of accuracy. When you store an integer as a 32-bit number, you can get nearly 10 digits of accuracy. In a 32-bit floating-point number, however, the accuracy is barely seven digits. To maintain the greatest possible accuracy, floating-point numbers are *normalized*, shifting the mantissa as far left as possible to include as many bits after the decimal point as possible. This assures that we have the greatest number of significant bits. In most cases (but not all) the mantissa is shifted until its leftmost bit is a 1. This fact can be used to gain a bit of precision: if you know that the leftmost bit is always 1, you don't need to store it. You can simply imagine it to be there and only store the bits one level down. This concept is called the *phantom bit*.

Note that the mantissa can never be greater than (or equal to) 1.0 if the decimal point is to the far left. Furthermore, if the high bit is always 1, it can never be less than 0.5. Thus, normalization forces the mantissa to the range to be between 0.5 and 1.0.

You need to know one more fact about the exponent field. Although I've seen some home brew implementations use a two's complement signed integer for the exponent, most practical implementations reserve the high bit (the sign bit) for the sign of the entire number, and they use an offset to keep the exponent field positive. For example, if I use seven bits (just under one byte) for the exponent, this number can range from –64 to +63 or, in hex, -40 to 3f. To keep the exponent positive, I add 0x40 to all exponents, so they now cover the range 0 to 7f. This concept is called the *split on nn* convention, and as far as I know, almost every industrial-strength format uses it.

A picture is worth a thousand words. Let's see how the method works with a hypothetical floating-point format. Assume seven bits of exponent, sixteen bits of mantissa, and no phantom bit. I'll write the number in the order: sign, exponent, mantissa (which is the usual order). I'll split the exponent on 0x40, or 64 decimal. Numbers in the format might appear as in Table 4.3.

Table 4.3   Floating-point representation.

Value	Sign	Exponent	Mantissa	Number in Hex
1.0	0	41	8000	41.8000
2.0	0	42	8000	42.8000
3.0	0	42	c000	42.c000
4.0	0	43	8000	43.8000
5.0	0	43	a000	43.a000
8.0	0	44	8000	44.8000
10.0	0	44	a000	44a000
16.0	0	45	8000	45.8000
0.5	0	40	8000	40.8000
0.25	0	3f	8000	3f.8000
0.333333333	0	3f	5555	3f.5555
π	0	42	c90f	42.c90f
2.71e–20		00	8000	00.8000
9.22e18		7f	ffff	7f.ffff
–1.0	1	41	8000	c1.8000

Note that the integers less than 65,536 can be represented exactly because the mantissa is big enough to hold their left-shifted value. Note also that the decimal point is assumed to be just left of the mantissa; I've shown it in the hex number column for emphasis. Finally, note that the high bit of the mantissa is always set, as advertised. For this reason, powers of two always have a mantissa of 0x8000 — only the exponent changes. This mantissa is equivalent to 0.5 decimal, which explains why a 1.0 needs one power of two in the exponent.

The last two rows express the smallest and largest number that can be represented in this format. You'll notice that the range is asymmetric; you can go farther in the small direction than in the large. To counter this, some designers, including Intel, use an offset that has more headroom on the high side.

I've only shown one negative number. As you can see, all you do is set the high bit for such numbers.

Back to Your Roots

You may be wondering why I chose this particular point to digress and talk about how floating-point numbers are encoded. The reason is simple: to take the next step in the process of defining a fast-converging square root algorithm, I must understand and be able to hack into the floating-point word.

There are almost as many floating-point formats as there are compilers; each vendor seemed to use its own until IEEE stepped in and defined the most complicated format humanly possible. Most compiler vendors and CPU designers try to adhere to the IEEE standard, but most also take liberties with it because the full IEEE standard includes many requirements that affect system performance.

Programs based on Intel 80x86 architecture and its math coprocessor allow for three flavors of floating-point notation, as shown in Table 4.4.

Table 4.4   The Intel formats.

Name	Exponent Bits	Mantissa Bits	Phantom Bit?	Total Bits
float	7	24–25 [a]	sometimes (!)	32
double	10	54	yes	64
long double	15	64	no	80

a. The float format uses an exponent as a power of four, not two, so the high bit may not always be 1.

Based on the size of the two fields, I can estimate both the dynamic range and precision of the number. Rough estimates are shown in Table 4.5.

Table 4.5   Range and precision of Intel formats.

Name	Range	Accuracy (decimal digits)
float	$10^{\pm 38}$	6–7
double	$10^{\pm 308}$	16
long double	$10^{\pm 4,032}$	19

The dynamic range of these numbers, even the simple float format, is hard to get your arms around. To put it into perspective, there are only about 10^{40} atomic particles in the entire Earth, which would almost fit into even the ordinary float format. One can only speculate what situation would need the range of a long double, but in the real world, you can be pretty sure that you need never worry about exponent overflow if you use this format.

The less said about the float format the better. It uses a time-dishonored technique first made popular in the less-than-perfect IBM 360 series and stores the exponent as a power of something other than two (the Intel format uses powers of four). This allows more dynamic range, but at the cost of a sometimes-there, sometimes-not bit of accuracy.

Think about it. If you have to normalize the mantissa, you must adjust the exponent as well by adding or subtracting 1 to keep the number the same. Conversely, because you can only add integral values to the exponent, you must shift the mantissa by the same factor. This means that the IBM format (which used powers of 16 as the exponent) required shifts of four bits at a time. Similarly, the Intel format requires shifts by two bits. The end result: you may have as many as one (Intel) or three (IBM) 0 bits at the top of the word. This means you can only count on the accuracy of the worst case situation. In the case of floats on an Intel processor, don't count on much more than six good digits.

One other point: Intel uses a split on the exponent that's two bits smaller than usual, to make the range for large and small numbers more symmetrical. They also use the phantom bit. Thus, in their double format, 1.0 is stored as

```
3ff.(1)00000 00000000.
```

Armed with an explanation of floating-point formats, I'm now in a position to hack the number to separate exponent and mantissa, which allows me to take the square root of the mantissa without worrying about the exponent. Because this leaves me with a mantissa in the range of 0.5 to 1.0, I can expect rapid convergence.

Listing 4.1 shows just such a hack in Borland C++, with the math coprocessor enabled. This code works for the type `double`. I won't address the other two formats, but `long doubles` are easier because they have no phantom bit.

The program of Listing 4.1 takes the input value as the initial guess and halves the exponent (note the little twist to get the exponent to round up). This puts the argument passed to the root finder always in the range 0.5 to 1.0.

Listing 4.1   Hacking the floating-point format.

```
typedef union{
    double fp;
    struct{
        unsigned long lo;
        unsigned long hi;
    };
} hack_structure;


hack_structure x;


void main(void){
    short expo;
    unsigned short sign;

    while(1){
        // get an fp number
        cin >> x.fp;

        // grab the sign and make it positive
        sign = ((long)x.hi < 0);
        x.hi &= 0x7fffffff;
```

Listing 4.1   Hacking the floating-point format. (continued)

```cpp
        // grab the exponent
        expo = (short)(x.hi >> 20);
        expo -= 0x3fe;

        // normalize the number
        x.hi &= 0x000fffff;
        x.hi += 0x3fe00000;

        // get square root of normalized number
        x.fp = sqrt(x.fp);     // using sqrt for now

        // force the exponent to be even
        if(expo & 1)
            x.fp *= factor;
            ++expo;
        }

        // halve the exponent
        if(expo < 0)
            expo = expo/2;
        else
            expo = (expo + 1)/2;
                        // note the round upwards here

        // put it back
        expo += 0x3fe;
        x.hi &= 0x000fffff;
        x.hi += ((unsigned long)expo << 20);

        // show the new number
        cout << "Sqrt is: " << x.fp << endl;
    }
}
```

As it stands, Listing 4.1 may be the most inefficient square root finder ever conceived, because it does all that work and still ends up calling the library function sqrt() in the end. I hope you understand that none of this work is necessary if you're programming in a language that already has the library function. The code shown is intended to be a guide for those of you who are programming in assembly language, or may otherwise, someday, have to write your own square root function.

To actually find the root I need another function, shown in Listing 4.2. Substitute the name of this function at the call to sqrt() in Listing 4.1. Later in this chapter, I'll discuss how to decide what the iteration loop count n should be.

Listing 4.2 Finding the root by iteration.

```
double my_sqrt(double a, int n){
    double x = a;
    if(x <= 0.0)
        return x;
    for(int i=0; i<n; i++){
        cout << "The guess is " << x << endl;
        x = 0.5 * (a/x + x);
    }
    return x;
}
```

A little experimenting with this code should convince you that I always get convergence in, at most, five iterations for all values of input. Whenever the input is an even power of two, convergence is immediate. Not bad at all for a quick and dirty hack. If you need a floating-point square root in a hurry, here it is.

Take special note that things get a little interesting when I put the exponent and mantissa back together if the original exponent is odd. For the even exponents, I simply divide by two. If the exponent is odd, I must multiply by the factor $\sqrt{2}$ to allow for "half a shift." Some folks don't like this extra complication. They prefer to use only even exponents. You can do this, but you pay a price in a wider range for the mantissa, which can then vary by a factor of four, from 0.5 to 2.0. This leads to one or two extra steps of iteration. Personally, I prefer the method shown because it gives me fewer iterations, but feel free to experiment for yourself.

Improving the Guess

In the preceding quick hack, I just left the mantissa alone, using it for the first guess. I could have also set it to some constant value, like 0.5. Question: Can I do better by choosing an optimal value of the mantissa? Answer: Yes, by quite a bit.

You've already seen that this method can handle a wide range of exponents. I'll set that part aside for the moment and assume that the input number will always be in the range 0.5 to 1.0 (the mantissa of a floating-point number). Figure 4.1 shows a graph of sqrt(x), with the range of interest highlighted. As you can see, the solution varies from 0.70711 to 1.0. Obviously, a good initial guess would be something in between.

Figure 4.1   Square root.

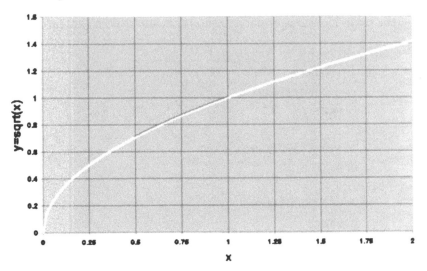

It turns out that the optimal first guess is the *geometric* mean of the two extreme values, because this value gives a relative error that's the same at both ends of the range. For this range, the value is

[4.8] $x_0 = \dfrac{1}{\sqrt{\sqrt{2}}} = 0.840896.$

Applying the iterator of Listing 4.2 with this starting value and an input of $1/_2$ (one extreme of the input range) gives the sequence shown in Table 4.6, which reaches convergence in four iterations in the worst case.

Table 4.6   **Constant initial guess.**

Trial	Guess	Error
1	7.1774998634e–01	1.0643205154e–02
2	7.0718569294e–01	7.8911751189e–05
3	7.0710678559e–01	4.4028638513e–09
4	7.0710678119e–01	0.0000000000e+00

Can I improve on this? Again, the answer is yes. I've seen a number of ways to do this. The most obvious is to use a table of starting values based on the value of the mantissa. Using a full table lookup would take up entirely too much space, but you can use the leading few bits of the mantissa to define a short table lookup. I won't go into detail on the method here, because I think I have a much better approach. However, the general idea of the table lookup approach is shown in the code fragment below.

```
switch((x.hi >> 14) & 3){
case 0:
    r = 0.747674391;
    break;
case 1:
    r = 0.827437730;
    break;
case 2:
    r = 0.900051436;
    break;
case 3:
    r = 0.967168210;
    break;
}
```

This code splits the range into four segments based on the first two bits in the range, then it uses the optimum starting value for that range.

The Best Guess

I said there was a better approach. The key to it can be found in Figure 4.1. For the initial guess, what you're really looking for is an approximation $r(x)$ to the curve

[4.9] $f(x) = \sqrt{x}$.

Using a constant value of 0.8409 is equivalent to assuming a straight horizontal line. That's a pretty crude approximation, after all. Surely I can do better than that.

Using a table of values, as suggested above, is equivalent to approximating the function by the staircase function shown in Figure 4.2. But even better, why not just draw a straight line that approximates the function? As you can see, the actual curve is fairly straight over the region of interest, so a straight line approximation is really a good fit.

Figure 4.2   Initial guesses.

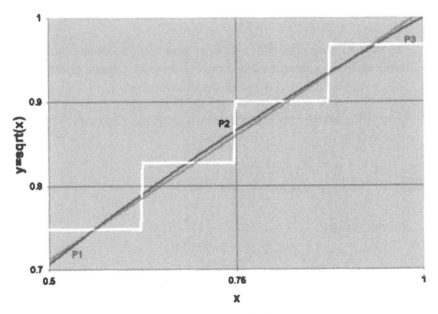

The general equation of a straight line is

[4.10] $r(x) = A + Bx$.

All I have to do is find the best values for the constants A and B. The derivation is much too tedious to give in all its glory here, but the concept is simple enough, so I'll outline it for you.

First, note that you don't want the straight line to actually touch the true function at the end points. That would make the error one-sided. Instead, try to locate the points so that they're higher than the curve at those extremes. Done properly, this makes the straight line drop below the curve in the middle, which tends to even out the error, plus and minus. To optimize the placement of the points, I'll require that the *relative* errors at the three extremes are equal in magnitude. I define this error to be

[4.11] $E(x) = \dfrac{f(x) - r(x)}{f(x)}$

and insist that $E(0.5) = E(1.0)$. Doing this gives a relation between A and B that turns out to be

[4.12] $B = \sqrt{2}A$.

The function then becomes

[4.13] $r(x) = A(1 + \sqrt{2}x)$.

To find A, note that there is some point P_2 where the curve rises above the straight line to a maximum. I don't know where this point is, but a little calculus reveals that the error is maximum at $1/\sqrt{2} = 0.70711$. This number is my old friend, the geometric mean. Forcing the error at this point to be the same (but in the opposite direction) as at the extremes, I finally arrive at the optimal value

[4.14] $A = \dfrac{1}{\sqrt{\sqrt{2}} + \dfrac{1}{2} + \dfrac{1}{\sqrt{2}}}$.

Numerically,

        A = 0.41731924
        and

[4.15]   B = 0.59017853.

Sure enough, the line

[4.16] $r(x) = 0.41731924 + 0.59027853x$

matches the sqrt() curve nicely with a relative error less than 1.0075 at the three extreme points. With Equation [4.16], the initial guess will never be off by more than 0.8 percent. Using this approach, I get convergence as shown in Table 4.7.

Table 4.7   **Optimal initial guess**.

Trial	Guess (r)	Error
1	7.0712650885e–01	1.9727663130e–05
2	7.0710678146e–01	2.7557689464e–10
3	7.0710678119e–01	0.0000000000e+00

This is a remarkable result, because by using this optimal guess, I get convergence in a *maximum* of three iterations. For 32-bit floating-point accuracy (equivalent to C float, roughly six digits), convergence is good enough in only one iteration. That's a result pretty hard to improve upon.

It's worth mentioning that any general-purpose square root function should be good to full machine accuracy, but there are times when speed is more important than accuracy (otherwise, you could simply use the C function). It's important to keep the accuracy issue in focus. The long string of zeros above is nice to see, but not always necessary. For a proper perspective, look at the accuracy expressed as a percent error in Table 4.8.

Table 4.8   **Accuracy using optimal initial guess**.

Iterations	Percent Error
0	0.75
1	0.003
2	0.00000004

When expressed as a percentage, the error after two iterations is truly small... roughly corresponding to measuring the distance from New York to Los Angeles with an error equal to the diameter of a human hair. For many cases (e.g., pixel graphics) even the error after one iteration is perfectly acceptable. For that matter, an error of less than 1 percent, which you get after *zero* iterations, is enough in some cases (within a control loop, for example, where the feedback keeps errors from growing). In the Z-8000 real-time program I mentioned earlier in this chapter, where I had to compute a lot of functions in a hurry, I needed only one iteration with the square root function by using the solution from the previous cycle as a starting value. That approach proved to be more than adequate for the problem.

Putting it Together

All that remains is to put the concepts together into a callable function, with proper performance over the exponent range. This is shown in Listing 4.3. Note that I've added the usual error check for a negative argument. I also have to check for zero as a special case. Finally, because I only need, at most, three iterations, I've taken out the testing loop and simply executed three steps inline.

Listing 4.3   Putting it together.

```
// Model for fast square root with optimal first guess

double my_sqrt(double a){
    short expo;
    double factor = root_2/2;
    const double A =  0.417319242;
    const double B = 0.590178532;
    hack_structure x;
    double root;

    // check for negative or zero
    if(a <= 0)
        return 0;
    x.fp = a;

    // grab the exponent
    expo = (short)(x.n.hi >> 20);
    expo -= 0x3fe;

    // normalize the number
    x.n.hi &= 0x000fffff;
    x.n.hi += 0x3fe00000;

    // get square root of normalized number
    // generate first guess
    root = A + B * x.fp;

    // iterate three times (probably overkill)
```

Listing 4.3   Putting it together. (continued)

```
        root = 0.5 * (x.fp/root + root);
        root = 0.5 * (x.fp/root + root);
        root = 0.5 * (x.fp/root + root);

        // now rebuild the result
        x.fp = root;

        // force the exponent to be even
        if(expo & 1){
            x.fp *= factor;
            ++expo;
        }

        // halve the exponent
        if(expo < 0)
            expo = expo/2;
        else
            expo = (expo + 1)/2;

        // put it back
        expo += 0x3fe;
        x.n.hi &= 0x000fffff;
        x.n.hi += ((unsigned long)expo << 20);

        return x.fp;
    }
```

I think you will find that Listing 4.3 handles all input values with accuracy and speed. This routine can serve as a template for a very fast assembly language function. If you need more speed and can tolerate less accuracy, leave out one, two, or even all three of the iterations.

Integer Square Roots

The algorithm given above for the square root of a floating-point number is hard to beat. But there are times when all you need or want is integer arithmetic. In many embedded systems, there is no floating-point processor, so floating-point arithmetic must be done via software emulation. For this and other reasons, it's often necessary to deal with the square root of an integer.

Can I apply Newton's method with equal success to the case of integers? Surprisingly, the answer is no, not quite. Using integer arithmetic naturally leads to the fast integer operations built into most CPUs, so in most cases doing things with integers will indeed speed things up. But there are some offsetting factors that make the integer algorithm a little more delicate.

There are two problems. The Newton formula translates easily enough. The integer version of Equation [4.1] still works. Unfortunately, it's less stable, because of the way the integer divide operation truncates the result. To convince yourself of this, try to find the square root of 65,535. You'll discover that the result oscillates between the two values 255 and 256.

One of these is just plain wrong. Unlike the floating-point case, I expect integer arithmetic to be *exact*. A reasonable definition of the square root of an integer x is the largest value whose square does not *exceed* the input value. By this definition, there is only one acceptable solution: 255. But in this case, applying Equation [4.1] using $x = 255$ drives me away from the solution.

This is a terrible result. In the floating-point case, I was always assured that each successive iteration drove me inexorably closer to the root, at least until the result got within the limits of machine precision. But that's the point. In integer arithmetic, the programmer expects every bit to count. If the algorithm sometimes forces the result in the wrong direction, the programmer must take measures to see that it's stopped at the right time. One ramification of this is that I can't just use a fixed number of iterations as I did in the floating-point square root calculation.

An algorithm that can be assured to work is shown in the code fragment below.

```
do{
    temp = a / x;
    e = (x - temp) / 2;
    x = (x + temp) / 2;
}
while(e != 0);
```

This is an iterative algorithm, and it will result in a variable number of iterations. Note that I've used the error, as defined by Equation [4.4], as a stop criterion, but that it does not really enter into the computation for x. This seems like a case of redundant code, and so it is. In fact, to get a reliable stop criterion, I've just about doubled the amount of work the CPU must do per iteration. What's more, there are many input values for which the algorithm takes an unnecessary extra iteration step.

I'd like to be able to tighten this algorithm up a bit, but every attempt that I've made to do so has failed. You're welcome to try, and I would welcome any improvement that's suggested, but my advice is simple: this algorithm is fragile; DON'T MESS WITH IT!

Another problem with the integer algorithm is that it has to cover an incredible range of numbers. In the floating-point case, I could limit the attention to the highlighted area of Figure 4.1, and optimize the initial guess for that area. For integers, though, the range is much broader, which generally leads to more iterations before convergence.

The Initial Guess

Even with this gloomy prospect, I can still improve the optimal guess using Equation [4.10]. I won't go through the derivation for the optimal coefficients, but it parallels that for the floating-point case. Given a range of input values from m to n, the coefficients of the straight line are found to be:

$$A = \frac{\sqrt{mn}}{\sqrt{\sqrt{mn}} + \frac{1}{2}(\sqrt{m} + \sqrt{n})}$$

and

[4.17] $$B = \frac{A}{\sqrt{mn}}.$$

That result looks pretty messy, but it's not quite as bad as it seems. Remember that the limits m and n are often powers of two — in fact, *even* powers of two. For example, m is often unity. For such cases, the formulas for the range of a 16-bit unsigned integer simplify to

$$A = \frac{512}{289}$$

and

[4.18] $$B = \frac{A}{256}.$$

In any case, they're just numbers after all is said and done. Who cares how nice the equation looks? In Table 4.9, I've shown the numerical values for some cases of special interest.

Table 4.9   Coefficients for optimal guess.

m	n	A	B
1	255	1.27949892	0.08012533
1	32,767	1.7328374	0.00957281074
1	65,535	1.7716255	0.00692046492
1	$2^{31} - 1$	1.98154732	4.2760179e–5
256	65,535	20.47996875	0.00500003052
256	$2^{31} - 1$	30.84313598	4.15981114e–5
65,536	$2^{31} - 1$	443.60684344	3.73932609e–5
2^{31}	$2^{32} - 1$	27,348.6708	9.00515849e–6

As a matter of interest, if you scale the last pair of coefficients by the appropriate power of two, you'll find that they're the same as for the floating-point case. Not surprising, since they correspond to the same range.

Approximating the Coefficients

There's just one problem with the coefficients in Table 4.9: they're clearly not integers! What good does it do to have an integer square root algorithm if you need a floating-point algorithm to start the process? The next challenge is to approximate the floating-point coefficients A and B using only integer arithmetic.

There are standard ways to convert floating-point numbers to rational fractions. However, it would be a diversion to focus on them now. I've made the conversions for some cases. In this particular case, I'm helped by the fact that I don't need super-accurate values. For example, if I let A be 444 in the seventh entry of Table 4.9, I can guess that the speed of convergence won't change much. Likewise for B, I can take the nearest reciprocal to get a divisor. An acceptable set of coefficients is shown in Table 4.10.

Table 4.10   Rational coefficients.

m	n	A	B
1	255	1	1/12
1	32,767	2	1/104
1	$2^{31} - 1$	2	1/23,386
256	65,535	21	1/200
65,536	$2^{31} - 1$	444	1/26,743

These numbers don't look bad at all. Figure 4.3 shows an example of the fit. It's not as elegant as the floating-point case, but it's not bad. Because the integer case must cover such a wide range, the curve of the square root function is not nearly as linear as for the floating-point case, and you can expect the straight line fit to be considerably less accurate. Even so, the largest relative error is less than 40 percent, which means the function should converge in three or four iterations.

Figure 4.3   Integer square root.

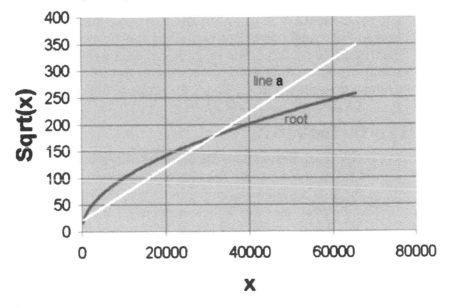

As a final tweak, you can try to limit the range a bit by looking at the values of the numbers. For example, if the high half of a double word is 0, I know that the integer can't be larger than 65,535, so I choose the coefficients for that range. A code fragment describing the process is shown below.

```
if ((a & 0xffff0000) != 0)
    x = 444 + a / 26743;
else if ((a & 0xff00) != 0)
    x = 21 + a / 200;
else
    x = 1 + a / 12;
```

Despite all my carping about the integer algorithm, the final results are pretty darned good. Within the range 1 to 256, convergence always occurs in, at most, two iterations, despite the sloppier fit. Over the entire range, the worst case takes five iterations. I think that's about the best you can hope for with integer arithmetic.

Tweaking the Floating-Point Algorithm

Ironically, perhaps the best place to use integer arithmetic is within the floating-point algorithm. Remember, I already handle the exponent myself, and the mantissa is always normalized. This means that I can still limit the range to the single power of two that's shown in Figure 4.1 if I treat the mantissa as an integer. For that case, the integer algorithm converges in two iterations. A final tweak to the routine in Listing 4.3 is to apply integer arithmetic to the mantissa. That's the ultimate floating-point algorithm.

Good-Bye to Newton

I've pretty much beaten poor Dr. Newton to death, and you now have two algorithms that work well. The integer algorithm is perhaps not as satisfying as the floating-point algorithm, but it gets the job done.

Which do you use? The answer depends on the application and the CPU hardware. If your system has a math coprocessor chip, my advice is to forget the integer algorithm and use floating-point math. Even if you're using integer arithmetic for everything else, it might be worthwhile to convert to floating point and back again.

On the other hand, if your CPU uses software floating point, the integer algorithm is definitely for you.

I've still not exhausted the subject of square roots, because there are other methods in addition to Newton's. I'll wrap up this chapter by taking a look at some of these other methods.

Successive Approximation

The Newtonian methods, especially using the optimal initial guess, are rather sophisticated as square root algorithms go. Now I'll take a trip from the sublime to the ridiculous and ask what is the *dumbest* algorithm I can cook up.

Why would I want a dumb algorithm? Because I might be using a dumb CPU. Newton's method is fine if the CPU has built-in operations for fast multiply and divide, but this book is about embedded systems programming, and many embedded systems still use simple eight-bit controllers that are not blessed with such luxuries. For such processors, I may not be interested in raw speed so much as simplicity.

For the case of integer square roots, I'll be specific about what I want: I want the largest integer x whose square is less than a. I can find such an integer with an exhaustive search by starting with $x = 1$, incrementing x until it's too big, then backing off by one. Listing 4.4 shows the simplest square root finder you will ever see. Note carefully the way the loop test is written. The <= sign is used, rather than < to guarantee that the program always exits the loop with x too large. Then I decrement x by one on the way out.

Listing 4.4   The simplest square root.

```
unsigned long sqrt(unsigned long a){
    unsigned long x = 1;
    while(x*x <= a)
        ++x;
    return --x;
}
```

The successive approximation method provides the ultimate in simplicity. Basically, it's no more than trial and error. Simply try every possible integer until you find the one that fits. The code in Listing 4.4 does just that. It's certainly neither pretty nor fast, but it sure gets the answer, and without any convergence concerns. It also does it in four lines of executable code, which surely must make it the smallest square root finder I can write.

I'll leave aside for a moment the obvious problem that this algorithm is going to use a lot of iterations, in fact x of them. A more fundamental problem is that the algorithm uses a multiply operator, which I had hoped to avoid. But this can be eliminated easily by the technique known as strength reduction.

Eliminating the Multiplication

Before I look at the math behind strength reduction, an example may make the idea more obvious and more credible. Consider the list of squares shown in Table 4.11.

Table 4.11   Integer squares.

n	n^2	Difference
0	0	
1	1	1
2	4	3
3	9	5
4	16	7
5	25	9
6	36	11

The key thing to notice is the last column, which is the difference between successive squares. Note that this difference is simply a progression of all odd integers. What's more, this difference is closely related to the root. Add one to the difference, divide by two, and you have the integer. The next equation will make the relationship obvious:

$$[4.19] \qquad (n+1)^2 = n^2 + 2n + 1 .$$

Suppose that I've already calculated both n and n^2 in a previous step. Then Equation [4.19] tells me that I can calculate the next square by adding one number, namely, the odd integer given by the term $2n+1$.

I can make one other improvement to the algorithm of Listing 4.4. That listing includes the test:

```
x*x <= a
```

Evaluating the test implies a subtraction. If the test is passed, I subtract again. To save time, I can simply subtract the numbers as I go and quit when the difference goes negative. The algorithm is shown in Listing 4.6, which provides an integer square root algorithm in only six lines of code. It uses only addition and subtraction, and thus can be used on the dumbest of CPUs. There are also no convergence worries. If you're creating an embedded system with a single-chip processor and don't have too many time worries, this is your algorithm. PIC programmers take note.

Listing 4.5   A more efficient version.

```
unsigned long sqrt(unsigned long a){
    unsigned long x = 1;
    unisgned long xsqr = 1;
    unsigned long delta = 3;
    while(xsqr <=a){
        ++x;
        xsqr += delta;
        delta += 2;
    }
    return --x;
}
```

Listing 4.6 The binary square root.

```
unsigned short sqrt(unsigned long a){
    unsigned long rem = 0;
    unsigned long root = 0;
    unsigned long divisor = 0;
    for(int i=0; i<16; i++){
        root <<= 1;
        rem = ((rem << 2) + (a >> 30));
        a <<= 2;
        divisor = (root<<1) + 1;
        if(divisor <= rem){
            rem -= divisor;
            root++;
        }
    }
    return (unsigned short)(root);
}
```

At this point, we have a very simple, but slow, algorithm.

I next ask if I can do something to make the algorithm run faster. The answer is a resounding yes.

The Friden Algorithm

The next improvement to the algorithm, making it run faster, requires some background, which is best given in the form of ancient history. Once upon a time, in a galaxy far, far away, people didn't have ready access to computers. They didn't even have electronic calculators. I was working at NASA in those days. When I needed more accuracy than I could squeeze out of a slide rule, I used an electromechanical calculator — a massive, noisy monster made by Friden. This thing was bigger than the largest typewriter, also had a moveable carriage, cost about $1,500 in 1960 dollars, and had hernia-inducing mass.

The Friden keyboard was huge, consisting of 16 columns of 10 keys each, representing the 10 digits. You punched in a number by picking one from column A, one from column B, and so on. On the carriage was a set of rotating wheels, which displayed the answer as digits in little windows. A set of operator keys on the right told the computer what to do, and it did so with a great gnashing of gear teeth.

When there was a Friden in use in the room, everyone knew it. The room lights dimmed, the motor roared, the entire desktop shook, and the wheels spun with a sound somewhere between a threshing machine and a car in dire need of Mr. Transmission. It sounded as though the machine couldn't possibly last the day, but in fact the Friden was quite a reliable machine. The basic Friden was strictly a four-function calculator, with no memory except the little wheels on the carriage. However, a square root version was available at considerable extra cost. Our office had only one square root Friden, so we all had to take turns with it when we needed square roots. Or so I thought.

One day I was busily thinking deep-space thoughts while my office mate was banging away on our common, non-square-root Friden. I heard a strange sound that went something like "punch-punch-cachunk, punch-punch-cachunk, punch-punch-cachunk-DING, punch-punch-DING-clang-clang" in a repeated rhythm. I thought my mate had either lost his marbles, or he was creating some new kind of computer game.

I asked him what the heck he was doing. He said, "finding a square root."

"But, um ...," said I, "this isn't a square root Friden."

"I know," he said, "that's why I have to do it this way."

It turns out that my office mate, Gap, was an old NASA hand and one who could make the Friden do all kinds of tricks, including the famous "Friden March" (a division of two magic numbers that made the thing sound like the "rah, rah, rah-rah-rah" of a football cheer).

Somewhere along the line, Gap had learned the Friden algorithm for finding a square root on a non-square-root Friden. It was a very similar algorithm, in fact, to the one programmed into the cam-and-gear "ROM" of the square root version. Gap taught it to me, and now I'll teach it to you. (Gap, if you're still out there and listening, thanks for the education.)

The method involves a minor rewrite to Equation [4.19].

[4.20] $(n+1)^2 = n^2 + n + (n+1)$

The keys on the Friden keyboard were "sticky," meaning they held whatever was punched into them. The nice part about the formulation in Equation [4.20] is that the last digit punched in was also the number whose square you were taking.

To see the algorithm work, watch the squares develop in Table 4.12.

Table 4.12 Friden square root.

Initial Number	Numbers Added	Result
0	0 + 1	1
1	1 + 2	4
4	2 + 3	9
9	3 + 4	16
16	4 + 5	25
25	5 + 6	36
36	6 + 7	49
49	7 + 8	64
64	8 + 9	81

As you can see, the number in the third column of this table is always the square of the last number added in.

Except Gap wasn't really adding, he was subtracting from the number whose root he was seeking. The DING sound I had heard occurred each time the subtraction caused an overflow. This was the signal to Gap that he'd gone too far, so he'd back up by adding the last two numbers back in the *reverse* order he'd subtracted. In that way, the number remaining in the keyboard was always the largest digit whose square didn't exceed the input number. This is precisely the number I seek for the integer square root.

Getting Shifty

This next part of the algorithm is critically important, both for the Friden solution and for my computerized solution. It's the notion of shifting in a manner similar to that used in long division.

Consider a two-digit number n, where the digits are a and b. I can write this as

[4.21] $n = am + b$,

where m is the modulus of the number (10 in the decimal system). Then n^2 is given by

[4.22] $n^2 = a^2 m^2 + 2abm + b^2$.

You can think of this as another number with the modulus equal to m^2 and the first digit equal to a^2. It's not immediately obvious that there are no "carries" from that next term, $2abm$, into the m^2 column, but that is in fact the case. Because b must be less than m, I can write

$$n^2 < a^2m^2 + 2am^2 + m^2n^2 < (a^2 + 2a + 1)m^2$$

or

[4.23] $$n^2 < [(a+1)m]^2.$$

In other words, no matter how large b is, it can never be large enough to change the value of a. This is very important, because it means I can solve for a independently from b.

Put another way, just as in long division, once you solve for a digit, it's solved forever. The only difference between the square root and the division algorithms is that you have to consider the digits of the operand two at a time (modulus m^2), instead of one at a time. In other words, if you work in decimal arithmetic, you operate on numbers in the range 0 to 99. That's the reason for the old "point off by twos" rule you may recall from high school (see "The High School Algorithm," page 78). This is best illustrated by an example, shown in Table 4.13. Here, I'll take the square root of the decimal integer 1,234,567,890.

The first step is to "point off," or group the digits into pairs.

---

12 34 56 78 90

Do this from the *right* starting with the decimal point. Note that this means the leftmost group can have either one or two digits. We do this because we'll be solving for a (in the formulas above) one digit at a time. Since the square of a one-digit number can produce a two-digit result, this requires dealing with the square two digits at a time.

Beginning at the left in Table 4.13, our first pair of digits is 12. Therefore, we begin by finding the largest integer whose square is less than 12. This much, we should be able to do in our heads, since we're supposed to know our multiplication tables. However, to stick to the algorithm, we'll find this first digit by subtraction, just as Gap taught me. The last digit subtracted, 3, is our first digit of the root.

After we find that first digit, the leading 3, we proceed to the next pair of digits. Note, however, that we don't forget about the 3. This is an important point, and quite unlike the case with long division. In long division, you'll recall, once a digit has been solved and subtracted, we don't use it anymore. The digits of each column are processed separately. In the square root algorithm, on the other hand, the first digits of the solution are retained, and appear in all subsequent subtractions.

You'll also note that, unlike division, our subtraction doesn't necessarily remove all the higher digits. In general, the length of the remainder gets longer and longer as we go, as does the number we're subtracting.

In following the development of Table 4.13, I urge you to not just look at it, but to actually perform the calculations yourself. That's the only way you can truly see how the algorithm works. In short, pretend you're a Friden.

Table 4.13   Decimal square root.

12	34	56	78	90
− 0 − 1				
11				
− 1 − 2				
8				
− 2 − 3				
3	334			
	− 30 − 31			
	273			
	− 31 − 32			
	210			
	− 32 − 33			
	145			
	− 33 − 34			
	78			
	− 34 − 35			
	9	956		
		− 350 − 351		
		255	25578	
			− 3510 − 3511	
			18557	

12	**34**	**56**	**78**	**90**
			− 3511 − 3512	
			11534	
			− 3512 − 3513	
			4509	450990
				− 35130 − 35131
				380729
				− 35131 − 35132
				310466
				− 35132 − 35133
				240201
				− 35133 − 35134
				169934
				− 35134 − 35135
				99665
				− 35135 − 35136
				29394

As you can see from the table, I'm still subtracting numbers in pairs, each larger by one digit than the other. Note, however, that I don't ignore the higher order digits, as I would in long division. Each of the subtracted numbers includes all the digits previously found, and I'm working my way up through the last digit, starting with zero. This process turned out to be just the right approach for the Friden calculator, which had sticky digits and thus remembered the leftmost digits already found. The nicest part of the whole approach from the point of view of utility was that the numbers left stuck in the keyboard after the last digit was found did, in fact, represent the square root — in this case

[4.24] $\sqrt{1234567890} = 35136 +$ fractional part .

It's worth mentioning that this may be the only case in computing history where the result of the computation is found entered into the keyboard, instead of the result being displayed.

You might feel that I'm still doing iterations here, but I'm not. This is no more an iterative process than its close cousin, long division, so it properly deserves the title of a closed-form algorithm.

With shifting, I get a dramatic improvement in the number of computations. Instead of r passes through the loop, I need only the sum of all the digits in the result number, which must be less than 10 times the number of digits. Now the algorithm is practical.

Another subtlety of the Friden algorithm is also worth mentioning: In Table 4.13, I shifted the "dividend" by two digits to tackle the next digit of the result. This meant copying the digits found thus far over one digit to the right. In the real Friden algorithm, I don't want to move the digits already found — they're stuck in the keyboard. To avoid moving them, I shift the carriage one place to the left instead of two, which leaves the digits of the root lined up correctly without having to move them. You'll see a similar behavior in the final binary version.

Finally, note that, unlike the situation in division, the number I'm subtracting accumulates digits as I go, growing from one digit to five digits. In general, the number of digits in the result, which is also the number I'm subtracting, is half that of the original argument. If I'm to get this multidigit result in the keyboard, it stands to reason that the number of digits involved in the subtractions must grow as the process unfolds.

Describing this process in words can't begin to evoke the sound that the Friden calculator generated. The sound of a square root Friden is the sound of a thrashing calculator whose decibels, not to mention complexity of sound, rise at an alarming rate, roughly comparable to the sound of a Yugo being seriously overrevved. The first time you hear it, you think it's a toss-up which will arrive first: the solution or the demise of a calculating engine. However, I never saw a Friden fail or give an incorrect result.

The High School Algorithm

The Friden algorithm may seem to be among the best kept secrets in Computer Science: that there exists a closed form, that is, noniterative, solution to the square root. You and I know, however, that we all learned a pencil and paper computation of the square root in high school — and promptly forgot it.

As fascinating and as valuable from a historical point of view as the Friden algorithm may be, it has little if any value when using other calculators that lack the sticky, matrix-like keyboard of the Friden calculator. Nevertheless, it is the basis for the high school algorithm and gives us enough insight into the process that, with any luck, you will be able to remember it this time.

I'll return to Equations [4.21] and [4.22] and set the modulus m to 10.

[4.25] $n = 10m + b$

[4.26] $n^2 = 100a^2 + 20ab + b^2$

I'll assume that I've already managed to find a by some method or other. I'm now looking for the next digit, b. Rewriting Equation [4.26] gives:

$$20ab + b^2 = n^2 - 100a^2$$
$$b(20a + b) = n^2 - 100a^2$$

or

[4.27] $b = \dfrac{n^2 - 100a^2}{20a + b}$.

   This formula gives a rule for finding b. Notice that the numerator of the right-hand side is the original number, less the square of a, shifted left so that the two line up. In short, the numerator is the *remainder* after I've subtracted the initial guess for the root. To get the denominator, I must double my guess a, shift it left one place, and add b.

   At this point, if you've been paying attention, you're asking how you can add b before you know what it is. The answer is that you can't. This makes the square root algorithm a bit of a trial and error process, just like the division algorithm. In division, you also must guess at the quotient digit by looking at the first couple of digits of the dividend and divisor. You can't be sure that this guess is correct, however, until you've multiplied it by the entire divisor and verified that you get a small, positive remainder.

   In the same way, in the square root process, assume a value for b based on a divisor of $20a$ then substitute the new value of b and make sure the division still works.

   It seems complicated, but it's actually no more complicated than division, and in some ways it's a bit easier.

   When I first wrote Equation [4.26], I treated a and b as though they were single digits. However, you'll note that there's nothing in the math that requires that. I can now let a be all the digits of the root that I've found so far and let b be the one I'm seeking next. In division, the trial quotient digit must be tested and often adjusted at each step. The same is true in the square root algorithm. However, in the latter, the larger a is, the less influence b will have on the square and the less likely you'll have to backtrack and reduce a. Therefore, although you must still always check to be sure, there's actually less backtracking in the square root. The only thing that makes the algorithm seem harder is that the roots, and therefore the differences, are getting larger as you go along, as shown in Table 4.13.

As an example, I'll use the same input value I used before: 1,234,567, 890. The first step is to point it off by twos.

---

12 34 56 78 90

Next, I write down the first digit of the root. I don't need a fancy algorithm for this. The first piece of the argument is only two digits long, and its root can only be one of nine numbers. I should be able to figure out, by inspection, which of the nine it is. In this case, four squared is 16, which is too large. So the first digit must be three. I write it down above the input value, square it, and subtract it to get the remainder.

$$
\begin{array}{r}
3 \\
\hline
12 \quad 34 \quad 56 \quad 78 \quad 90 \\
9 \\
\hline
3 \quad 34
\end{array}
$$

As you can see, I've brought down the next digits, just as I would in division, except I bring them down by twos, so the next dividend is 334. This is the top half of the division in Equation [4.27].

The tricky part is remembering that the bottom half is not $10a + b$, but $20a + b$. Before I look for b, I must double the current root then tack on a zero (bet that's the part you forgot). At this point I have

$$
\begin{array}{r}
3 \\
\hline
12 \quad 34 \quad 56 \quad 78 \quad 90 \\
9 \\
\hline
60 \mid 3 \quad 34
\end{array}
$$

The division is 334/60, which yields five and some change. Before I write down the new digit, however, I must make sure that the division still works when I stick the five into the divisor. Now I have 334/65, which still yields a five. I'm OK, and I don't need to backtrack. I write this next digit down and the trial root is now 35.

It's very important for you to see that the trial root and the number you divide with are not the same because of the factor of two. The last digit is the same, but the rest of the divisor is double that of the root.

At this point, I have

```
     3   5
   ┌─────────────────
   │12  34  56  78  90  .
   │ 9
 65│ 3  34
```

What next? How do I subtract to get the new remainder? I can't just square 35 and subtract it, because I've already subtracted the square of 3. Again, Equation [4.26] provides our answer. I have

$$n^2 = 100a^2 + 20ab + b^2,$$

which I have already written in the form

$$b(20a + b) = n^2 - 100a^2.$$

I've also already performed the subtraction on the right; that's how I got the remainder to divide with, to find b. Now that I've found it, I must complete the subtraction by subtracting out the left-hand side. That is, the remainder I now must obtain is

[4.28] $\text{rem} = n^2 - 100a^2 - b(20a + b).$

In this example, $b = 5$, and $20a + b = 65$, so I subtract 325. This step is easier to do than to explain. Note that it's identical to a division operation; that is, I multiply the last digit found by the "divisor" (65 in this example), and subtract it. After the subtraction, I have

```
     3   5
   ┌─────────────────
   │12  34  56  78  90
   │ 9
 65│ 3  34
   │ 3  25
   └──────────
        9  56
```

At this point, you can begin to see how the algorithm works.

1. Point off the argument by twos.
2. Write down the first square root digit by inspection.
3. Subtract its square from the first two digits.
4. Obtain the remainder by drawing down the next two digits.

5. Double the square root and append a zero.
6. Estimate the next root digit by dividing the remainder by this number.
7. Verify the digit by substituting it as the last digit of the divisor.
8. Multiply the last digit by the divisor.
9. Subtract to get the new remainder.
10. Repeat from step 5 until done.

By now, you should be able to complete the process, which looks like this:

	3	5	1	3	6	—	
	12	34	56	78	90		initial argument
	9						square first digit
65	3	34					draw down, divide by 20*root
	3	25					subtract 5*65
701		9	56				draw down, divide by 20*root
		7	01				subtract 1*701
7023		2	55	78			draw down, divide by 20*root
		2	10	69			subtract 3*7023
70266			45	09	90		draw down, divide by 20*root
			42	15	96		subtract 6*70266
			2	93	94		remainder

Note again that the last digit of the divisor must always be the same as the multiplier when I subtract the next product. That's because they're both b.

Is the result correct? Yes. If you square 35,136, you get 1,234,538,496, which is smaller than the original argument. If you square the next larger number, 35,137, you get 1,234,608,969, which is too large. Thus the root I obtained, 35,136, is indeed the largest integer whose square is less than the original argument.

You've just seen the high school algorithm in all its glory. Now that you've seen it again, you probably recognize at least bits and pieces of it. If, like me, you've tried to apply it in later years and found that you couldn't, it's probably because you forgot the doubling step. The thing that you're dividing to get the next digit (the divisor) is not the root, nor is it exactly twenty times the root. You get it by doubling the root, then appending a zero. Also, before finally multiplying this divisor by the new digit, don't forget to stick that new digit into the divisor in place of the original zero. When you do, there's always the chance that the new product is too large. In that case, as in division, you must decrement the latest digit and try again.

Doing It in Binary

I have now described how to find square roots by hand using decimal arithmetic. However, the algorithm is a lot more useful and universal than that. In deriving it, I made no assumptions as to the base of the arithmetic, except when I put the 10 in Equation [4.25]. Replace that 10 by any other base, and the method still works. As in the case of division, it's particularly easy when I use base 2, or binary arithmetic. That's because I don't have to do division to get the next digit. It's either a 1 or a 0, so I either subtract or I don't. I know how to test for one number greater than another, and that's as smart as I need to be to make this method work in binary.

In the decimal case, you'll recall that I made the first estimate of the quotient digit by using the divisor with a zero as the last digit. Then, once I had the new digit, I stuck it into the divisor in place of the trailing zero and tried again. Using binary arithmetic, I don't even have to do this step. If the division succeeds, it must succeed with the 1 as the last bit, so I might as well put it there in the first place.

The example below repeats the long-form square root in binary. The input argument is:

$$45,765 = \text{0xb2c5} = \text{0b1011001011000101}$$

I perform exactly the same process as in decimal arithmetic, except that I don't have to do division. I merely need to look at the magnitude of the two numbers and either subtract or not.

```
              1   1   0   1   0   1   0   1
            ┌─────────────────────────────────
            10  11  00  10  11  00  01  01
             1                                    first square is always 1
      101    1  11                                times 100, append 1
             1  01                                it fits, subtract
     1101       10  00                            doesn't fit, digit is 0
               10  00  10                         bring down next
    11001        1  10  01                        now it fits, subtract
                10  01  11                         bring down next
   110101       11  01  01                        doesn't fit, digit is 0
                10  01  11  00                     bring down next
  1101001        1  10  10  01                     now it fits, subtract
                11  00  11  01                      bring down next
 11010101       11  01  01  01                      doesn't fit, digit is 0
                11  00  11  01  01                   bring down next
110101001        1  10  10  10  01                   now it fits, subtract
                 1  10  00  11  00                   remainder
```

The root is 0b11010101, which is 0xd5, or decimal 213. This result tells me that the square root of 45,765 is 213, which is true within the definition of the integer root. The next larger root would be 214, whose square is 45,796, which is too large. Any doubt you might have about the method is expelled by looking at the remainder, which is

    0b110001100 = 0x18c = 396,

which is also $45,765 - (213)^2$.

Implementing It

Now that you see how the algorithm works, I'll try to implement it in software. The major difference between a hand computation and a computerized one is that it's better to shift the argument so that I'm always subtracting on word boundaries. To do this, I'll shift the input argument into a second word, two bits at a time. This second word will hold partial remainders. I'll build up the root in a third word and the divisor in a fourth. Remember, the divisor has an extra bit tacked on (the multiplication by two) so I need at least one more bit than the final root. The easiest thing to do is make both the divisor and remainder words the same length as the input argument. Listing 4.6 shows the result. Try it with any argument from zero through the maximum length of an unsigned long word (4,294,967, 295), and you'll find it always gives the correct root. It gives it quickly, too, in only sixteen iterations, with nothing in the loop but a test and a subtraction.

Note that I don't test for the error of a negative argument as I do in floating-point implementations. Because the argument is assumed to be unsigned, no such test is necessary.

Better Yet

The code in Listing 4.7 looks hard to beat, but it uses a lot of internal variables. It needs one to hold the remainder, into which I shift the input argument, one to hold the root the program is building, and one to hold the divisor. Can I do better?

I can. The secret is in noting the intimate relationship between the root and the divisor. They differ only in that the root ends in 1 or 0, whereas the corresponding divisor has a 0 tucked in front of the last bit and so ends in 01 or 00. You can see this relationship by comparing the two, as I've shown in Table 4.14 for the example problem. The divisors for successful as well as unsuccessful subtractions are shown.

Table 4.14   Root versus divisor.

Root	Divisor	Successful?
1	101	yes
11	1101	no
110	11001	yes
1101	110101	no
11010	1101001	yes
110101	11010101	no
1101010	110101001	yes

Whether the subtraction is done or not, you can see that the divisor is always four times the root plus one. The success or failure of the subtraction only determines whether or not the bit inserted into the next root is a 1 or a 0.

Because there's always a predictable relationship between the root and the divisor, I don't need to store both values. I can always compute the final result from the last divisor. The binary algorithm, modified to use a single variable for the divisor, is shown in Listing 4.7. In this implementation, I basically carry the desired information in the divisor. A single division by two at the end gives the desired root.

Listing 4.7   An improved version.

```
unsigned short sqrt(unsigned long a){
    unsigned long rem = 0;
    unsigned long root = 0;
    for(int i=0; i<16; i++){
        root <<= 1;
        rem = ((rem << 2) + (a >> 30));
        a <<= 2;
        root ++;
        if(root <= rem){
            rem -= root;
            root++;
        }
        else
            root--;
    }
    return (unsigned short)(root >> 1);
}
```

Why Any Other Method?

Now that you know there's a closed-form algorithm that requires no iteration, you may be wondering why you should ever use the Newtonian iterative method. The answer is that it can be faster, depending on the available hardware. Remember, the Newton method can converge in one iteration, whereas this closed-form method always requires as many passes (and shift operations) as there are bits in the result. So even though using Newton's method requires a divide operation, it may still be faster than the straightforward subtract-and-shift.

On the other hand, some processors use a lot of cycles to divide, and some require software to do so. In the latter case, there is no contest: the subtract-and-shift method is the way to go because it can find the square root in about the same time it takes to do the same subtract-and-shift operations as the divide alone. The bottom line: look at your hardware and choose the method that's fastest for that processor.

Conclusions

In this chapter, I described a super algorithm for fast floating-point square roots that hacks the exponent to narrow the range and uses a linear function to give a first guess. This reduces the iterative process to only two or three iterations. In the process, I discussed how to optimize functions over their range of operation. This chapter also described three approaches to the integer square root, each with its own strengths and weaknesses.

As you can see, there's much more to this "simple" function than meets the eye. It's little wonder that it is poorly implemented so often. As you go through the other algorithms in this book, you'll find that they often have much in common with the square root function. In particular, there's a lot left unsaid when someone just hands you an equation, such as Equation [4.1], and assumes that the problem has been solved. A proper implementation calls for a lot of attention to detail and consideration of things like the kind of application, the needs for speed and accuracy, and the capabilities of the target CPU. If the method is iterative, as Equation [4.1] requires, you must concern yourself with how to perform the iteration, how to select the starting value, and how to decide when to stop. Even if the method is a closed form, there are still a number of implementation details to work out. If there's any one lesson to be learned from this chapter, it is this:

Never trust a person who merely hands you an equation.

Chapter 5

Getting the Sines Right

In the last chapter, I looked at the square root function and studied it in laborious detail. I'll be doing similar things with the sine and cosine functions in this chapter and with other fundamental functions a bit later. You'll find a few common threads that run through these chapters.

- The solution can be found in any computer handbook.
- To get the best results, the range of the input variable must be limited.
- Iterative solutions require starting values and stopping criteria.

In a nutshell, blindly following the handbook formula will get you in trouble. The purpose of this book is to show you how to stay *out* of trouble.

The Functions Defined

A good way to start with any function is by defining it. The sine and cosine functions are defined using the geometry of Figure 5.1. If I have a circle with unit radius, as shown, a line drawn at any angle will intersect the circle at a point with coordinates (i.e., projections on the x- and y-axes) of

[5.1] $\qquad x = \cos(\theta) \qquad y = \sin(\theta)$.

Figure 5.1   Sine/cosine definition.

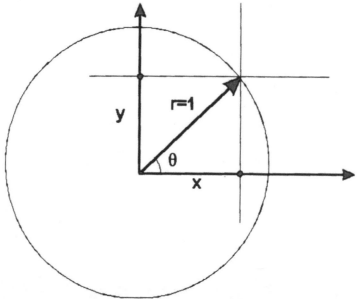

It should be pretty obvious that the coordinates will change in proportion if you change the radius of the circle to some other value, r. In fact, that's the main attraction of these trig functions, because I can then write

$$x = r\cos(\theta)$$
and

[5.2] $$y = r\sin(\theta)$$

If I advance θ through its full range and plot the values of the two functions defined in Equation [5.1], I get the familiar sine waves shown in Figure 5.2. I need go no further than this figure to see a few important features of the functions.

- Both functions are bounded by values of +1.0 and –1.0.
- Both functions repeat with a period of 360°.
- The sine function is odd, meaning that it's reversed as it crosses zero.

  [5.3] $\sin(x) = -\sin(-x)$

Figure 5.2   The sine/cosine functions.

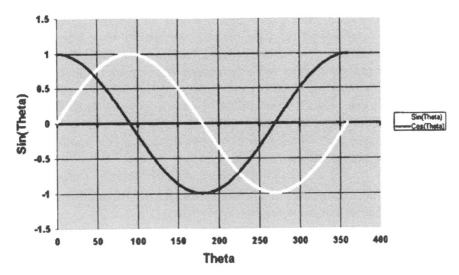

- The cosine function is even, meaning that it's reflected through the y axis.

    [5.4] $\cos(x) = \cos(-x)$

- The cosine and the sine functions are identical except for a phase angle.

    [5.5] $\cos(x) = \sin(x + 90)$

If you've ever worked on a problem involving geometry, I don't have to sell you on the usefulness of these functions. It's impossible to avoid them. The question is how to compute them.

The Series Definitions

The principal values of the trig functions are those for angles from 0° through 45°; all others can be computed from these using formulas like Equations [5.3] through [5.5], and one other pair:

$$\sin(180 - x) = \sin(x)$$
[5.6] $$\cos(180 - x) = -\cos(x).$$

When I was in high school, my textbook included trig tables with the values tabulated for all angles in steps of 0.1°. I was taught to interpolate between tabular values to get the results for the input angles needed. Because I only needed values from 0° through 90°, the tables had 900 entries

— about 15 pages. When I graduated from college and was working for NASA, I received a book of four-place tables, about 150 pages. One fellow had a book of five-place tables, which he guarded as though it were the Holy Grail. Without those tables, we would have been in deep trouble, because there were no scientific calculators in those days.

I used to wonder who generated those tables and how. Did someone actually perform very accurate measurements with a protractor, then record them in the book? That couldn't be it, because it would be very difficult, as a practical matter, to measure dimensions to five or six digits of accuracy. You might also wonder how that pocket marvel, the modern scientific calculator, computes them.

As it turns out, calculus gives a way of deriving a representation for the sine function using an infinite series. (Ditto for e^x, ln x, and many other functions). The series are shown below.

[5.7]
$$\sin x = x - \frac{x^3}{3!} + \frac{x^5}{5!} - \frac{x^7}{7!} + \frac{x^9}{9!} - \frac{x^{11}}{11!} + \cdots$$

[5.8]
$$\cos x = 1 - \frac{x^2}{2!} + \frac{x^4}{4!} - \frac{x^6}{6!} + \frac{x^8}{8!} - \frac{x^{10}}{10!} + \cdots$$

You will find these definitions for these functions in any textbook or handbook on computer math. However, I hope you learned from the square root example that a formula alone is not always enough to serve you in practical applications.

I'm sure you can see the patterns in the series. The signs of the terms alternate, the powers are successive odd powers for the sine and successive even ones for the cosine, and the denominator is the factorial of the integer exponent in the numerator. (For those of you not familiar with the notation, the ! means "factorial," and $n!$ is the product of all integers from 1 through n.) The factorial function $n!$ grows very rapidly as n increases, which is a Good Thing, because it means that the series converge rapidly. As you'll soon see, however, "rapidly" can be a relative term, and not necessarily satisfactory in the real world.

For those who like to see things come to a satisfactory end, once you see the pattern, it's easy enough to write the general terms of the series (Equations [5.9]).

$$\sin{}_n x = -(1)^n \frac{x^{2n+1}}{(2n+1)!}$$

[5.9] $(n = 0,1,...)$

$$\cos{}_n x = -(1)^n \frac{x^{2n}}{(2n)!}$$

It is these series that were used — and still are — to calculate the tables in trig books. It may interest you to know that the generation of tables such as these were the purpose behind Charles Babbage's invention of the mechanical marvel he called the Difference Engine back in 1823. The Difference Engine, which was never built, was supposed to compute functions to 20-digit accuracy.

If you had a sufficiently large memory, you could take the path of those old trig textbooks (and Babbage) and simply build a table of the functions. But modern calculators and computers compute the functions, not to four or five decimal places, but to anywhere from 10 to 20. That would take a table of 45,000,000,000,000,000,000,000 entries — not exactly practical, even in today's climate of RAM extravagance. Instead, programmers are faced with computing the values in real time as they are needed. This chapter explores the practical ramifications of this need.

Why Radians?

There's a small "gotcha" in the series above: the angle x must be expressed in *radians*. The radian, in turn, is defined such that there are 2π in a circle. In other words,

1 circle = 2π radians = 360°

so

1 radian = $360/(2\pi) = 180/\pi° = 57.29578°$.

Why do mathematicians use radians instead of degrees? I wish I could give you a good answer. People tend to think in degrees, not radians, and since the first computer was built, programmers have been converting angles from degrees to radians and back. There are a few (but very few) cases where the radian is the more natural measure. But usually, I find myself wishing the trig functions had been implemented for degrees in the first place [DEC's VAX FORTRAN does include these functions, SIND() and COSD()].

The only real excuse for using radians is that the radian is the natural unit for which the series were derived. The original implementations used radians, and programmers have been stuck with them ever since. Sorry. Even computer science has its traditions.

Cutting Things Short

As mentioned, the factorial is a function that increases very rapidly as its argument increases. Its presence in the denominators of the terms in the series guarantees that the series will converge (i.e., home in on a finite value) rapidly for any argument x.

The problem with infinite series is that they're... well... infinite. If I had an infinite computer and infinite time, I could wrap this chapter up now and go on to the next. In the real world, you can't wait to evaluate an infinite number of terms. Fortunately, you also don't have or need infinite accuracy. Any real number in a computer is only an approximation anyway, limited by the computer word length. As the computer calculates successively higher order terms in the series, the terms eventually become smaller than the bit resolution of the computer. After that, computing more terms is pointless. A word for purist mathematicians: just because all terms in a series are small, doesn't necessarily mean that the sum of those terms is also small, or even finite. A counter example is the harmonic series,

$$S = \frac{1}{2} + \frac{1}{3} + ... \frac{1}{n} + ...$$

As n increases in this series, the terms get smaller and smaller, but the sum is infinite. Mathematicians must, and do, worry about the convergence of power series; however, be assured that if someone gives you a series for a function like $\sin(x)$, they've already made sure it converges. In the case of the sine and cosine functions, the alternating signs make sure that small terms do not add up to large numbers.

If higher order terms become negligible, you can *truncate* the series to obtain polynomial approximations. You must always do this with any infinite series. Sometimes you can estimate the error that has been introduced, but most often programmers simply keep enough terms to be sure the error stays within reasonable bounds.

It might be helpful to take a look at an example. Table 5.1 computes the sine and cosine for $x = 30°$ or $\pi/6$ radians. Within the limits of accuracy shown, both series have converged after the fifth term (the cosine is always a little slower because its terms have one less power of x in them, term for term). As promised, the sizes of the terms decrease rapidly thanks to the factorial in the denominator.

Term Limits

The number of terms required for the sine or cosine series to converge depends on the size of the angle. The infinite series are guaranteed to converge for all values of x, but there are no guarantees as to *when*. In my example, I deliberately chose a fairly small angle, 30°, for the input argument. That's the reason for rapid convergence. Suppose, instead, I try the value $x = 1,000$. Then the numerators of the terms in Equation [5.7] will be 1000, 10^9, 10^{15}, 10^{21}, and so on. It will take a very large number of terms before the factorials begin to assert their authority over numbers like that. In the practical sense, it will never happen, because I'm likely to get floating-point overflows (using single-precision floating point, at least). Even if I didn't, the final answer would be buried under round-off error, which is generated by subtracting nearly equal terms.

Table 5.1   Sample computation: $x = 30°$.

Term	Sine Term	Sine	Cosine Term	Cosine
1	0.523598776	0.523598776	1.000000000	1.000000000
2	–.023924596	0.499674180	–.137077839	0.862922161
3	0.000327953	0.500002133	0.003131722	0.866053883
4	–.000002141	0.499999992	–.000028619	0.866025264
5	0.000000008	0.500000000	0.000000140	0.866025404
6	0.000000000	0.500000000	0.000000000	0.866025404

The end result for any practical computation of the trig functions is that I must limit the range of x, just as I did for the square root function. Clearly, the smaller the range, the fewer terms I'll need. I haven't shown you yet how to limit the range, but assume for the moment that it can be done. How many terms will be required?

In this case, it's fairly straightforward to compute how many terms I'll need. Because the terms in the series have alternating signs, I needn't worry about sneaky effects such as lots of small numbers summing up to a larger one. As you saw in Table 5.1, the intermediate results oscillate around the final result. Because of this, I can be assured that the error will always be smaller (sometimes a lot smaller) than the size of the first neglected term. All I need to do is to compute the values of each term in the equations for representative values of the angular range. I can then compare the results with the resolution of the least significant bit (LSB) in the numeric representation

I'm using. Because I can't know in advance the numeric format you might be using, it's best to separate the two steps. The term values are shown in Tables 5.2 and 5.3. Notice how dramatically the error decreases for smaller ranges.

In Tables 5.4 and 5.5, I've used these results to compute how many terms would be required for typical computer representations. Usually, one would expect to see mantissas in the range of 24 bits (for 32-bit floating point) or 56 bits (for the eight-byte format used for C doubles). The smaller numbers are sometimes useful in special-purpose embedded systems, whereas the largest ones correspond to double precision and numeric coprocessor word lengths. Again, it should take only one look at these tables to convince you that it's worth spending some computing power in an effort to reduce the range of the input parameter.

Table 5.2   Size of last term (sine function).

				x			
Term	180°	90°	45°	30°	22.5°	15°	11.25°
1	3.1411	1.5708	0.7854	0.5236	0.3927	0.2618	0.1963
2	5.1677	0.6460	0.0807	0.0239	0.0101	0.0030	0.0013
3	2.5502	0.0797	0.0025	0.0003	7.8e–5	1.0e–5	2.4e–6
4	0.5993	0.0047	3.7e–5	2.1e–6	2.9e–7	1.7e–8	2.2e–9
5	0.0821	1.6e–4	3.1e–7	8.2e–9	6.1e–10	1.6e–11	1.2e–12
6	0.0074	3.6e–6	1.8e–9	2.0e–11	8.6e–13	9.9e–15	4.2e–16
7	4.7e–4	5.7e–8	4.3e–12	3.6e–14	8.5e–16	4.4e–18	1.0e–19
8	2.2e–5	6.7e–10	1.3e–14	4.7e–17	6.2e–19	1.4e–21	1.9e–23
9	8.0e–7	6.1e–12	2.9e–17	4.7e–20	3.5e–22		
10	2.3e–8	4.4e–14	5.1e–20				
11	5.4e–10	2.6e–16					
12	1.1e–11	1.3e–18					
13	1.7e–13	5.2e–21					
14	2.4e–15						
15	3.0e–17						
16	3.1e–19						
17	2.9e–21						

Little Jack's Horner

Now that I know how many terms I need to include in the truncated series, I still have to design code to evaluate them. A naive programmer might program Equation [5.7] just as it's written, computing x^n and $n!$ over and over again for each term. A good stopping criterion might be just to compare the value of each term with some error criterion (in which case, Tables 5.2 to 5.5 aren't necessary).

As a matter of fact, this was precisely the way I first implemented the function, just to prove that I could. The sidebar "A Simple-Minded Sine Function" presents a wonderful example of how *not* to generate a sine function.

It is a terrible implementation because it requires the repeated raising of x to higher and higher powers, not to mention the overhead of calling two time-consuming functions for each term. Yes, the function gives the correct answer for any input, no matter how large, but the price you pay in run-time performance is horrible.

A Simple-Minded Sine Function

The following naive implementation of Equation [5.4] is a terribly impractical approach to the sine function. Note the repeated raising of x to higher and higher powers and the two time-consuming function calls for each term. *Do not use this code.* Read the rest of this chapter for a better approach.

```
double factorial(unsigned long n){
    double retval = 1.0;
    for(; n>1; --n)
        retval *= (double)n;
        return retval;
}


double sine(double x){
    double first_term;
    double next_term = x;
    double retval = x;
    double sign = 1.0;
```

```
double eps = 1.0e-16;
int n = 1;
do{
    cout << next_term << endl;
    first_term = next_term;
    n += 2;
    sign *= -1.0;
    next_term = sign * pow(x, n)/factorial(n);
    retval += next_term;
}
while((abs(next_term) > eps));
return retval;
}
```

Table 5.3   Size of last term (cosine function).

Term	180°	90°	45°	30°	22.5°	15°	11.25°
1	1.0000	1.0000	1.0000	1.0000	1.0000	1.0000	1.0000
2	4.9348	1.2337	0.3084	0.1371	0.0771	0.0343	0.0193
3	4.0578	0.2537	0.0159	0.0031	0.0010	2.0e–4	6.2e–5
4	1.3353	0.0209	3.3e–4	2.9e–5	5.1e–6	4.5e–7	8.0e–8
5	0.2353	9.2e–4	3.6e–6	1.4e–7	1.4e–8	5.5e–10	5.5e–11
6	0.0258	2.5e–5	2.5e–8	4.3e–10	2.4e–11	4.2e–13	2.3e–14
7	0.0019	4.7e–7	1.2e–10	8.9e–13	2.8e–14	2.2e–16	6.9e–18
8	1.0e–4	6.4e–9	3.9e–13	1.3e–15	2.4e–17	8.1e–20	1.5e–21
9	4.3e–6	6.6e–11	1.0e–15	1.5e–18	1.5e–20		
10	1.4e–7	5.3e–13	2.0e–18	1.4e–21			
11	3.6e–9	3.4e–15	3.3e–21				
12	1.1e–11	1.8e–17					
13	7.7e–11	8.2e–20					

x

Term	180°	90°	45°	30°	22.5°	15°	11.25°
				x			
14	2.1e–14						
15	2.7e–16						
16	3.1e–18						
17	3.1e–20						

I can improve things considerably by noting that each term can be computed from the preceding one. I'll go back to the general term for the sine function given in Equation [5.9].

$$\sin_n x = (-1)^n \frac{x^{2n+1}}{(2n+1)!}$$

The next term in the series is

[5.10]
$$\sin_{n+1} x = (-1)^{n+1} \frac{x^{2(n+1)+1}}{(2(n+1)+1)!}$$

$$= (-1)^{n+1} \frac{x^{2n+3}}{(2n+3)!}$$

Table 5.4   Terms needed (sine function).

Bits	180°	90°	45°	30°	22.5°	15°	11.25°
				x			
8	6	4	2	2	2	2	1
16	7	5	4	3	3	2	2
24	9	7	5	4	4	3	3
40	12	9	7	6	6	5	4
48	13	10	8	7	6	6	5
56	15	11	9	8	7	6	6
64	16	12	10	9	8	7	7

Taking the ratio gives

$$\frac{\sin_{n+1} x}{\sin_n x} = \frac{(-1)^{n+1} \dfrac{x^{2n+3}}{(2n+3)!}}{(-1)^n \dfrac{x^{2n+1}}{(2n+1)!}}$$

$$= \frac{-x^2 (2n+1)!}{(2n+3)!}.$$

Let this ratio be R_{n+1} given by

[5.11] $$R_{n+1} = \frac{-x^2}{(2n+3)(2n+2)}.$$

An improved version of the code in the sidebar eliminates calls to both pow() and factorial(). This version is shown in Listing 5.1.

The code in Listing 5.1 is reasonably efficient, and many folks might be tempted to stop there. However, this function is still a pretty awful approach. I can improve on things a bit by computing x^2 once, but there's a much better way, and the relation in Equation [5.11] gives the clue: every term is a factor of the next term. This includes, in fact, the value x, which is present in every term. Thus, I can factor Equation [5.7] and [5.8] as

$$z = x^2$$

[5.12] $$\sin x = x \left[1 - \frac{z}{6} \left(1 - \frac{z}{20} \left(1 - \frac{z}{42} \left(1 - \frac{z}{72} \left(1 - \frac{z}{110}(1 - \dots) \right) \right) \right) \right) \right].$$

Similarly,

[5.13] $$\cos x = 1 - \frac{z}{2} \left[1 - \frac{z}{12} \left(1 - \frac{z}{30} \left(1 - \frac{z}{56} \left(1 - \frac{z}{90}(1 - \dots) \right) \right) \right) \right].$$

Table 5.5   Terms needed (cosine function).

Bits	180°	90°	45°	30°	22.5°	15°	11.25°
8	6	4	4	2	2	2	2
16	8	6	4	4	3	3	3
24	10	7	5	4	4	4	4

Bits	180°	90°	45°	30°	22.5°	15°	11.25°
40	13	9	7	7	6	5	5
48	14	11	8	7	7	6	6
56	15	11	9	8	8	7	7
64	16	13	10	9	8	7	7

This formulation is called *Horner's method*, and it's just about the optimal way to calculate the series. Note that the denominators at each step are the products of two successive integers — in the sine, $2*3$, $4*5$, $6*7$, and so on. Once you spot this pattern, you'll be able to write these functions from memory. If you don't have access to a canned sine function, and you need one in a hurry, direct coding of these two equations is not bad at all.

There are still a couple of details to iron out. First, I have a small problem: the method as shown uses a lot of stack space because of all the intermediate results that must be saved. If your compiler is a highly optimizing one, or if you don't care about stack space, you have no problem. But if you're using a dumb compiler and an 80x87 math coprocessor, you will quickly discover, as I did, that you can get a coprocessor stack overflow.

Listing 5.1   Avoiding function calls .

```
double sine(double x){
    double last = x;
    double retval = x;
    double eps = 1.0e-16;
     long n = 1;
    do{
      n += 2;
      last *= (-x * x/((double)n * (double)(n-1)));
      retval += last;
    }
    while((abs(last) > eps));
    return retval;
}
```

The old Hewlett-Packard (HP) RPN (reverse Polish notation) calculators had only four stack locations. Those of us who used them heavily learned to always evaluate such complicated expressions from the inside out. As HP was always happy to demonstrate, if the expression is written that way, no more than four stack entries are ever required.

For example, the HP way of computing Equation [5.12] would be

[5.14] $\sin x = -\left[-\left(-\left(-\left(-\left(\cdots \frac{z}{110}-1\right)\frac{z}{72}-1\right)\frac{z}{42}-1\right)\frac{z}{20}-1\right)\frac{z}{6}-1\right]x.$

This is the best form to use when using a pocket calculator. It's easy to remember, and it makes effective use of the CHS (+/-) key. But for a computer implementation, I can eliminate half of the sign changes by yet another round of transformations. It may not be obvious, but if I carefully fold all the leading minus signs into the expressions in parentheses, I can arrive at the equivalent forms.

[5.15] $\sin x = \left[\left(\left(\left(\left(\cdots \frac{z}{110}+1\right)\frac{z}{72}-1\right)\frac{z}{42}+1\right)\frac{z}{20}-1\right)\frac{z}{6}+1\right]x$

[5.16] $\cos x = \left[\left(\left(\left(\cdots \frac{z}{90}+1\right)\frac{z}{56}-1\right)\frac{z}{30}+1\right)\frac{z}{12}-1\right]\frac{z}{2}+1$

This is about as efficient as things get. The equations are not pretty, but they compute really fast. Just as I can add more terms on the right in Equations [5.12] and [5.13], I can add them on the left in Equations [5.15] and [5.16]. Just remember that this time the signs alternate, beginning with the plus signs shown at the right end.

A direct implementation of these equations is shown in Listing 5.2. The number of terms used are appropriate for 32-bit floating-point accuracy and ±45° range. Because most computers perform multiplication faster than division, I've tweaked the algorithms by storing the constants as their reciprocals (Don't worry about the expressions involving constants; most compilers will optimize these out). Notice the prepended underscore to the names. This is intended to underscore (pun intended) the fact that the functions are only useful for a limited range. I'll use these later in full-range versions.

Listing 5.2   Horner's method.

```
// Find the Sine of an Angle <= 45
double _sine(double x){
    double s1 = 1.0/(2.0*3.0);
    double s2 = 1.0/(4.0*5.0);
    double s3 = 1.0/(6.0*7.0);
```

Listing 5.2   Horner's method.

```
    double s4 = 1.0/(8.0*9.0);
    double z = x * x;
    return (((((s4*z-1.0)*s3*z+1.0)*s2*z-1.0)*s1*z+1.0)*x;
}


// Find the Cosine of an Angle <= 45
double _cosine(double x){
    double c1 = 1.0/(1.0*2.0);
    double c2 = 1.0/(3.0*4.0);
    double c3 = 1.0/(5.0*6.0);
    double c4 = 1.0/(7.0*8.0);
    double z = x * x;
    return ((((c4*z-1.0)*c3*z+1.0)*c2*z-1.0)*c1*z+1.0);
}
```

The second detail? I've quietly eliminated a test for deciding how many terms to use. Since you must compute Horner's method from the inside out, you must know where "inside" is. That means you can't test the terms as you go; you must know beforehand how many terms are needed. In this case, the extra efficiency of Horner's method offsets any gains you might achieve by skipping the computation of the inner terms, for small x. This result parallels an observation from the discussion of square roots in Chapter 4: sometimes it's easier (and often faster) to compute things a fixed number of times, rather than testing for some termination criterion. If it makes you feel better, you can add a test for the special case when x is very nearly zero; otherwise, it's best to compute the full expression.

You may be wondering why I've included functions for both the sine and cosine, even though I've already shown In Equation [5.5] that the cosine can be derived from the sine. The reason is simple: you really need to be able to compute both functions, even if you're only interested in one of them. One approximation is most accurate when the result is near zero and the other when it's near one. You'll end up using one or the other, depending on the input value.

Home on the Range

At this point, I've established that I can get a nice, efficient implementation of the sine function if the range of the input parameter can be limited. It remains to be shown that I can indeed limit the range.

Again, it might be useful to spend a few moments discussing a naive implementation before I go on to a better one. The general idea is straightforward enough. First, recognize that no one should call the functions with larger values because the functions repeat after 360° (2π radians). But I can't guarantee that someone won't, so I'll perform a modulo function

[5.17] $\sin(x) = \sin(\text{mod}(x, 2\pi))$.

This at least protects me from foolish inputs like 10,000π. Note that the result of this operation should always yield a positive angle, in the range 0 to 2π.

Second, the sine function is odd, so

[5.18] $\sin(x + \pi) = -\sin(x)$.

If the angle is larger than 180° (π radians), I need only subtract π then remember to change the sign of the result.

Third, the sine and cosine functions are complementary (see Equation [5.5]), so

[5.19] $\sin(x + \pi/2) = \cos(x)$.

Using this gets me the range 0° to 90° ($\pi/2$ radians). Finally, I can use the complementary nature again in a little different form to write

[5.20] $\sin(x) = \cos\left(\dfrac{\pi}{2} - x\right)$.

If the angle is larger than 45° ($\pi/4$ radians), I subtract it from 90°, yielding an angle less than 45°. This transformation serves two purposes: it reduces the range to 0° to 45° and it guarantees that I use whichever of the two series (Equation [5.15] or [5.16]) that is the most accurate over the range of interest.

All of this is best captured in the pseudocode of Listing 5.3. It's straightforward enough, but rather busy. I've actually seen the sine function implemented just as it's written here, and even done it that way myself when I was in a hurry. But the nested function calls naturally waste computer time. More than that, the code performs more sign changes and angle transformations than are really needed. Even the real-variable mod( ) function is nontrivial. (You'd be surprised how many implementations get it wrong. See jmath.cpp for one that gets it right). I can eliminate the nested function calls, but I get instead either messy nested if statements, or a few flags that must be set and tested. None of these approaches are very satisfying.

Listing 5.3   A naive range reduction.

```
double sine(double x) {
    x = mod(x, twopi);
    if( x > pi){
        return - sin1(x - pi);
    else
        return sin1(x);
}

double sin1(double x) {
    if (x > halfpi)
        return cos2(x - halfpi);
    else
        return sin2(x);
}

double sin2(double x) {
    if (x > pi_over_four)
        return _cosine(halfpi - x);
    else
        return _sine(x);
}

Function cos2(x) {
    if (x > pi_over_four)
        return _sine(halfpi - x);
    else
        return _cosine(x);
}
```

Note the name change in the base functions; to distinguish them from the full-range functions, I've prepended an underscore to their names.

A Better Approach

Fortunately, there's a very clean solution that eliminates all the problems, even the need for mod( ). Part of the solution hinges on the fact that I don't really have to limit the angle to positive values. A look at Equations [5.7] and [5.8] show that they're equally valid for negative arguments.

Figure 5.3   Range reduction.

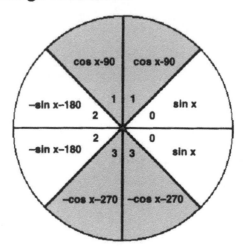

For the rest of the algorithm, look at Figure 5.3. It might seem at first glance that the unit circle is divided into its eight octants. However, a closer look reveals only four 90° quadrants, each centered about the x and y axes. The quadrants are rotated 45° counterclockwise from their usual positions. Within each quadrant, the formula needed to compute the sine of the angle is shown.

I'll assign each quadrant a number, zero through three, as shown. This number can be very quickly calculated if I use rounding instead of truncation to get the integer.

$$n = \text{Round}(x/90)$$

I can then reduce the angle to its essence by subtracting the appropriate value.

$$x = x - 90n$$

Note that this formula will reduce any angle, positive or negative, to the range −45° to 45°. The quadrant number may then be found by taking n mod 4.

A simple case statement now gives the result I need for every case. The resulting code is shown in Listing 5.4. Note that I don't need a separate set of computations to compute the cosine. I merely add 90° ($\pi/2$ radians) to the angle and call the sine. Using this approach, I end up handling both mod( ) and the angle reduction in a single step and with a minimum of sign changes and logic.

Listing 5.4   The sine function, the right way.

```
double sine(double x){
  long n = (long)(x/halfpi + 0.5);
  x -= n * halfpi;
  n = mod(n, (long)4);
  switch(n){
    case 0:
        return _sine(x);
    case 1:
        return _cosine(x);
    case 2:
        return - _sine(x);
    case 3:
        return  - _cosine(x);
  }
}


double cosine(double x){
  return sine(x + halfpi);
}
```

Tightening Up

That range reduction to an angle less than 45° was pretty easy. From Tables 5.4 and 5.5, you can see that I'll need nine terms to get double-precision accuracy. I'd like to reduce the number of terms even further, but the obvious next question is: Can I? Answer: Yes, but not without a price.

All of the range reductions can be derived from a single trig identity.

[5.21] $\sin(A + B) = \sin(A)\cos(B) + \cos(A)\sin(B)$

If I let A = x and B = $90n$, I'll get the relations embodied in Listing 5.4. To carry the same process further, I'll try other values of B; Tables 5.6 through 5.9 give the results.

Table 5.6 Six segments (±30° range).

n	60n	cos(60n)	sin(60n)	sin(x + 60n)
0	0	1.00000	0.00000	$\sin(x)$
1	60	0.50000	0.86603	$0.5 \sin(x) + 0.86603 \cos(x)$
2	120	–0.50000	0.86603	$-0.5 \sin(x) + 0.86603 \cos(x)$
3	180	–1.00000	0.00000	$-\sin(x)$
4	240	–0.50000	–0.86603	$-0.5 \sin(x) - 0.86603 \cos(x)$
5	300	0.50000	–0.86603	$0.5 \sin(x) - 0.86603 \cos(x)$

The good news from these tables is that, at worst, I have to store and use a couple of multiplicative constants. With steps of 30°, one of the constants is 0.5, which can be accomplished by a shift if I'm using integer arithmetic. The 45° case (±22.5°) is especially appealing because the same constant multiplies both parts when there are two parts to compute.

The bad news is I have two parts to compute in half the cases. This means I must calculate *both* the sine and cosine via the series. Because the range has been reduced, I need fewer terms in the series (see Tables 5.4 and 5.5). Still, it turns out that the total number of multiplies required will always be higher than for the four-quadrant case. As long as I only need a sine or a cosine, I gain nothing in speed by limiting the range to less than ±45°.

Table 5.7 Eight segments (±22.5° range).

n	45n	cos(45n)	sin(45n)	sin(x + 45n)
0	0	1.00000	0.00000	$\sin(x)$
1	45	0.70711	0.70711	$0.70711(\sin(x) + \cos(x))$
2	90	0.00000	1.00000	$\cos(x)$
3	135	–0.70711	0.70711	$-0.70711(\sin(x) - \cos(x))$
4	180	–1.00000	0.00000	$-\sin(x)$
5	225	–0.70711	–0.70711	$-0.70711(\sin(x) + \cos(x))$
6	270	0.00000	–1.00000	$-\cos(x)$
7	315	0.70711	–0.70711	$0.70711(\sin(x) - \cos(x))$

Table 5.8   **Twelve segments (±15° range).**

n	30n	cos(30n)	sin(30n)	sin(x + 30n)
0	0	1.00000	0.00000	$\sin(x)$
1	30	0.86603	0.50000	$0.86603 \sin(x) + 0.5 \cos(x)$
2	60	0.50000	0.86603	$0.5 \sin(x) + 0.86603 \cos(x)$
3	90	0.00000	1.00000	$\cos(x)$
4	120	–0.50000	0.86603	$-0.5 \sin(x) + 0.86603 \cos(x)$
5	150	–0.86603	0.50000	$-0.86603 \sin(x) + 0.5 \cos(x)$
6	180	–1.00000	0.00000	$-\sin(x)$
7	210	–0.86603	–0.50000	$-0.86603 \sin(x) - 0.5 \cos(x)$
8	240	–0.50000	–0.86603	$-0.5 \sin(x) - 0.86603 \cos(x)$
9	270	0.00000	–1.00000	$\cos(x)$
10	300	0.50000	–0.86603	$0.5 \sin(x) - 0.86603 \cos(x)$
11	330	0.86603	–0.50000	$0.86603 \sin(x) - 0.5 \cos(x)$

There is a very important "but" implied in that statement, though. Many cases, such as coordinate rotations, require both the sine and cosine of the same angle. In the cases where I've already computed both trig functions of the reduced angle for, say, the sine function, the cosine comes virtually free. In that case, using smaller ranges makes great sense. Furthermore, by limiting the range, I can use fewer terms in the series. Listing 5.5 shows a superfast (for single-precision floating point) implementation based on Table 5.8 (15° range). Recall that for this range and precision, I can reduce the series to three terms for the sine and four terms for the cosine. The two expressions then take on very simple forms.

[5.22] $$\sin x = \left(\left(\frac{z}{20} - 1\right)\frac{z}{6} + 1\right)x$$

[5.23] $$\cos x = \left(\left(\frac{z}{30} + 1\right)\frac{z}{12} - 1\right)\frac{z}{2} + 1$$

This is starting to look much simpler. Because these functions are so short, I can speed things up even more by inserting them as inline functions.

Listing 5.5 Both at once.

```
// Find both the sine and cosine of an angle

void sincos(double x, double *s, double *c){
    double s1, c1;
    long n = (long)(x/pi_over_six + 0.5);
    x -= (double)n * pi_over_six;
    n = n % 12;
    if(n < 0)
        n += 12;
    double z = x*x;
    double s1 = ((z/20.0-1.0)*z/6.0+1.0)*x;
    double c1 = ((z/30.0+1.0)*z/12.0-1.0)*z/2.0+1.0;

    switch(n){
    case 0:
        s = s1;
        c = c1;
        break;
    case 1:
        s =  cos_30 * s1 + sin_30 * c1;
        c = -sin_30 * s1 + cos_30 * c1;
        break;
    case 2:
        s =  sin_30 * s1 + cos_30 * c1;
        c = -cos_30 * s1 + sin_30 * c1;
        break;
    case 3:
        s =  c1;
        c = -s1;
```

```
       break;
    case 4:
       s = -sin_30 * s1 + cos_30 * c1;
       c = -cos_30 * s1 - sin_30 * c1;
       break;
    case 5:
       s = -cos_30 * s1 + sin_30 * c1;
       c = -sin_30 * s1 - cos_30 * c1;
       break;
    case 6:
       s = -s1;
       c = -c1;
       break;
    case 7:
       s = -cos_30 * s1 - sin_30 * c1;
       c =  sin_30 * s1 - cos_30 * c1;
       break;
    case 8:
       s = -sin_30 * s1 - cos_30 * c1;
       c =  cos_30 * s1 - sin_30 * c1;
       break;
    case 9:
       s = -c1;
       c =  s1;
       break;
    case 10:
       s =  sin_30 * s1 - cos_30 * c1;
       c =  cos_30 * s1 + sin_30 * c1;
       break;
```

```
case 11:
    s =  cos_30 * s1 - sin_30 * c1;
    c =  sin_30 * s1 + cos_30 * c1;
    break;
    }
}
```

Table 5.9   Sixteen segments (±11.25° range).

n	22.5n	cos(22.5n)	sin(22.5n)	sin(x + 22.5n)
0	0	1.00000	0.00000	$\sin(x)$
1	22.5	0.92388	0.38268	$0.92388 \sin(x) + 0.38268 \cos(x)$
2	45	0.70711	0.70711	$0.70711 \sin(x)$
3	67.5	0.38268	0.92388	$0.38268 \sin(x) + 0.92388 \cos(x)$
4	90	0.00000	1.00000	$\cos(x)$
5	112.5	–0.38268	0.92388	$-0.38268 \sin(x) + 0.92388 \cos(x)$
6	135	–0.70711	0.70711	$-0.70711(\sin(x) - \cos(x))$
7	157.5	–0.92388	0.38268	$-0.92388 \sin(x) + 0.38268 \cos(x)$
8	180	–1.00000	0.00000	$-\sin(x)$
9	202.5	–0.92388	–0.38268	$-0.92388 \sin(x) - 0.38268 \cos(x)$
10	225	–0.70711	–0.70711	$-0.70711(\sin(x) + \cos(x))$
11	247.5	–0.38268	–0.92388	$-0.38268 \sin(x) - 0.92388 \cos(x)$
12	270	0.00000	–1.00000	$-\cos(x)$
13	292.5	0.38268	–0.92388	$0.38268 \sin(x) - 0.92388 \cos(x)$
14	315	0.70711	–0.70711	$0.70711(\sin(x) - \cos(x))$
15	337.5	0.92388	–0.38268	$0.92388 \sin(x) - 0.38268 \cos(x)$

The Integer Versions

You've now seen implementations that are just about optimal using floating-point arithmetic. But as is usual in embedded systems, programmers often need to use integer arithmetic.

Except for a couple of gotchas, the sine and cosine functions are uniquely suited to the use of integers because both the input parameter and the output values are bounded. I'll illustrate the procedures below assuming 16-bit arithmetic, but the results can be extended easily to other word lengths.

BAM!

For starters, there's a neat trick you can use to represent the input angle. Instead of using degrees or radians, you can use a different scale factor. The best choice is the so-called Binary Angular Measure (BAM), in which the angle is measured in "pirads." This quantity, as the name implies, stands for an angle of π radians; that is,

1 BAM = 1 pirad = π radians = 180°.

Imagine the behavior of an angle as it increases from 0°. It reaches a maximum at 180°, which is also equal to –180°. Beyond that, you can think of it as continuing to increase to 360°, or you can think of it as a decreasing negative number, until it finally arrives at 360 = 0°.

This is precisely the same behavior you see in an integer represented in a computer using two's complement arithmetic. You can either think of it as a two's complement signed (–180 to +180) or unsigned (0 to 360) number, but in either case the number wraps around to 0 when incremented far enough. If I can pick the right scale factor, I can use the natural behavior of a two's complement integer to take care of a lot of the problems. The mod( ) function, for example, doesn't even come up, because any angle larger than 360° is truncated automatically when it's loaded.

To see how this works, I've shown a few angles in the BAM representation in Table 5.10. I've also shown the segment number for both four and sixteen segments — I'll need this segment number later.

Table 5.10   Angles in binary angular measure (pirads).

Angle (°)	Hex	Segment (4)	Segment (16)
0	0000	0	0
30	1555	0	1
45	2000	0	2
60	2AAB	1	3
90	4000	1	4
120	5555	1	5
135	6000	2	6
150	6AAB	2	7
180	8000	2	8
210	9555	2	9
225 = –135	A000	3	10
240 = –120	AAAB	3	11
270 = –90	C000	3	12
300 = –60	D555	3	13
315 = –45	E000	0	14
330 = –30	EAAB	0	15
360 = 0	0000	0	0

The use of BAM presents another huge freebie: I no longer have to use modulo tricks to identify the quadrant numbers. As long as I stick to dividing the circle into 4, 8, or 16 segments — that is, powers of two — the quadrant number is easy to get from the high bits of the angle. As you can see in Table 5.10, the segment number for the 16-segment case is simply the high hex digit, rounded. Similarly, for four segments, it's the high two bits, rounded. This makes the range reduction logic easy. After computing the segment number n, I can reduce the angle simply by subtracting n again, appropriately shifted left. Because of the rounding, the result could be either positive or negative.

More specifically, the algorithm for finding the segment numbers is

$$n = (x + 0x2000) \gg 14 \qquad \text{for four segments}$$
or
$$n = (x + 0x800) \gg 12 \qquad \text{for 16 segments}$$

Representing the function result is a little bit trickier. In floating point, the sine has a range of −1.0 to +1.0. My 16-bit integer has a range of −32,768 to +32,767. Notice that it is not symmetrical; I am one bit short of being able to represent the result with maximum resolution. In other words, if I define the binary point to be at the far left of the word, I can represent the −1, but not the +1.

I have two options: I can say "that's close enough," and simply let 7FFF be "one," or I can bite the bullet and lose one bit of accuracy to get the proper scaling. I prefer to take the former choice, to give me as many bits of accuracy as possible. With short integers, I need all the bits I can get. To keep me out of trouble with asymmetries, I'll resolve to limit the function symmetrically for both positive and negative integers. Thus I'll define the integer sine function as:

[5.24] $i\sin(x) = 32,768\sin(x)$

But I'll artificially limit the extreme values to ±1. This function gives 15 bits of resolution.

If I were using 32-bit integers, I'd probably take the opposite tack, and waste one bit to get an exact representation of the value 1. That is, I'd let 1.0 be represented by 0x40000000, and not worry about the next-to-highest bit that never gets used.

Only the very critical issue of scaling remains. When you're working with integers, it's important that all coefficients and intermediate results are scaled to get the maximum resolution at every step, without overflow. Also, don't forget that the argument in the series is supposed to be expressed in radians, but input is measured in pirads. To get things right again, you have to fold the conversion factor π into each term. You could simply convert the argument back to radians before you start, but it turns out that you can get better scaling by taking a different approach. For integer arithmetic, things work better if all the coefficients are roughly the same size, and you can get that effect by a different factoring.

Let's go back to the form of Equations [5.7] and [5.8]:

[5.7] $\sin x = x - \dfrac{x^3}{3!} + \dfrac{x^5}{5!} - \dfrac{x^7}{7!} + \dfrac{x^9}{9!} - \dfrac{x^{11}}{11!} + \ldots$

[5.8] $\cos x = 1 - \dfrac{x^2}{2!} + \dfrac{x^4}{4!} - \dfrac{x^6}{6!} + \dfrac{x^8}{8!} - \dfrac{x^{10}}{10!} + \ldots .$

Let

[5.25] $x = \pi y,$

where y is the angle measure in pirads. This time, let

$z = y^2,$

and factor out only the powers of z, not the coefficients. You'll get

$$\sin(x) = y\left[\pi - z\left(\frac{\pi^3}{3!} - z\left(\frac{\pi^5}{5!} - z\left(\frac{\pi^7}{7!}\right)\right)\right)\right]$$

and

$$\cos(x) = 1 - z\left[\frac{\pi^2}{2!} - z\left(\frac{\pi^4}{4!} - z\left(\frac{\pi^6}{6!}\right)\right)\right]$$

or

[5.26] $\sin(x) = y(s1 - z(s2 - z(s3 - z \cdot s4)))$

and

[5.27] $\cos(x) = 1 - z(c1 - z(c2 - z \cdot c3)),$

where the coefficients are computed in Table 5.11.

Table 5.11   Coefficients for integer algorithm.

Term	Numeric Value	Hex Value
$s1 = \pi$	3.141592654	$6488
$s2 = \pi^3/3!$	5.167712782	$2958
$s3 = \pi^5/5!$	2.550164042	$051A
$s4 = \pi^7/7!$	0.599264530	$004D
$c0 = 1$	1.000000000	$7FFF
$c1 = \pi^2/2!$	4.934802202	$4EF5
$c2 = \pi^4/4!$	4.058712128	$103E
$c3 = \pi^6/6!$	1.335262770	$0156

Notice how nearly equal these coefficients are. In this case, that's a blessing because it helps me avoid most of the shifts normally needed to keep enough significant digits. What shifting I need is pretty much taken care of by the multiplications and fixed-point scaling. This scaling is based on a couple of assumptions. First, the input value (0x2000 pirads at its largest) will be shifted left one bit for more significance. Second, the value of z will be shifted left two bits after it's computed.

If I put all this together and "Hornerize" it one more time, I get the algorithm shown in Listing 5.6. Here I'm showing the code in C, but you should think of this as mere pseudocode. Since the goal of the integer versions is speed, I'm assuming that you'd actually implement the functions in assembly language.

The many right shifts and type-casts are a consequence of using integer arithmetic to represent left-adjusted real numbers.

Normally, in fixed-point arithmetic, we try to adjust the scale of all parameters to maintain as many non-zero bits as possible. In short, we tend to normalize them to the left. In Table 5.11, you will see that I've violated this rule. In this case, the exception is justified because the higher-order terms only contribute slightly to the final result. I've chosen the scaling of the coefficients very carefully, to ensure maximum accuracy while retaining efficiency. The final algorithms are within ±1 bit over the entire range, which is about all one can hope for using integer arithmetic. If you're willing to accept the extra bother of rounding after each multiplication (shifting only truncates), you can make the algorithms accurate within $\pm^1/_2$ bit, which is *exact* within the limits of our chosen word length. To round, shift right one less bit than shown in the code, add one to that last bit, then shift right one more time.

Note that, even though the input argument, y, is never larger than 0x2000, I've made no attempt to normalize it left. Remember that y is measured in pirads, which uses all 16 bits to measure angles. We already have all the precision the format allows. Shifting left buys you nothing; you'd only be shifting in meaningless zeros.

On the other hand, the square $z = y^2$ has more significant bits, and we can retain a few of these by shifting the product right a few bits fewer than normal. The scaling of z is reflected in that of the coefficients. One happy consequence of my choices of scaling is that most of the right shifts are 16 bits. This won't make much difference if your CPU has a barrel-shifter, but it makes a world of difference for the small CPU's that don't, because it means that you can effect the "right shift" simply by addressing only the upper half of the 32-bit product.

The same approach (and scaling) works for the 16-segment case. The only difference is that, with the reduced range, I can use lower order, and therefore much simpler, expressions for the functions. Use only the s1 and s3 coefficients in the sine function, and c0 and c2 in the cosine.

```
sin: return -y*(z*s2-s1) << 1;
cos: return 0x4000 + z(z*c2-c1);
```

Listing 5.6   Integer sine/cosine.

```
short function _sin(short y){
    static short s1 = 0x6488;
    static short s3 = 0x2958;
    static short s5 = 0x51a;
    static short s7 = 0x4d;
    long z, prod, sum;
    z = ((long)y * y) >> 12;
    prod = (z * s7) >> 16;
    sum = s5 - prod;
    prod = (z * sum) >> 16;
    sum = s3 - prod;
    prod = (z * sum) >> 16;
    sum = s1 - prod;

    // for better accuracy, round here
    return (short)((y * sum) >> 13);
}


short function _cos(short y){
    static short c0 = 0x7fff;
    static short c2 = 0x4ef5;
    static short c4 = 0x103e;
    static short c6 = 0x156;
    long z, prod, sum;
    z = ((long)y * y) >> 12;
    prod = (z * c6) >> 16;
    sum = c4 - prod;
```

```
    prod = (z * sum) >> 16;
    sum = c2 - prod;

    // for better accuracy, round here
    prod = (z * sum) >> 15;                         ; note, not 16

    return (short)(c0 - prod);
}
```

Chebyshev It!

An article about sines and cosines wouldn't be complete without some mention of the use of Chebyshev polynomials. Basically, the theory of Chebyshev polynomials allows the programmer to tweak the coefficients a bit for a lower error bound overall. When I truncate a polynomial, I typically get very small errors when x is small, and the errors increase dramatically and exponentially outside a certain range near $x = 0$. The Chebyshev polynomials, on the other hand, oscillate about zero with peak deviations that are bounded and equal. Expressing the power series in terms of Chebyshev polynomials allows you to trade off the small errors near zero for far less error near the extremes of the argument range. I will not present a treatise on Chebyshev polynomials here; for now, I'll only give the results of the process.

You don't need to know how this is done for the purposes of this discussion, but the general idea is to substitute every power of x by its equivalent in terms of Chebyshev polynomials, collect the terms, truncate the series in that form, and substitute back again. When all the terms have been collected, you'll find that you are back to a power series in x again, but the coefficients have been slightly altered in an optimal way. Because this process results in a lower maximum error, you'll normally find you can drop one term or so in the series expansion while still retaining the same accuracy.

Just so you don't have to wait for the other shoe to drop, the Chebyshev version of the four-quadrant coefficients is shown in Table 5.12. Note that I need one less term in both functions.

Table 5.12   Chebyshev coefficients for integer algorithm.

Term	Numeric Value	Hex Value
s1	3.141576918	$6487
s3	5.165694407	$2953
s5	2.485336730	$04F8
c0	0.999999842	$7FFF
c2	4.931917315	$4EE9
c4	3.935127589	$0FBD

The implementations shown here are about as efficient as you can hope to get. As you can see by what I went through to generate practical software, there's a lot more to the process than simply implementing the formulas in Equations [5.7] and [5.8] blindly. As was the case with square roots, the devil is in the details.

What About Table Lookups?

Remember those trig tables in your high school textbooks? Some of you may still be wondering why I don't just store the tables in the computer and interpolate between table entries, just as we did in school. Won't this work? Yes, as a matter of fact, it does, and it can be far faster than computing the functions via a formula. Generally the problem with a table lookup is accuracy.

Most tables are based on linear interpolation — same as the high school method. That's fine for accuracies of three to four digits, but it's very difficult to get accuracies of six to 20 digits. To do so, you must make the tables very large. How large? We can get a feel for the trade-offs by comparing the actual value of a function with an approximation obtained by linear interpolation.

Consider the function $y = \sin(x)$, and assume that you've tabulated it at intervals of Δx. Let two successive points of the table be at the values x_1 and x_2. Because they're neighboring points, the distance between them must be

[5.28] $x_2 - x_1 = \Delta x$.

If you're seeking $\sin(x)$ for a value in between these two, the interpolation formula is

$$y(x) = y(x_n) + (x - x_1)\frac{y_{n+1} - y_n}{x_{n+1} - x_n}$$

[5.29]

$$= y_1 + \frac{(x + x_1)}{\Delta x}(y_2 - y_1).$$

The error in the estimate is

[5.30] $e(x) = f(x) - y(x) = f(x) - \left[y_1 + \frac{(x - x_1)}{\Delta x}(y_2 - y_1) \right].$

So far, things look pretty messy. The nature of the error function is certainly not intuitively obvious at this point. However, consider the error at the two endpoints.

$$e(x_1) = f(x_1) - \left[y_1 + \frac{(x_1 - x_1)}{\Delta x}(y_2 - y_1) \right]$$

$$= f(x_1) - y_1$$

Similarly,

$$e(x_2) = f(x_2) - \left[y_1 + \frac{(x_2 - x_1)}{\Delta x}(y_2 - y_1) \right]$$

[5.31]

$$= f(x_2) - [y_1 + (y_2 - y_1)]$$

$$= f(x_2) - y_2.$$

If the tabular values are correct, the value for y_1 is in fact $f(x_1)$, and ditto for y_2, so

[5.32] $e(x_1) = e(x_2) = 0.$

When you stop and think about it, this result should hardly be a surprise. I've chosen the tabular points to yield values of $f(x)$ at the tabular values of x, so the interpolation should give exact results for any value that happens to lie on a tabular point. As I progress from one tabular point to another, I must assume that the error climbs from zero to some peak value then shrinks back to zero again. Where is the peak value? I don't know for sure — only calculus can answer that question — but a very good guess would be that it's close to the midpoint.

$$x = \frac{1}{2}(x_1 + x_2)$$

[5.33]
$$= \frac{1}{2}(x_1 + x_1 + \Delta x)$$

$$x = x_1 + \frac{\Delta x}{2}$$

At this point, the error is

$$e_{max} = f\left(x_1 + \frac{\Delta x}{2}\right) - \left[y_1 + \frac{\left(x_1 + \frac{\Delta x}{2} - x_1\right)}{\Delta x}(y_2 - y_1)\right]$$

[5.34]
$$= f\left(x_1 + \frac{\Delta x}{2}\right) - \left[y_1 + \left(\frac{1}{2}\right)(y_2 - y_1)\right]$$

$$e_{max} = f\left(x_1 + \frac{\Delta x}{2}\right) - \frac{1}{2}(y_1 + y_2)$$

or

[5.35]
$$e_{max} = f\left(x_1 + \frac{\Delta x}{2}\right) - \frac{1}{2}[f(x_1) + f(x_1 + \Delta x)].$$

So far, I have not made use of the nature of $f(x)$; the formula above is good for any function. Now I'll specialize it to the sine function.

[5.36]
$$e_{max} = \sin\left(x_1 + \frac{\Delta x}{2}\right) - \frac{1}{2}[\sin(x_1) + \sin(x_1 + \Delta x)]$$

Clearly, the value of this error depends on the choice of x; I can expect the maximum error between tabular points to vary over the range of x. I'm now going to assert without proof that the "maximum of the maximum" error — call it E — occurs at 90°, or $\pi/2$ radians. It's easy enough to prove this rigorously, but an arm-waving argument is satisfying enough here; it stands to reason that the error in a linear interpolation is going to be largest when the curvature of the function is greatest. For the sine function, this is the case when $x = \pi/2$.

The math will come out easier if I assume that x_1 and x_2 straddle this value, so that

$$x_1 + \frac{\Delta x}{2} = \frac{\pi}{2}$$

[5.37]
$$x_1 = \frac{\pi}{2} - \frac{\Delta x}{2}$$

$$x_2 = \frac{\pi}{2} + \frac{\Delta x}{2}.$$

For this case,

[5.38]
$$\sin\left(x_1 + \frac{\Delta x}{2}\right) = \sin\left(\frac{\pi}{2}\right) = 1.$$

Also, a trig identity gives

[5.39]
$$\sin(A + B) = \sin A \cos B + \cos A \sin B.$$

Letting $A = \pi/2$ and $B = -\Delta x/2$ gives

$$\sin\left(\frac{\pi}{2} - \frac{\Delta x}{2}\right) = \sin\left(\frac{\pi}{2}\right)\cos\left(\frac{\Delta x}{2}\right) - \cos\left(\frac{\pi}{2}\right)\sin\left(\frac{\Delta x}{2}\right)$$

[5.40]
$$= (1)\cos\left(\frac{\Delta x}{2}\right) - (0)\sin\left(\frac{\Delta x}{2}\right)$$

$$\sin(x_1) = \cos\left(\frac{\Delta x}{2}\right).$$

Similarly,

[5.41]
$$\sin(x_2) = \cos\left(\frac{\Delta x}{2}\right).$$

Equation [5.36] now gives the maximum error:

$$E = 1 - \frac{1}{2}\left[\cos\left(\frac{\Delta x}{2}\right) + \cos\left(\frac{\Delta x}{2}\right)\right]$$

or

[5.42]
$$E = 1 - \cos\left(\frac{\Delta x}{2}\right).$$

To go further, I must consider the nature of the cosine function. If Δx is small, as it surely ought to be, I can approximate the cosine by its first two terms:

$$\cos\left(\frac{\Delta x}{2}\right) \approx 1 - \frac{1}{2}\left(\frac{\Delta x}{2}\right)^2,$$

so

$$E = 1 - \left[1 - \frac{1}{2}\left(\frac{\Delta x}{2}\right)^2\right]$$

or

[5.43] $E = \dfrac{\Delta x^2}{8}.$

I now have the relationship I need to estimate what the error will be for any spacing, Δx, between tabular points. More to the point, now I can compute how many table entries I'll need, using linear interpolation, to get any desired accuracy. To be useful, any table must cover the range 0° to 45°, or 0 to $\pi/4$ radians. (You can get the rest of the range using the angle conversion formulas in Equations [5.17] to [5.20]). The number of table entries must be roughly

[5.44] $N = \dfrac{\pi}{4\Delta x}.$

Reverting Equation [5.43] gives

[5.45] $\Delta x = \sqrt{8E},$

and Equation [5.44] now becomes

$$N = \frac{\pi}{8\sqrt{2E}}$$

or approximately

[5.46] $N = \dfrac{0.278}{\sqrt{E}}.$

I now tabulate N for various allowable errors E. (See Table 5.13.)

You can see in the table that the number of entries needed for the lower precision formats makes table lookup eminently practical, and I recommend such an approach over computing the power series in such cases. Conversely, you can also see that the precision of 32-bit floating-point numbers

(24-bit mantissa) is already approaching the practical limits of table size, and for the higher precision cases, the table size is completely impractical.

I can always reduce the error between tabular points and therefore use a more widely spaced grid by using a higher order interpolation formula. A quadratic interpolation formula might allow me to extend the method a bit. On the other hand, if I get too fancy with high-order interpolation formulas, I'll use so much computer time that I might as well compute the power series. As a practical rule of thumb, I'd have no qualms about recommending the table lookup approach for short-integer and 32-bit floating-point implementations, but I'd use the power series when more accuracy is required.

Table 5.13   Entry numbers for table lookup.

Precision	E	N
short integer	2^{-15}	50
long integer	2^{-31}	12,882
24-bit float	10^{-5}	50
32-bit float	10^{-7}	880
36-bit float	10^{-8}	2,780
40-bit float	10^{-10}	27,800
64-bit float	10^{-17}	8.8e7
80-bit float	10^{-19}	8.8e8

One last point remains to be made. Remember those little boxes in your trig books titled "proportional parts"? In case you've forgotten what they were, these were the slopes of the function in the range covered by that page. In short, they were the values of the expression

[5.47] $$m = \frac{y_2 - y_1}{\Delta x}.$$

The proportional parts allow you to skip computation of this term, which makes the interpolation go faster. You can use the same approach in a computerized table lookup by storing the value of m associated with each pair of tabular points. You can store these values in a second table. This trick reduces the interpolation to a single multiply and add, yielding a very fast implementation.

Because the curvature of the sine function is greater near 90° and small near 0°, you might be tempted to implement a table lookup using points that are not evenly spaced. DON'T DO IT! The beauty of using evenly spaced tabular values is that you can find the index of the entry you need almost instantaneously, via a single division:

[5.48] $n = \text{floor}\left(\dfrac{x}{\Delta x}\right)$

What's more, the fractional part $x - x_n$ is given by the modulo function

[5.49] $u = \text{mod}(x, \Delta x)$.

Using this approach, you can find both the integer (index) part and the fractional part of the input argument in two simple calculations, and if you've also stored the slope as suggested, you end up with a very fast algorithm.

The foregoing comments are even more appropriate if you use the BAM protocol for storing the angle. In this case, you can get both parts (integer and fraction) of the angle simply by masking the binary number representing it. The high bits give the index into the table and the low bits the part to be used in interpolation. The end result is an algorithm that is blazingly fast — far more so than the power series approach. If your application can tolerate accuracies of only 16 bits or so, this is definitely the way to go.

Chapter 6

Arctangents: An Angle–Space Odyssey

When you're working with problems that involve trigonometry, you'll find that about 80 percent of your work will be occupied with computing the fundamental functions sine, cosine, and tangent of various angles. You'll find the infinite series that give these transformations in almost any handbook. From my totally unbiased point of view, one of the more comprehensive treatments for the first two functions was just given in Chapter 5. The third function, the tangent, is relatively easy because it's related to the first two by the simple relation

[6.1] $\tan(x) = \dfrac{\sin x}{\cos x}.$

This function works for all angles except those for which the cosine is zero. At that point, the definition blows up, and the magnitude of the tangent goes to infinity. Note that this problem cannot be solved with tricky coding; the function itself blows up, so there's no way fancy coding will stop it. However, you can keep the "explosion" within bounds, which is what I did in Listing 3.2 (see the section "Is It Safe?").

But the ordinary trig functions like sine and cosine represent only one side of the coin. Having computed a trig function as part of my calculations, I almost certainly will need to be able to compute the angle that generates it. In other words, I need the inverse functions as well. It's no good to be able to compute trig functions from angles if I can't go back home to angles again. The key to this process is the inverse tangent, or arctangent, function, because all the other inverse functions can be computed from it.

By now, you must be getting used to the notion that implementing formulas as they appear in handbooks is not going to work. Nowhere is this awful truth more evident than in the case of the arctangent. In fact, here's a case where the typical power series is all but worthless. Instead, you must use a different and probably surprising form.

I'm sure there must be a magic formula that describes how to derive one form from the other, but I originally got my formula by dint of hard work, trial and error, and more than a little good luck. I could show you the result of the method in its final form, but in this case, getting there was more than half the fun, and it seems a shame not to let you share the adventure. So instead of just presenting the final result, I've decided to let you follow the evolution of the solution in its roughly chronological form.

Going Off on an Arctangent

Because the most fundamental trig functions are the sine and cosine, it seems reasonable to use their inverse functions, arcsine and arccosine directly. Just as in the "forward" case, it's possible to write infinite series for both of these functions. The series don't converge as rapidly as the sine and cosine series, but they get the job done. The old IBM FORTRAN intrinsic functions were implemented this way, and my first attempts used the same series, too.

Later, I found a better way: use the arctangent. As it turns out, the series for the arcsine and arccosine are really rather worthless, because it's both easier and faster to compute them based on the arctangent (see the formulas for arcsine and arccosine in Chapter 3, "The Functions that Time Forgot"). As you will see in a moment however, it's not immediately obvious that using the arctangent for *anything* can be a good idea.

As usual, it's good to begin with a definition. The tangent (not the arctangent) is defined in Figure 6.1. Like the sine and cosine, it's based upon the idea of a unit circle. But where the sine and cosine are given by projections of the arrow tip onto the x- and y-axes, the tangent is given by extending the arrow to the vertical line shown.

As you can see in the graph of the function shown in Figure 6.2, it's not exactly well behaved or bounded. Whereas the sine and cosine can have values only within the range of ±1.0, the tangent exceeds 1.0 for angles greater than 45° and, in fact, wanders off the graph to infinity as θ approaches 90°, only to show up again coming in from *negative* infinity beyond 90°. Still, as strange as the function is, you can find its value for any angle except ±90°. You can conceptually find its tangent by looking it up on the graph: draw a vertical line for the desired angle, and the tangent is the corresponding value where this line crosses the curve.

Figure 6.1   Tangent definition.

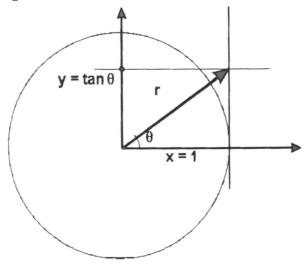

The same approach can be used in reverse: to find the arctangent, draw a horizontal line corresponding to a given value of the tangent; the desired angle is given by the point where this line crosses the tangent curve. (It might help if you turn the page sideways!) The good news is that the arctangent is well behaved and defined even for those infinite values. The bad news is that you have to deal with those infinities! You can't use a graphical approach with a computer. You need an equation, and you'll find it in the handbooks in the form of yet another infinite series, as given in Equation [6.2].

[6.2] $\tan^{-1}x = x - \dfrac{x^3}{3} + \dfrac{x^5}{5} - \dfrac{x^7}{7} + \dfrac{x^9}{9} - \dfrac{x^{11}}{11} + \dots$

Notice the notation: because the arctangent is the inverse function of the tangent, it is often written as tan^{-1} in mathematical formulas.

The series in Equation [6.2] is interesting. It's easy to remember because, except for one teensy little difference, it's identical to the series for sin(x), which is shown below for comparison.

[6.3] $$\sin x = x - \frac{x^3}{3!} + \frac{x^5}{5!} - \frac{x^7}{7!} + \frac{x^9}{9!} - \frac{x^{11}}{11!} + \ldots$$

Figure 6.2   The tangent function.

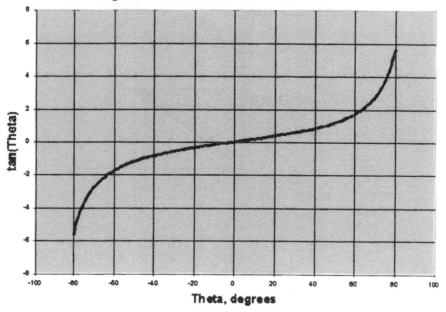

That "teensy" difference, however, makes *all* the difference. The exclamation point stands for the factorial function, and as I'm sure you know, the value of this function increases extremely rapidly as its argument increases. Because it appears in the denominators of the fractions in Equation [6.3], it guarantees that the series converges rapidly for all values of x. Even when x is huge and its powers are increasing rapidly with each higher term, the factorial will eventually overwhelm the numerator, and the terms approach zero.

By contrast, the convergence of the series for the arctangent is painfully slow, and it doesn't converge at all for values of |x| > 1. A picture is worth a thousand words, and in this case, you certainly can't grasp the enormity of the effect of that factorial sign without seeing it in action. Table 6.1 shows the value of each term in the arctangent series through the first 20 terms for the limiting value of 45°, where the tangent is equal to 1.0. The correct answer should be 45° expressed in radians, or 0.785398163.

At first glance at Table 6.1, the series appears to work well. After the first few terms, the answer seems to be closing in on the exact value. However, by the 20th term, you can see that something is badly wrong, because things are not improving very much. The error in the 20th term is only half that of the 10th. You can see that although the error is indeed decreasing (the series, after all, *does* converge), it is decreasing much too slowly. For a dramatic comparison, see the performance of the sine series shown in Table 6.2 and Figure 6.3.

Table 6.1   Arctangent series performance.

n	Term	Sum	Error
1	1	1	−0.2146
2	−0.333333333	0.666666667	0.118731
3	0.2	0.866666667	−0.08127
4	−0.142857143	0.723809524	0.061589
5	0.111111111	0.834920635	−0.04952
6	−0.090909091	0.744011544	0.041387
7	0.076923077	0.820934621	−0.03554
8	−0.066666667	0.754267954	0.03113
9	0.058823529	0.813091484	−0.02769
10	−0.052631579	0.760459905	0.024938
11	0.047619048	0.808078952	−0.02268
12	−0.043478261	0.764600691	0.020797
13	0.04	0.804600691	−0.0192
14	−0.037037037	0.767563654	0.017835
15	0.034482759	0.802046413	−0.01665
16	−0.032258065	0.769788349	0.01561
17	0.03030303	0.800091379	−0.01469
18	−0.028571429	0.77151995	0.013878
19	0.027027027	0.798546977	−0.01315
20	−0.025641026	0.772905952	0.012492

Table 6.2   Sine series performance.

n	Term	Sum	Error
1	1	1	–0.15853
2	–0.166666667	0.833333333	0.008138
3	0.008333333	0.841666667	–0.0002
4	–0.000198413	0.841468254	2.73e–06
5	2.75573e–06	0.84147101	–2.5e–08
6	–2.50521e–08	0.841470985	1.6e–10
7	1.6059e–10	0.841470985	–7.6e–13
8	–7.64716e–13	0.841470985	2.78e–15
9	2.81146e–15	0.841470985	0

Figure 6.3   Error in function approximations.

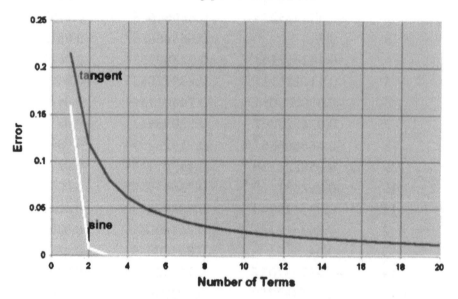

As you can see, the difference is dramatic. In fact, the error curve for the sine function in Figure 6.3 is rather dull; the error is not even visible after the third term because of the all-important factorial in the denominator. By contrast, the error behavior of the tangent in Figure 6.2 displays the classic $1/n$ behavior, as it should, because the denominators of Equation [6.2] are proportional to n.

Lest I paint the convergence of Equation [6.2] to be worse than it really is, I should point out that it does have one saving grace — the terms alternate in sign. To get a true picture of the error behavior, I should consider two terms at a time. The value of the nth and $(n + 1)$th terms, taken together, is:

[6.4]
$$v_n = \frac{x^{2n+1}}{2n+1} - \frac{x^{2(n+1)+1}}{2(n+1)+1}$$

$$= \frac{x^{2n+1}}{2n+1} - \frac{x^{2n+3}}{2n+3}$$

Factoring out the power of x and collecting terms gives

$$v_n = (x^{2n+1})\left[\frac{1}{2n+1} - \frac{x^2}{2n+3}\right]$$

$$= (x^{2n+1})\left[\frac{2n+3-x^2(2n+1)}{(2n+1)(2n+3)}\right].$$

For the case where $x = 1$,

$$v_n = \frac{2n+3-(2n+1)}{(2n+1)(2n+3)}$$

$$= \frac{2}{(2n+1)(2n+3)}.$$

Except for the smallest values of n, you can ignore the additive terms 1 and 3 and write

[6.5] $$v_n \approx \frac{1}{2n^2}.$$

So the error is not really proportional to $1/n$, but to $1/n^2$, which is still bad. How bad? Suppose you require eight significant digits. That means you will need

$$\frac{1}{2n^2} = 1.0 \times 10^{-8},$$

which gives

$$2n^2 = 1 \times 10^8$$

$$n^2 = \frac{1}{2} \times 10^8,$$

which leads to

[6.6] $n = 7,071.$

If you agree that calculating more than 7,000 terms is not practical, you will see that something radically different must be done to get a practical algorithm.

If the series for the arctangent converges so slowly, why use it? Maybe you'd be better off using the arcsine or arccosine series, after all. Fortunately, and perhaps surprisingly, you can convert this series to a different form that converges very fast, which is what this chapter is all about. I'll use some of the same tricks that I used in previous chapters — limiting the range of the input argument and using identities to compute the values for other arguments. Most importantly, however, I'll use a trick that you haven't seen before and one that you would be well advised to add to your toolbox. The trick is to simply recognize that the infinite series is not the only way to represent a function. If the series doesn't do the job, it's time to look for another, completely different representation, which often is dramatically better than any tweaked form of the original series.

Necessity is the Mother of Invention

At this point, it's time for a little personal history. In 1976 I was working on an embedded system (using an 8080) that required the computation of all the fundamental functions, including the arctangent. I didn't need very high accuracy, but I did need speed. Certainly I couldn't wait for the computation of 7,000 terms. The project also had a very short fuse, so there was no time to do lots of library research. I needed a fast algorithm, and I needed it quickly.

My reasoning went something like this: The conventional series is too "linear"; I needed a function that was much more nonlinear. Here's a good rule to remember: if you need a function that's inherently nonlinear, put something in the denominator of a fraction. In other words, whereas

$$f(x) = x$$

is the ultimate linear function, its inverse

$$g(x) = 1/x$$

is decidedly *non*linear.

At that point in my career I had heard of a functional form called "continued fractions," but I didn't know much about them and didn't have time to learn. But this tidbit of knowledge inspired me to try a solution in the form of a ratio of polynomials:

[6.7] $$f(x) = \frac{P(x)}{Q(x)},$$

where $P(x)$ and $Q(x)$ are polynomials.

This approach worked beautifully. I found that I could represent the arctangent of x to the desired accuracy using only three multiplications and one division with the following form.

[6.8] $$f(x) = \frac{x(1 + ax^2)}{1 + x^2(b + cx^2)}$$

I don't know about you, but where I come from, five terms is better than 7,000 times. Such is the power of rational polynomials.

Although the form I chose to fit the function was purely empirical, it was based on a certain amount of logic. First, I knew that for small x, the formula had to reduce to

$$\tan^{-1}(x) = x,$$

so I knew that the numerator must have a leading value of x. For the same reason, the rest of the function must reduce to 1.0 at $x = 0$. The higher powers of x let me tweak the detailed shape of the curve.

But what values should I use for the three coefficients a, b, and c? Because time was a factor, I took the coward's way out. I simply forced the function to be exact at four points: 0°, 45°, and two angles in between. This gave me three linear equations to solve for the coefficients. I won't give you the values, because they are not optimal, but the approach gave me a practicable formula in a short time, and the resulting implementation met all the needs of the program. Ah, the beauty of empiricism.

Rigor Mortis

After all the dust had settled, I knew what I had accomplished, but I wasn't quite sure *how*, or what the mathematical basis for the formula was. A year or two later the subject came up again when I needed an arctangent function for my then-new Radio Shack TRS-80 (Level I BASIC had no trig functions). This time, because I had some time to spend, I resolved to derive the formula with a lot more rigor and see if I could find a more general solution. Because long constants in BASIC take up valuable character space in the source program (which was not tokenized), I was also hoping to find some equivalent to Equation [6.2], in which the coefficients are all nice, short integers. The derivation that led me to a general solution represents the main thrust of this chapter, as well as an odyssey into the world of continued fractions. Follow along with me and you'll see it unfold.

I'll begin at the beginning, with Equation [6.2], which is reproduced here.

$$\tan^{-1}x = x - \frac{x^3}{3} + \frac{x^5}{5} - \frac{x^7}{7} + \frac{x^9}{9} - \frac{x^{11}}{11} + \dots$$

It's easy enough to see that x is present in every term, so a good beginning would be to factor it out.

$$\tan^{-1}(x) = x(\text{something}).$$

You'll also note that the "something" must reduce to 1 for small x. We're on the right track. The trick was to figure out what that something was. Just replacing it by the factored version of Equation [6.2] is no good — I'd still have slow convergence. Remembering my rule about nonlinearity, not to mention my past success with a rational fraction, I decided to try the form

[6.9] $$\tan^{-1}x = \frac{x}{1 + P(x)},$$

where $P(x)$ is some polynomial to be determined. I don't know what it is yet, but I can solve for it with a little algebra:

$$1 + P(x) = \frac{x}{\tan^{-1}x}$$

or

[6.10] $$P(x) = \frac{x}{\tan^{-1}x} - 1.$$

I already have a representation of $\tan^{-1}(x)$ in the slowly converging series. I can divide it into x by synthetic division. It's not an easy task — the process is tedious in the extreme, and it's terribly easy to make a mistake — but it's a feasible one. As my major professor in college liked to say, "Conceptually, there's no problem." It helps to have a computer program like Maple, Mathcad, or Mathematica that can do symbolic algebra, but I didn't have access to one at the time, and I was still able to solve the puzzle through brute force and tenacity.

After some work, I found that $P(x)$ was not a polynomial, but the infinite series

$$[6.11] \qquad P(x) = \frac{x^2}{3} - \frac{4x^4}{45} + \frac{44x^6}{945} - \frac{438x^8}{14,175} + \frac{10,196x^{10}}{467,775} - \frac{1,079,068x^{12}}{638,512,875} + \cdots .$$

Yes, the coefficients look messy, and no, I didn't have a clue how to write the general term. However, I was greatly encouraged by the form of the series. First, I saw that it only contained even powers — always a good omen. Second, I saw that the coefficient of the first term was the simple fraction, $1/3$. Finally, I noted that the terms have alternating signs, which aids convergence.

Thus encouraged, I decided to see if I could find a form that would simplify the remaining coefficients in the series. Therefore, I continued with the form

$$[6.12] \qquad P(x) = \frac{x^2}{3(1 + Q(x))} .$$

Another round of tedious algebra gave me the form for $Q(x)$.

$$[6.13] \qquad Q(x) = \frac{4x^2}{15} - \frac{12x^4}{175} + \frac{93x^6}{2,635} - \frac{7,516x^8}{336,875} + \cdots$$

Proceeding for one more cycle, I assumed the form

$$[6.14] \qquad Q(x) = \frac{4x^2}{15(1 + R(x))}$$

and got the following form for $R(x)$.

$$[6.15] \qquad R(x) = \frac{9x^2}{35} - \frac{16x^4}{245} + \frac{624x^6}{18,865} - \cdots$$

At this point, I began to wonder if I was really going in the right direction or simply being led on a wild goose chase. Yes, the leading terms of each series were still fairly simple, but subsequent terms were getting more complicated at every step, and there seemed to be no end to this trend. Performing the tedious algebra had long since ceased being fun, and I began to wonder if I was getting anywhere or if I was merely digging myself into a deeper hole. The answer came when I wrote the results to date, substituting each result into the previous equation. I got the form

[6.16]
$$\tan^{-1}x = \cfrac{x}{1+\cfrac{x^2}{3\left(1+\cfrac{4x^2}{15\left(1+\cfrac{9x^2}{35(1+\langle\text{something}\rangle)}\right)}\right)}} .$$

By canceling each leading term with the denominator of the next term, I got an even simpler form.

[6.17]
$$\tan^{-1}x = \cfrac{x}{1+\cfrac{x^2}{3+\cfrac{4x^2}{5+\cfrac{9x^2}{7(\langle\text{something}\rangle)}}}}$$

Although my algebra skills had just about run out of gas, I saw that I already knew enough to see the pattern. The leading constants in each denominator comprise the class of successive odd integers: 1, 3, 5, 7, ... Similarly, the numerators are the squares of the successive integers: 1, 2, 3, ... This gave me just enough information to guess that the pattern continues in the same way.

[6.18]
$$\tan^{-1}x = \cfrac{x}{1+\cfrac{x^2}{3+\cfrac{4x^2}{5+\cfrac{9x^2}{7+\cfrac{16x^2}{9+\cfrac{25x^2}{11+...}}}}}}$$

With a certain amount of trepidation, I tried this formula on the arctangent of 1. It worked! Subsequent analysis with Mathematica has proven that Equation [6.18] is, indeed, the right equation for the function.

The form in Equation [6.18] is called, for obvious reasons, a *continued fraction*. Although it is not seen as often as the power series, it has much in common with it. Where the power series calls for the sum of successive product terms, the continued fraction requires division of successive sums. In practice, as with the power series, the continued fraction must be truncated somewhere and computed from the bottom up. This makes the computations a bit more messy than for a power series, but sometimes the extra degree of difficulty is well worth the effort. This is one such time.

How Well Does It Work?

One thing that had not yet been established was whether or not the continued fraction converged any faster than the power series. From my success with the ratio of polynomials, I was pretty sure I already knew the answer, but I still had to verify it. So I computed $\tan^{-1}(1)$ for various orders (depths) of the continued fraction. The results are shown in Table 6.3. For comparison, I've shown the values of the original power series, using the same number of terms.

At 10 levels, the continued fraction has an error of only 0.000003 percent, or 25 bits of accuracy, which is good enough for single-precision floating-point arithmetic. At the same level, the power series is still trying to determine the first digit and has only five bits of accuracy. The continued fraction seems to be a roaring success. To emphasize the difference, the error behavior of the power series and the continued fraction are compared in Figure 6.4. As you can see, the rapid convergence of the continued fraction makes it far more practical than the power series.

Bear in mind that a continued fraction is not the solution to all problems. For example, the power series for the sine function is about as good a solution as you can get. However, in cases where the power series converges slowly, you can expect the continued fraction to converge rapidly. In general, the more slowly the series converges, the more rapidly its equivalent continued fraction will converge. That's a handy rule to remember.

Table 6.3 **Comparison of accuracy.**

Terms	Power Series	Continued Fraction
1	1.0	1.0
2	0.666666667	0.750000000
3	0.866666667	0.791666667
4	0.723809523	0.784313725
5	0.834920634	0.785585585
6	0.744011544	0.785365853
7	0.820934621	0.785403726
8	0.754267954	0.785397206
9	0.813091483	0.785398238
10	0.760459904	0.785398135

Figure 6.4 **Error, continued fraction.**

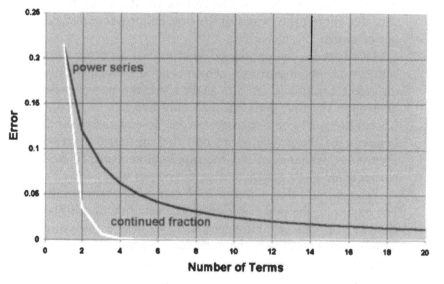

Is There a Rule?

Since doing the conversion described above, from series to continued fraction, I've done a fair number of others. It might interest you to know that you needn't confine yourself to series; you can apply the same technique to ordinary numbers. If the numbers are rational (meaning they can be expressed as the ratio of two integers), don't bother calculating the continued fraction. The interesting cases are the irrational numbers. I've shown a few examples in Table 6.4.

Table 6.4   **Continued fractions for constants.**

Define a sequence... $\langle a_0, a_1, a_2, ..., a_n \rangle$.

Represent the continued fraction as $a_0 + \cfrac{1}{a_1 + \cfrac{1}{a_2 + ... \cfrac{1}{a_n}}}$.

Then:

$\sqrt{2} = \langle 1, 2, 2, 2, 2, 2, 2, ... \rangle$

$\sqrt{3} = \langle 1, 2, 1, 2, 1, 2, 1, 2, ... \rangle$

$\sqrt{5} = \langle 2, 4, 4, 4, 4, 4, 4, ... \rangle$

$\sqrt{7} = \langle 2, 1, 1, 1, 4, 1, 1, 1, 4, 1, 1, 1, 4, ... \rangle$

$\phi = \langle 1, 1, 1, 1, 1, 1, 1, 1, 1, ... \rangle$ (golden ratio)

$e = \langle 2, 1, 2, 1, 1, 4, 1, 1, 6, 1, 1, 8, 1, 1, 10, 1, 1, 12, ... \rangle$

$\pi = \langle 3, 7, 15, 1, 292, 1, 1, 1, 2, 1, 1, 14, 2, 1, 1, 2, 2, 1, 4, ... \rangle$

As you can see, all except the pesky π turn out to have simple forms when expressed as continued fractions. Note that this doesn't make the numbers any less irrational: because both the decimal and continued fraction forms go to infinity, neither can be converted to a ratio of two integers. However, in the continued fraction form, the coefficients do have a pattern, which is not true for decimal digits. Only π has no repetitive form in this representation. Later, however, you'll see that you can give it a regular form also.

The continued fraction form delivers one great advantage over the decimal form of the number. By truncating, you can get successively better rational approximations to the number. The approximations for the square root of two are shown below as an example. Each successive ratio represents a better approximation to the actual value than the previous one. Contrast this to the process of finding rational approximations by trial and error. In that case, increasing the denominator by no means guarantees that the next approximation is an improvement over the previous one.

$$\sqrt{2} \approx 1, \frac{3}{2}, \frac{7}{5}, \frac{17}{12}, \frac{41}{29}, \frac{99}{70}$$

For the record, the last ratio happens to be that used by ancient mathematicians, who wrote it in the form

$$\sqrt{2} \approx \frac{1}{2}\left(\frac{10}{7} + \frac{14}{10}\right).$$

It is in error by only 0.007 percent, which is accurate enough to build pyramids.

You may well ask if there is an algorithm that lets one compute the coefficients of the continued fraction from those of the power series. If there is, I don't know about it. I've studied continued fractions for quite a while, and I have a few books on the subject. They have rules for how to compute the *values* of the fractions using recurrence relationships; however, I have not seen a formula for generating the continued fraction in the first place. As far as I know, the only way to do it is by using brute force, as in the method I've shown here.[1]

From Continued Fraction to Rational Fraction

There's only one problem with the continued fraction formulation: it involves successive divisions, which are almost always much slower than additions or even multiplications. You can eliminate this problem by rewriting the fractions in such a way as to "rationalize the denominator."

The infinite continued fraction of Equation [6.18] can't be rationalized, precisely because it's infinite. However, in practice, you must always truncate the fraction to a fixed number of terms. Once you do that, the fraction is no longer infinite and you can simplify the expression to the ratio of polynomials that I found empirically years ago.

---

1. Since this chapter was written, the author found two methods, due to Wallis and Euler, that give general ways to convert a power series to its equivalent continued fraction form. See: Battin, Richard H., *An Introduction to the Mathematics and Methods of Astrodynamics*, AIAA Education Series, American Institute of Aeronautics and Astronautics, 1999, pp.44–68.

For example, truncating at the level of $P(x)$ in my derivation I can write

[6.19]
$$\tan^{-1}x = \cfrac{x}{1 + \cfrac{\frac{x^2}{3}}{}}$$
$$= \frac{3x}{3 + x^2}.$$

The math to rationalize the continued fraction is not as tedious as that for the synthetic division, but it's tedious enough. Fortunately, programs like Maple, Mathcad, or Mathematica handle such problems with ease. In Table 6.5 I've given the equivalent rational fractions for all orders through Order 11. For the record, the formula I used in my original empirical solution was equivalent to that shown as Order 7. Using Mathematica, I've worked out the formulas through Order 21, but it's highly unlikely that you'll ever need them.

Table 6.5   Equivalent rational fractions.

$\tan^{-1}(x)$, where $y = x^2$

x	Order 1
$\dfrac{3x}{3+y}$	Order 3
$\dfrac{x(15 + 4y)}{15 + 9y}$	Order 5
$\dfrac{x(105 + 55y)}{105 + 90y + 9y^2}$	Order 7
$\dfrac{x(945 + 735y + 64y^2)}{945 + 1050y + 225y^2}$	Order 9
$\dfrac{x(1155 + 1190y + 231y^2)}{1115 + 1575y + 525y^2 + 25y^3}$	Order 11

To underscore the performance of the continued fraction, substitute the values $x = y = 1$ into the equations of Table 6.5, and you'll get the fractions shown in Table 6.6. These correspond to the first six "terms" in Table 6.3, so their errors are also the same.

Table 6.6   **Rational fractions: accuracy.**

Order	Fraction	Error
1	1/1	−0.214601836
3	3/4	0.035398163
5	19/24	−0.006268503
7	160/204	0.001084437
9	1,744/2,220	−0.000187422
11	2,576/3,280	0.000032309

Back Home on the Reduced Range

If you look carefully, you can see two things from the formulas and results in Tables 6.3 through 6.6 and Figure 6.4. First, as I've noted before, the performance of the continued fraction is dramatically better than the power series representation. At this point, you should be persuaded that you can attain any desired accuracy, simply by increasing the order of the approximation. If you were to plot the error characteristic logarithmically, you'd find that the error at $x = 1$ decreases by roughly a factor of six for each step (increasing the order by two). At lower values of x, it's even more dramatic. At $x = \frac{1}{2}$, the error is reduced by a factor of 10 at each step. Clearly, if you need more digits of accuracy, you need only increase the order.

On the other hand, as good as the performance of the continued fraction is, it's still not good enough. To get even 25 bits of accuracy (1 part in 3×10^7), you'd need 10 "terms" in the continued fraction, corresponding to Order 19. Because each order adds two multiplications and one addition, you'd need 18 multiplications, 9 additions, and a division to get even this single-precision result, and even more for higher accuracy. This is not an outrageously high number, as was the case for the series, but we'd still like to avoid the complexity that goes with such a high order.

Furthermore, you saw me say earlier that I found the Order 7 equation to be adequate. How could that be? The answer becomes clear when you look at the error curve for Order 11, which is shown in Figure 6.5.

As you can see, the error stays extremely low until near the end of the range. At $x = \frac{1}{2}$, the error is still barely more than 1×10^{-8}. It only begins to grow and it does that quickly, as x nears 1.0. The error curves for other orders are similar. As the order of approximation increases, the error decreases across the range, but the improvement is more dramatic at the lower values of x. To put it another way, the error curve becomes flatter, with a sharper "knee" near $x = 1$.

This characteristic of the error curve brings us back to the guiding principle learned in previous chapters: to get even more accuracy, you don't need to improve the formula, you simply need to limit the range of the input argument.

You've already seen two implicit reductions that are very powerful, although I didn't make much of them at the time. The tangent of an angle ranges all the way from $-\infty$ to $+\infty$. However, note that the error curves I've drawn and the calculations I've made are always in the range 0 to 1. Even more importantly, the power series and the continued fraction derived from it are not even valid for values of x outside the range -1 to 1.

Figure 6.5   Arctangent error function.

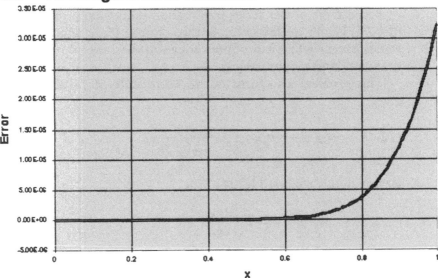

I can reduce the range in two steps. First, the tangent function is an odd function — odd in the mathematical, not the usual, sense.

[6.20] $\tan(-x) = -\tan(x)$

Using this identity reduces the range to only positive values. That's a shame because the formula is equally valid (and equally accurate) for negative values. However, eliminating negative values leaves room to make further reductions in range. Listing 6.1 is a "wrapper" that handles this bit of book-keeping.

Listing 6.1   Arctangent wrapper.

```
/* First wrapper for arctangent
 * Assures positive argument
 */
double arctan(double x){
    if(x < 0)
        return -arctan1(-x);
    else
        return arctan1(x);
}
```

Because the power series in Equation [6.2] is only valid for input values of $x \le 1$, I can't use the series, or the continued fraction derived from it, as it stands except when it produces angles in the range 0° to 45°. So how do I deal with larger input values?

This problem also turns out to be trivially easy. A second trig identity gives

[6.21] $\tan\theta = \cot\left(\dfrac{\pi}{2}-\theta\right) = \dfrac{1}{\tan\left(\dfrac{\pi}{2}-\theta\right)}.$

Reverting this equation leads to

[6.22] $\tan^{-1}x = \pi - \tan^{-1}\left(\dfrac{1}{x}\right).$

To solve for the arctangent of arguments greater than one, simply take the reciprocal of the input, find its arctangent, then subtract the result from π. In this way, those seemingly impossible cases where x approaches infinity become trivial. In fact, the closer I get to infinity, the faster the convergence. If I could represent a true infinity in the machine, its reciprocal would be zero and so would the output angle of the arctangent function.

Listing 6.2 is a second wrapper that deals with input values larger than 1.0. Between this wrapper and that in Listing 6.1, I can get an argument that is always in the range 0 to 1, as I've assumed in my analysis and plotted in Figure 6.5.

Listing 6.2   Reducing the range.

```
/* Second wrapper for arctangent
 * Assures argument <= 1
 */
double arctan1(double x){
    if(x <= 1.0)
        return arctan2(x);
    else
        return halpi - arctan2(1/x);
}
```

The Incredible Shrinking Range

We've seen that even this limitation is not enough — an argument of 1.0 still takes too many terms. Can we limit the range even further? The answer, as you might have guessed, is a resounding "yes." The mechanism for doing so comes from yet another trig identity, one of the double-angle formulas:

[6.23] $$\tan(a + b) = \frac{\tan a + \tan b}{1 - \tan a \tan b}$$

However, don't forget that I'm seeking the arctangent, not the tangent. It's sometimes difficult to remember what the input is and what the result is. Some changes in notation will help me keep my goal more easily in view.

Assume that the input value x is

[6.24] $x = \tan(a + b)$.

Also, let

[6.25] $\tan a = z$
              and
[6.26] $\tan b = k$,

where k is an unknown constant. Conversely,

[6.27] $a = \tan^{-1}(z)$
              and
[6.28] $b = \tan^{-1}(k)$.

Equation [6.23] now becomes

[6.29] $x = \frac{z + k}{1 - zk}$.

Solving this equation for z gives

[6.30] $z = \frac{x - k}{1 + kx}$.

Going the other way, from Equation [6.24],

$x = \tan(a + b)$,

which, reverted, gives

$\tan^{-1}x = a + b$.

Going back to arctangents gives

[6.31] $\tan^{-1}x = \tan^{-1}k + \tan^{-1}z$.

Remember, k is a simple constant; therefore, so is $b = \tan^{-1}k$, which I can precompute. Equation [6.31] says that I compute a new argument, z, from Equation [6.30], solve for its arctangent, then add it to the value of b.

It's important to note here that I have not broken the input argument into segments to reduce the range. I've simply introduced an offset, k, which is always applied. You can see how this helps by looking at the extreme values of z.

[6.32] $z(0) = \frac{0 - k}{1 + 0k} = -k$

[6.33] $z(1) = \frac{1 - k}{1 + k}$

The first value is clearly negative, with magnitude less than one, if k is itself less than one. The second value is clearly positive if k is less than one, and will also be less than one. Thus, by applying the offset, I've gotten back the negative side of my input argument. In doing so, I've reduced the maximum excursion to something less than one. Figure 6.6 shows the relationship between x and z for the case k = $^1/_2$. Note that the curve passes through zero when $x = k$; this must always be true, as is clear from Equation [6.30].

Figure 6.6   *x–z* **relationship.**

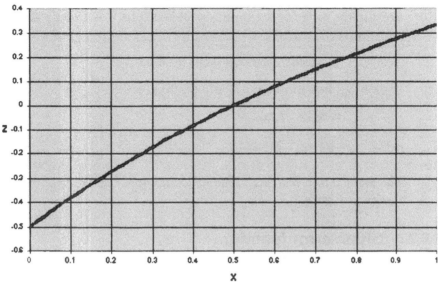

Note that I got this reduction without having to split the range up into regions and use different values of k (and therefore b) for each region. I still have only one region; I always apply the offset. Perhaps even more importantly, the same method opens the possibility of further range reductions, either by applying this one recursively or by breaking the input range into regions with a different k and *b* for each region. Either approach will let me reduce the range to an arbitrarily small region.

Choosing the Range

Now that I know how to reduce the range by introducing the offset k, the obvious next problem is what value(s) to assign to k. Over the years, I've tried a couple of methods. In my first few implementations, I was still thinking angles, so I tried to design an algorithm that would limit the range of the output variable to a fraction of 90°; say, 22.5° or 11.25°. Later, I decided that this made things too difficult. The input to the function is a numerical value for the tangent, not an angle, and it turns out to be easier to use simple fractions of the full range and let the angles fall where they may.

Most recently, for reasons that will become obvious, I've swung back the other way and now favor the angles again. Because the choice is not completely obvious, I'll give present both methods and the trade-offs for each. You are free to choose the one that best meets your needs.

Equal Inputs

Because the range of the input variable is 0 to 1, it seems reasonable to split the difference and let $k = \frac{1}{2}$. For this value, the arctangent is

[6.34] $b = 0.463647609$ radians or $26.5650518°$.

For $k = \frac{1}{2}$, Equations [6.32] and [6.33] give

$$z(0) = -1/2$$

[6.35] $z(1) = \dfrac{2-1}{2+1} = \dfrac{1}{3}$

The error curve for this value of k and an 11th-order approximation are shown in Figure 6.7.

Figure 6.7   Offset error function.

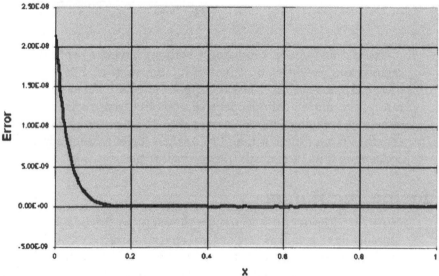

One look at this error curve gives graphic evidence of both the good and bad sides of the equal-value approach. On the one hand, I've reduced the maximum error, which now occurs at $x = 0$, from 0.000035 to barely 2×10^{-8} — a dramatic reduction by more than a factor of 1,500. On the other hand, the errors at the two extremes are horribly unbalanced; the error at $x = 1$ is not even visible: it's 130 times smaller than at $x = 0$. This is a sure sign that I have chosen a value of k that is far from optimal. Intuition and experience tell me that I am most likely to get the best fit, that is, the lowest maximum error, when the errors at the two extremes are equal in magnitude. This is the "minimax" principle, and it is a powerful one.

Folding Paper

You can gain wonderful insight into the nature of the range reductions using the following thought experiment. Imagine that you cut a round circle out of paper. This circle represents the full range of possible angles, from 0° to 360°. If it helps you to visualize the situation, imagine drawing horizontal and vertical lines on the circle, dividing it into four quadrants.

Although the angle can clearly fall into any of the four quadrants, the arctangent algorithm sees only two — the first and fourth. That's because the values of the tangent function for angles in the second quadrant are identical to those in the fourth, just as the values in the first quadrant are identical to those in the third. Because the arctangent function has no way to distinguish between the two pairs of quadrants, it traditionally returns values only in the first and fourth quadrant. To mirror this behavior, fold your circle about the vertical diameter into a semicircle, as in Figure 6.8-1.

Figure 6.8   Reducing the range.

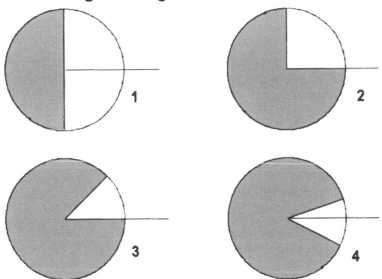

The first range reduction, as implemented in Listing 6.1, restricted the range to positive arguments and their corresponding angles. This restricted the angular range to the first quadrant — from 0° to 90°. Fold your semicircle vertically to reflect this reduction, as in Figure 6.8-2.

The second reduction used the reciprocals of arguments greater than one, which corresponded to angles greater than 45°. To show this reduction, fold your paper one more time to get an octant ranging from 0° to 45°, as in Figure 6.8-3.

The last reduction involved introducing an offset value, k, which corresponded to a rotation of the remaining octant. The rotation in Figure 6.8-4 shows a rotation through the angle $b = 26.565°$, which I got by letting $k = \frac{1}{2}$. You can now see that this was the wrong angle, which explains why the error curve is so greatly unbalanced. To get both a balanced error curve and the minimum excursion from zero, which should also yield the minimum largest error for a given range, balance the active area around zero by rotating the paper clockwise by 22.5°.

This method of viewing the range reductions not only illustrates a simple way of understanding the process, it suggests how to do further range reductions easily. For example, you needn't stop at the rotation through 22.5°. You could fold the paper again and rotate through 11.25° and repeat the process ad infinitum, reducing the angular range by a factor of two at each step. The only limiting factor is how many regions you're willing to accept.

Balanced Range

Now you can see that for optimal error performance you should use an offset that balances the output angle, not the input argument. Accordingly, let k be chosen so that

[6.36] $\tan^{-1}k = \dfrac{\pi}{8} = 22.5°$.

The value of k, then, is simply

[6.37] $k = \tan(22.5°)$.

It's easy enough to calculate this value on a pocket calculator, but you'll gain a little more insight into the situation by applying Equation [6.23], with $a = b = \pi/8$.

$$\tan(a+b) = \frac{\tan a + \tan b}{1 - \tan a \tan b}$$

If $a = b = \pi/8$, then $a + b = \pi/4$, and its tangent is equal to 1. Letting $q = \tan a$, gives

[6.38]
$$1 = \frac{2q}{1-q^2}$$
$$1 - q^2 = 2q$$
$$q^2 + 2q - 1 = 0.$$

This is a quadratic equation, so it has two roots. Applying the quadratic formula gives

$$q1 = \sqrt{2} - 1$$
$$q2 = -\sqrt{2} - 1.$$

Because the second root gives a negative value, I'll accept the positive one and write

[6.39] $\qquad q = \tan\frac{\pi}{8} = \sqrt{2} - 1.$

My calculator verifies that

$$k = \tan 22.5 = 0.414213\ldots,$$

which is the fractional part of the square root of two.

Now look at the value of z for the two extreme points. You've already seen from Equation [6.32] that

$$z(0) = -k.$$

In this case,

[6.40] $\qquad z(0) = -(\sqrt{2} - 1).$

To see the value at $x = 1$, I begin with Equation [6.33].

$$z(1) = \frac{1-k}{1+k}$$

and substitute my value for k, which gives

$$z(1) = \frac{1 - (\sqrt{2} - 1)}{1 + (\sqrt{2} - 1)}$$

$$= \frac{2 - \sqrt{2}}{\sqrt{2}}$$

$$= \frac{2\sqrt{2} - 2}{2}$$

$$= \sqrt{2} - 1$$

or

[6.41] $z(1) = k$.

Using equal output angles does indeed balance the range. The value of z, which is the argument for the continued fraction, now ranges between –k and k, which will always be less than one. Because the error is a function of z, the errors at the two endpoints will also be equal, and the minimax condition will be achieved.

[6.42] $|\text{error}(z(0))| = |\text{error}(z(1))|$

An Alternative Approach

I should pause here to mention that I could have approached this solution from the opposite end. I arrived at the balanced-angle solution by looking at the paper-folding example of Figure 6.8, and I rightly reasoned that equal arguments, z, would result in equal magnitudes for the errors at the two extremes. Suppose, however, I hadn't been able to figure this out. I could still have arrived at the same result by requiring the minimax condition, Equation [6.42], and working backward to determine k. The errors at the extremes will be equal in magnitude if

[6.43] $z(0) = -z(1)$.

Proceeding from there and using the definition for z given in Equation [6.30], I get

$$k = \frac{1 - k}{1 + k}$$

or

$$k(1+k) = 1-k$$

[6.44] $k^2 + 2k - 1 = 0$.

This is a quadratic equation in k. The positive root is

[6.45] $k = \sqrt{2} - 1$,

which is the same result I got by assuming k was the tangent of 22.5°.

Either way, this value of k is optimal for the range $x = 0$ to 1. A function to reduce the range using these values is shown in Listing 6.3. The optimized error curve is shown in Figure 6.9. (Note the call to the library function _atan() in Listing 6.3. I'm assuming that this function is provided by the programmer using one of the continued fraction forms. Remember, the whole purpose of this exercise is to *provide* the library function. If your compiler already has a satisfactory one, I'm wasting your time.)

Listing 6.3   Reduction by rotation.

```
/* Third wrapper for arctangent
 * Reduces argument via rotation
 */
double arctan3(double x){
  static const double b = pi/8;
  static const double k = root_2 - 1;
  return(b + _atan((x - k)/(1 + k*x)));
}
```

I've now reduced the maximum error another order of magnitude compared to Figure 6.7 and a full factor of 15,000 less than in Figure 6.5. Clearly, reducing the range by including an offset makes a profound difference in error behavior — especially if the offset is optimized — and that, in turn, allows you to use a lower order approximation. The degree of improvement can be seen clearly in Table 6.7, where I've given the maximum error for each order of approximation.

Figure 6.9   Error with optimal offset.

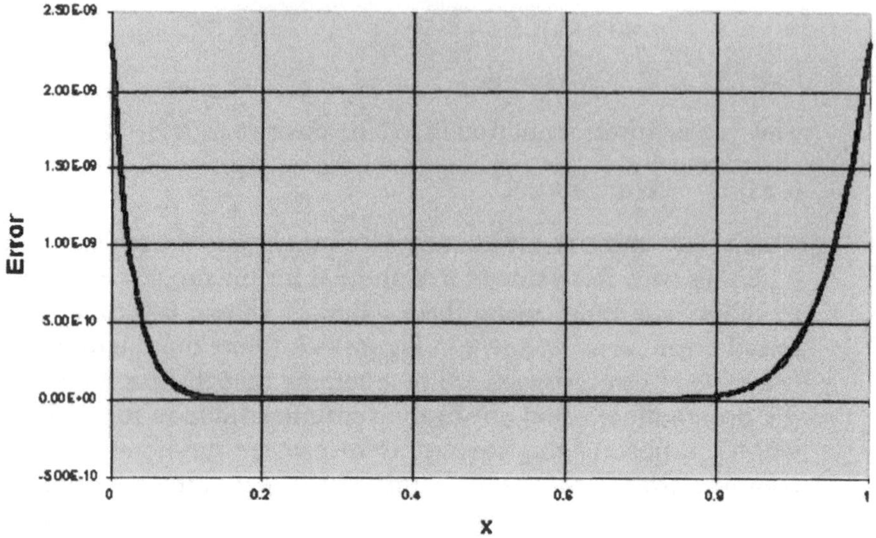

It's worth noting again that I achieved this range reduction *without* having to test the argument. I simply applied Equation [6.31] in all cases, producing an effective rotation of the octant by 22.5°.

However, I could easily perform further reductions by repeating the process, using splits every 11.25°. The last column of Table 6.7 shows the errors if I do this. Just as in the paper-folding exercise, you can continue this process as long as you like.

Table 6.7   Effect of range reduction.

Order	Full Range	One Reduction	Dual Reduction
1	0.214602	0.021514	0.002563
3	0.035398	0.000893	2.64e–05
5	0.006269	3.63e–05	2.63e–07
7	0.001084	1.46e–06	2.6e–09
9	0.000187	5.81e–08	2.54e–11
11	3.23e–05	2.31e–09	2.48e–13

The two-reduction case involves only a simple range test. The code to implement this reduction is shown in Listing 6.4. The errors for this two-stage reduction are shown in the third column of Table 6.7.

Listing 6.4   Optimal two-segment reduction.

```
/* Two-stage reduction
 * Uses equal-angle approach
 */
double arctan5(double x){
  static const double b1 = pi/16;
  static const double b2 = 3*pi/16;
  static const double b_mid = pi/8;
  static const double k1 = tan(b1);
  static const double k2 = tan(b2);
  static const double k_mid = tan(b_mid);
  double k, b;
  if(x <= k_mid){
    k = k1;
    b = b1;
    }
  else{
    k = k2;
    b = b2;
    }
  return(b + _atan((x - k)/(1 + k*x)));
}
```

As you can see from the table, the reduction in errors attributable to using the optimal offset is quite dramatic. Where before, even Order 11 was inadequate over the full range, I can now get single-precision accuracy with Order 7 using one reduction and with Order 5 using two. With two reductions, I can get accuracy much better than one part in 10^8, using only Order 7 — the same order I used in my first attempts. This kind of accuracy is a far cry from that given by the series approximation I started with. To grasp the full effect of the improvement, take another look at Table 6.5 and note the extreme simplicity of the equations for Orders 5 and 7.

More on Equal Inputs

Now that I've shown that equal-angle approach is optimal for the single-reduction case and you now have a mnemonic device for visualizing further reductions using equal angles, you may wonder why I bothered to mention the equal-value option at all. It clearly is inferior to the equal-angle technique in the single-reduction case.

The answer becomes clear if you look at higher numbers of reductions. The relationship between an angle and its tangent is clearly a nonlinear one. This is hardly surprising, but the implication for the equal-angle case is that if you divide the input range into regions defined by equal angles, the values of x separating the regions will not be equally spaced. This means you cannot use such tricks as masking and looking at high bits to determine which is the correct region.

The required process amounts to a table lookup. Although you can speed it up using a binary search or other similar mechanism, you can't escape the fact that the larger the number of segments, the slower the search process. Optimization of the algorithm, then, depends on trading off complexity in the continued fraction for lower numbers of segments against the time required for the search process for larger numbers of segments.

The situation is quite different if you use the equal-value approach. If the segments are equally spaced in the input parameter, you can use techniques similar to those used for the sine function in Chapter 5. Compute the segment number using a simple modulo computation and use that segment number as an index into a table of k and b values. For this case, the time required to locate the segment is a fixed (and small) time, regardless of how many segments are used.

This, then, is the real (and only) advantage of the equal-value method: It excels when the number of segments is large. If you make the number large enough, you can think of the algorithm not as a computation at all, but more as a table lookup with perhaps a more sophisticated and smart interpolation algorithm. Taken to the limit, the arctangent formula becomes the first-order formula and a pure table lookup with linear interpolation. The error characteristic of the equal-value method is bound to be worse than the equal-angle method because, as you've seen, the errors are highly unbalanced at the two extremes of each segment. However, if you use enough segments, the errors will be small in any case, so the issue might be moot.

To illustrate the differences between the two approaches, Listing 6.5 shows an example using four equal-value segments split on the fractions $^1/_8$, $^3/_8$, $^5/_8$, and $^7/_8$. The largest magnitude of the argument for this case is –k or

–0.125. At this magnitude, you can expect the error to be around 1.0e–8, using the fifth-order formula, or 4.0e–11, using seventh order. Figure 6.10 shows the error behavior for the equal-value fifth-order case. In this figure, I've left off the absolute value computation so you can see the discontinuities in the errors from segment to segment.

Listing 6.6 shows the code for an equal-angle approach, also using four segments.

Listing 6.5   Four equal-value segments.

```
/* Four-stage reduction
 * Uses equal-input approach
 */
double arctan4(double x){
  static const double atan_125 = 0.124354994;
  static const double atan_375 = 0.358770670;
  static const double atan_625 = 0.558599315;
  static const double atan_875 = 0.718829999;
  double k, b;
  int n = 4*x;
  switch(n){
    case 0:
      k = 0.125;
      b = atan_125;
      break;
    case 1:
      k = 0.375;
      b = atan_375;
      break;
    case 2:
      k = 0.625;
      b = atan_625;
      break;
    case 3:
      k = 0.875;
      b = atan_875;
      break;
  }
  return(b + atan((x - k)/(1 + k*x)));
}
```

Listing 6.6 Four equal-angle segments.

```
/* Four-stage reduction. Uses equal-angle approach */
double arctan7(double x){
  static const double b0 = pi/32;
  static const double b1 = 3*pi/32;
  static const double b2 = 5*pi/32;
  static const double b3 = 7*pi/32;
  static const double k0 = tan(b0);
  static const double k1 = tan(b1);
  static const double k2 = tan(b2);
  static const double k3 = tan(b3);

  static const double b_mid = pi/8;
  static const double b_hi = 3*pi/16;
  static const double b_lo = pi/16;
  static const double k_mid = tan(b_mid);
  static const double k_hi = tan(b_hi);
  static const double k_lo = tan(b_lo);
  double k, b, z;

  if(x <= k_mid)
    if(x <= k_lo){
      k = k0;
      b = b0;
      }
    else{
      k = k1;
      b = b1;
      }
  else
    if(x <= k_hi){
      k = k2;
      b = b2;
      }
    else{
      k = k3;
      b = b3;
      }
  return(b + atan((x - k)/(1 + k*x)));
}
```

For purposes of comparison, I've also shown the error behavior for the equal-angle approach in Figure 6.10. Comparison of the two curves thoroughly illustrates the superior accuracy of the equal-angle case. However, you can expect the search process needed to locate the correct segment to be more time consuming than for the equal-value case. More importantly, the equal-input approach can be extended much more easily to larger numbers of segments. The bottom line is that optimality of the errors becomes less and less important if they're well within bounds.

Figure 6.10   Error with multiple input regions.

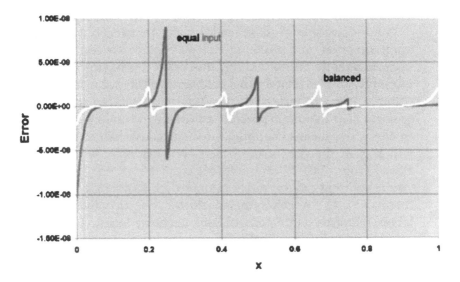

Which approach is best? As I said at the outset, the answer is not obvious, which is why I'm giving you both methods. However, in the remainder of this chapter, I'll assume the equal-angle method because of its superior error behavior.

Problems at the Origin

At this point, I've reduced a seemingly impossible problem — computing the arctangent from a slowly convergent power series — to an algorithm that's both fast and accurate. I accomplished this in two steps by

- converting the power series into an equivalent continued fraction and
- applying range reductions to limit the input argument.

In spite of these steps, the error curves in Figure 6.10 still leave a lot to be desired. First, there are sharp discontinuities with every move from one segment to another. This kind of behavior is generally not acceptable in function approximations. If, for example, you apply divided differences to the outputs to get a rate of change, you will see spikes in this rate at the transitions between segments. This is not good. The syndrome is very difficult to cure. It's almost built into the approximation formulas. However, in a moment you'll see one approach that does eliminate this effect.

Perhaps more worrisome is the fact that the largest error is at the origin. It's worrisome because for small angles

[6.46] $\tan(x) \approx x$.

Your "customers" have every right to expect this behavior in your approximation, and anything else will not be welcome. As they stand, the approximations do not exhibit this desired behavior because of the use of angle offsets. Fortunately, this problem is easily solved: simply do not use an offset near $x = 0$. Instead of dividing the range into equal angles, make the segment near the origin special and use the formula verbatim with no offset. In effect, this means chopping off the negative half of the one segment so it will be half the size of the others. For example, with two segments, one ranges from 0° to 15° and another from 15° to 45°, split on 30°. In effect, you're splitting the 45° range into an odd number of segments (i.e., an odd fraction of 45°, or $\pi/4$ radians), all but one of which will be paired into larger segments with positive and negative lobes. The first few possible arrangements are listed below.

- Three segments of 15° each ($\pi/12$)
- Five segments of 9° each ($\pi/20$)
- Seven segments of 6.42857° each ($\pi/28$)
- Nine segments of 5° each ($\pi/36$)
- Eleven segments of 4.0909091° each ($\pi/44$)
- Thirteen segments of 3.461538462° each ($\pi/52$)
- Fifteen segments of 3° each ($\pi/60$)

From the information you have already, you can calculate the maximum error associated with each of these cases. This is shown in Table 6.8 for various orders. You should now be confident that you can not only generate the arctangent function to any desired degree of accuracy, but you can do so quickly and with a reasonable number of segments and approximation orders.

Table 6.8   **Error for odd segment counts.**

Order	Segments			
	1	**3**	**5**	**7**
1	0.2146	6.15e–3	1.30e–3	4.73e–4
3	0.0354	1.13e–4	8.60e–6	1.59e–6
5	6.27e–3	2.01e–6	5.48e–8	5.16e–9
7	1.08e–3	3.54e–8	3.45e–10	1.65e–11
9	1.87e–4	6.19e–10	2.16e–12	5.26e–14
11	3.23e–5	1.08e–11	1.34e–14	—
13	5.56e–6	1.88e–13	—	—
15	9.57e–7	3.28e–15	—	—
	9	**11**	**13**	**15**
1	2.22e–4	1.22e–4	7.36e–5	4.79e–5
3	4.51e–7	1.65e–7	7.17e–8	3.50e–8
5	8.85e–10	2.17e–10	6.73e–11	2.47e–11
7	1.71e–12	2.81e–13	6.24e–14	1.72e–14
9	3.32e–15	—	—	—

Getting Crude

Now that you've seen the theoretically "correct" approximations, perhaps it's time to ask what the *least* complicated algorithm is that you can find. The obvious first choice must be the first-order approximation

$$\tan^{-1}(x) = x,$$

but this is crude indeed. The main problem is that the fit gets worse and worse as x gets larger, although the formula is exact for values of x near zero. But all is not lost. With a little tweaking, you can still use the idea of a linear equation but trade some of the error within the range for better accuracy near the end. Accordingly, assume that

[6.47] $\tan^{-1}(x) = ax.$

You now have a coefficient that you can play with to improve the fit. (You might be wondering why I didn't generalize further and use an offset constant. The reason is that whatever algorithm I use, I still want it to be exact at $x = 0$.)

With a value of $a \neq 1$, you will see the error begin to grow as x increases. However, with a proper choice for a, you can make the error pass through zero somewhere in the range, and then go in the opposite direction. The optimal value is one where the two errors are equal and opposite. It takes a bit of calculus to find the values, but the solution is fairly straightforward. Figure 6.11 shows the first-order function with no adjustment and optimized for the range 0 to 1. Figure 6.12 shows the error curves for the same two cases. The optimization gives an error reduction of about a factor of four. The error now amounts to less than 0.05 radians, or less than 3°. One can hardly argue that a function giving this much error is practical in real-world problems, although there may be cases (computer graphics, for example) where it is. However, the general approach for minimizing error certainly seems to hold promise.

Figure 6.11   Linear approximation.

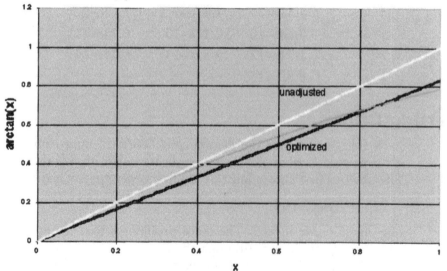

Figure 6.12 Error, linear approximation.

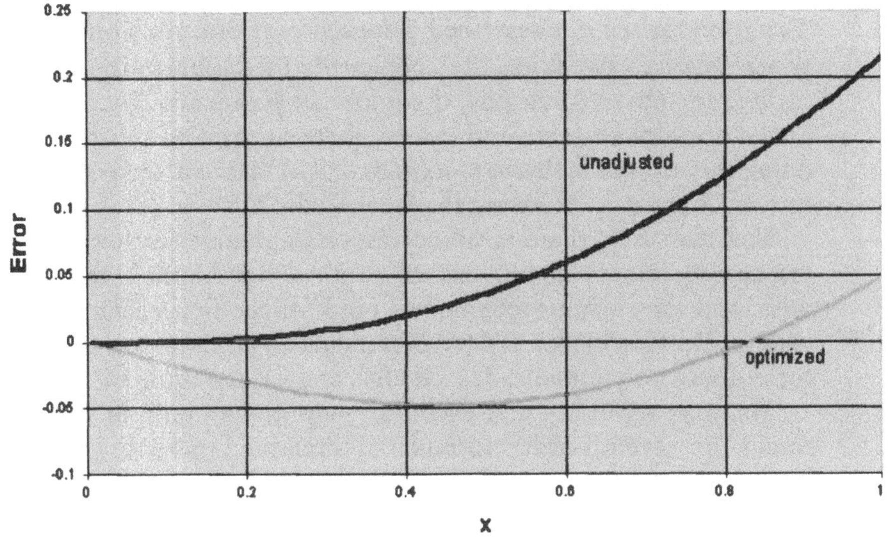

Table 6.9 shows the optimized coefficient and the maximum error for four choices of segmentation.

Table 6.9   Optimized first-order function.

Segments	a	Maximum Error
full (0 to 1)	0.83327886	0.0478807008
1	0.96073345	5.2497432649e-3
3	0.98417502	1.3335152673e-3
5	0.99401079	3.0956013149e-4
7	0.99690021	1.1516239849e-4

These results are nothing short of remarkable. With only three "segments," which are really only two (0° to 15° and 15° to 45°), and with only a simple first-order approximation, I got an error less than 0.8°. This kind of error is too great for high-precision work, but for many real-time embedded applications, the accuracy of this simple algorithm is more than sufficient. With seven "half-segments," you achieve an accuracy three times better than the 7th-order continued fraction was able to provide over the full range, 0 to 1.

If you're looking for a quick-and-dirty algorithm that requires very little arithmetic, look no further. This is it.

Don't Chebyshev It — Minimax It!

The previous section described a special case of a much more general one, which involves tweaking the coefficients to minimize the maximum error (called, for obvious reasons, the minimax problem). For power series such as the ones for the sine and cosine, there's a well-established technique for doing this using Chebyshev polynomials. I showed the results of that process in Chapter 5, "Getting the Sines Right."

Unfortunately, there is no corresponding analytical technique for minimaxing the errors in the case of rational polynomials that we're dealing with, and the calculus technique I used for the first order gets out of hand very quickly for higher orders. Nevertheless, minimaxing can be done, even for rational polynomials. The results can be quite dramatic.

You may recall my first encounter with the arctangent function, where I found the seventh-order formula of Equation [6.8] to give me adequate accuracy, even over the full range of 0 to 1. From what you've seen so far, it's difficult to believe that such an algorithm would be sufficient. I said that I didn't need high accuracy — 16 bits was enough — but Table 6.7 says I could only expect an accuracy of about 0.001 radians, or 0.06°. That's pretty crude and is only 12 bits of accuracy. So where did the required accuracy come from? The answer was the I had obtained a mimimax of sorts.

The general subject of the minimax process is far too deep to go into here; entire books have been written on that subject alone. However, I'll outline the process using that same seventh-order method, so you can get the general idea. Following this, I'll give the results of my efforts for many cases of practical interest.

The error curve for the seventh-order formula is shown in Figure 6.13. As usual, the error is essentially invisible below $x = \frac{1}{2}$. From there to the limiting value of $x = 1$, the error takes off and rises sharply.

You should ask yourself whether you can reduce the sharp peak near $x = 1$ by allowing more error in the midrange values. This philosophy is the same one just used to optimize the first-order case.

Figure 6.13   Error, seventh-order approximation.

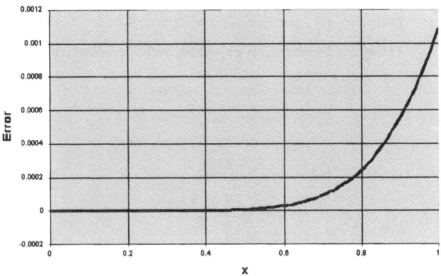

The solution is also the same: minimax. To optimize the error, the error curve must cross zero a certain number of times (the more, the better). You'll recall I got my first empirical formula by forcing zeros at the endpoint of $x = 1$ (output $= 45°$) plus two points in between. To see the way the method works, I'll revisit it precisely as I did it 22 years ago. I'll begin by repeating Equation [6.8],

$$f(x) = \frac{x(1 + ax^2)}{1 + x^2(b + cx^2)},$$

then define the three values of x for which the formula is exact.

[6.48] $$\tan^{-1}(x_i) = \frac{x_i(1 + ax_i^2)}{1 + x_i^2(b + cx_i^2)} \qquad (i = 1, 2, 3)$$

Although it may not look like it, this equation represents a linear equation in the three unknowns a, b, and c (remember, here the x's are not the variables — we're using them as "constant" input variables, and seeking to find the corresponding values of the "constants" a, b, and c). To see this more clearly, multiply the denominator to get

$$\tan^{-1}(x_i)[1 + x_i^2(b + cx_i^2)] = x_i(1 + ax_i^2).$$

In standard form, this becomes

[6.49] $ax_i^3 - b\,\tan^{-1}(x_i)x_i^2 - c\,\tan^{-1}(x_i)x_i^4 = x_i - \tan^{-1}(x_i)$.

I now have a linear algebra problem, which I can cast into matrix form:

[6.50] $A = \begin{bmatrix} x_0^3 & -x_0^2\tan^{-1}(x_0) & -x_0^2\tan^{-1}(x_0) \\ x_1^3 & -x_1^2\tan^{-1}(x_1) & -x_1^2\tan^{-1}(x_1) \\ x_2^3 & -x_2^2\tan^{-1}(x_2) & -x_2^2\tan^{-1}(x_2) \end{bmatrix}$

and

[6.51] $U = \begin{bmatrix} x_1 - \tan^{-1}(x_1) \\ x_2 - \tan^{-1}(x_2) \\ x_3 - \tan^{-1}(x_3) \end{bmatrix}$.

The desired coefficients can be found by inverting the matrix to get $Ak = U$, where k is the vector of coefficients {a, b, c}, in that order. Now I can compute the desired coefficients by inverting the matrix and writing the linear algebra solution.

[6.52] $k = \begin{bmatrix} a \\ b \\ c \end{bmatrix} = A^{-1}U$

It remains only to choose the x_n. In my original analysis, I wanted the error at the endpoint to be zero (a very good idea, by the way), so I chose one x to be 1.0. For the other two, I chose values by dividing the 0° to 45° region into three equal parts.

[6.53] $x_1 = \tan(15°) = 0.26795$
$x_2 = \tan(30°) = 0.57735$
$x_3 = \tan(45°) = 1.0$

For these values, I found the coefficients.

[6.54] $a = 0.4557085$
$b = 0.78896303$
$c = 0.06450259$

The resulting error curve is shown in Figure 6.14. As you can see, the error function does indeed cross zero at the points specified. The error is zero at $x = 0$, $x = 1$, and the two points in between. Between these zeros are hills and valleys as expected. The amplitudes of these hills and valleys, however, are necessarily affected by the need to pass through the zeros.

Figure 6.14   **Empirical formula**

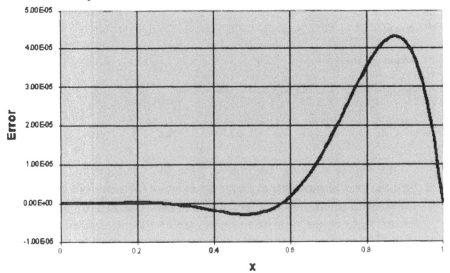

You won't fully grasp the significance of this error curve until you compare the maximum error with that of Figure 6.13. In that graph, the maximum error was about 0.001. The new error curve never exceeds a value of about 4.2×10^{-5} — an improvement by a whopping factor of 24. This error is equivalent to an angle of 0.002°, which was more than adequate for my needs at the time. Figure 6.15 shows the two error curves together for easy comparison. Shown on the same scale, the lump in Figure 6.14 doesn't loom nearly so large.

Figure 6.15 Error comparisons.

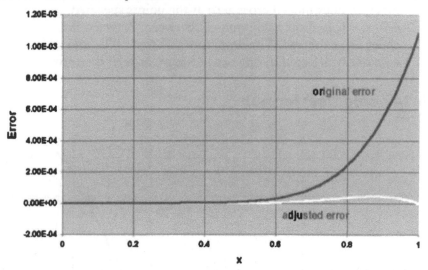

But this is only the beginning. One look at Figure 6.14 should also tell you that my choices for x_1, x_2, and x_3 were far from optimum. The strategy inherent in the minimax approach is to seek coefficients such that the height of the highest peak (positive or negative) was minimized. A little reflection should convince you that this is best accomplished when all the peaks have the same magnitude. If that isn't true, if one peak is higher than the others, you would be able to reduce it by shifting the values of x around a bit. This shifting would adversely affect the other peaks in turn. Therefore, you're going to get the smallest worst case error when all the peaks are the same magnitude. This optimum situation, which represents the minimax solution, occurs when all the peaks have equal amplitudes and alternating signs.

I've done just that kind of tinkering for the seventh-order function given in Equation [6.48]. The optimum values are as follows.

[6.55]
$$x_1 = 0.542$$
$$x_2 = 0.827$$
$$x_3 = 1.0$$

[6.56]
$$a = 0.42838816$$
$$b = 0.76119567$$
$$c = 0.05748461$$

Figure 6.16   Minimaxed error.

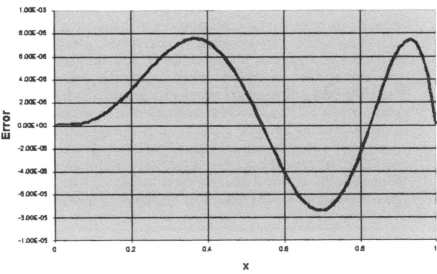

The new error curve is shown in Figure 6.16. You can see how the minimax process balances the heights of all the peaks. The maximum error is now down to less than 8×10^{-6}, which is equivalent to $0.00046°$, or only 1.65 seconds of arc. That's a small angle in anybody's book (it's about the apparent diameter of a dime, one mile away), and more than ample for almost any embedded application short of the Hubble Space Telescope. Through minimaxing, I've managed to increase the accuracy of the seventh-order formula by a whopping 125 times.

You may be wondering how I found the magic values of the coefficients. Twenty years ago, when I was seeking a true minimax solution for Equation [6.8], I found the coefficients by brute force. I programmed my old Tandy TRS-80 to search all combinations of *a*, *b*, and *c* until the maximum error was minimized. Time was not an issue here; I needed to compute the numbers only once, and since the computer didn't have much else to do, I was content to let it grind all day, all night, or even all week, if necessary. I did give the program a few smarts, in that it started with a fairly course step size and reduced it as it got near the optimal solution, but mostly it performed an exhaustive search. More sophisticated computer programs use general-purpose optimization algorithms to find the solutions in much less time.

However, the approach you just saw, in which you specify the values of *x* at the zero crossings and use linear algebra to compute the coefficients, is a more indirect, but far more effective, way of optimizing the coefficients. The great advantage of this approach is that every error curve generated has the required number of zero crossings. If you begin with Equation [6.8] and try to guess the values of the coefficients by trial and error, you will find it difficult to get even one zero crossing, much less three. Using the indirect method described in Equations [6.48] to [6.52], you are guaranteed to get those crossings and are left with the much simpler job of adjusting the heights of the error peaks. If you have access to a program like Mathcad, you'll find that about a half hour spent twiddling the x_n is sufficient to get the peaks equal within the limits of a graph. Such a program lets you see the error curve as you alter the values of x, and the effect of a change is readily apparent. Thus, it's not difficult to optimize even a trial-and-error fairly high-order formula, using the tried and true eyeball method. Except when the number of zero crossings is quite large, such an approach is adequate.

Mathcad

For some years now, I've strongly recommended Mathcad, the math analysis program from Mathsoft, Inc. My usual comment is, "If you don't have Mathcad now, get it now!"

Considering that background, you can imagine my embarrassment, not to mention disgust, when I upgraded to Mathcad v7.0, only to find that it was seriously broken. I'm not talking about minor inconveniences here; I'm talking major errors.

I have discussed this problem with the folks at Mathsoft, who have always been very nice and helpful, and they explained that the Mathcad user interface had been heavily modified based on feedback from users. Apparently, this also required changes in the interface between Mathcad and Maple, which Mathcad uses as its symbolic engine.

The Mathsoft support person told me that they were aware of some of the problems I'd found, and she supplied patches to fix them. Other problems, she tried unsuccessfully to convince me, were not bugs, but features. I must admit that the patched version is much more reliable and doesn't give me OS-crashing errors as before, but I still don't like the new user interface much, and those features are still bugs. More recently, I received Mathcad 8.0 which was worse yet and required three more patches to be usable. Mathcad 2000 is more of the same.

Someone pointed out that the problem with listening to complaints from users, in regard to user interfaces, is that the people you're most likely to hear from are the beginners who have just started using the program and can't figure out how to do things. Those of us who have been using the program for a while have learned — sometimes easily, sometimes not — how to deal with the interface that's there. We don't tend to call in to complain.

Although it's nice to have a program that's easy for new users to learn, it's even more important to have one that "power users" can use effectively. Sometimes, listening only to the complaints of new users can lead one down the garden path.

In my personal opinion, Mathcad version 7.0 and beyond, patched or not, is a distinct step downward from version 6.0, which was itself a step downward from version 5.0. Everything takes longer to do and is less reliable than in the older versions.

Because I've recommended Mathcad so strongly to my readers for a few years now, this puts me in an awkward position. I still like the program, but in all conscience, I can't recommend version 7.0 or later, patches or no patches. My best advice to you is, if you have version 6.0 or earlier, do not upgrade! If you haven't bought a copy yet, try to find a copy of Mathcad v5.0.

Smaller and Smaller

At this point, you may think you've come about as far as you can go in terms of optimization of the algorithm. Certainly I've squeezed all the performance I can get out of the seventh-order formula, right?

Wrong. For starters, recall that I forced the error function to be zero at the limit $x = 1$. This is not a bad idea: if I plan to split the range into separate regions later, it's nice to have zero error where the regions stitch together; otherwise, the approximations at those boundaries will be discontinuous, which is never a good idea.

Nevertheless, if minimum error is your goal, you can reduce it by relaxing that requirement. The optimum parameters for that case are shown below, and the error curve is shown in Figure 6.17.

$$[6.57] \quad \begin{aligned} x1 &= 0.533 \\ x2 &= 0.808 \\ x3 &= 0.978 \end{aligned}$$

[6.58]
$$a = 0.43145287$$
$$b = 0.7643089$$
$$c = 0.05828781$$

This step doesn't reduce the error much, but it's now about 6×10^{-6}, a reduction of 25 percent. Is a reduction of error by 25 percent worth accepting a discontinuity in the error behavior at the boundaries between regions? My gut feel is that it's not. I'd rather be sure that the error remains continuous at the boundaries. On the other hand, if the error is small enough, perhaps nobody cares what its value is. You are free to decide for yourself.

Figure 6.17 Relaxing the endpoint.

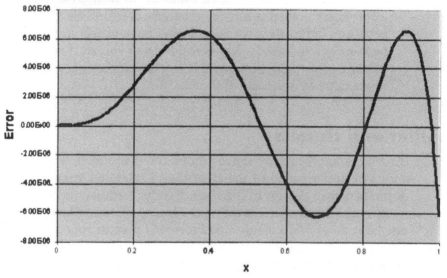

Sure, I'm finished with the example formula now, right? No, there's still more performance to squeeze out of it. Remember when I optimized the error in the first-order case by introducing a scale factor? I have not yet done that for this seventh-order formula. This is reflected by the zero slope of the error curve when $x = 0$. Adding the extra scale factor frees up this constraint, which affords a better chance to reduce error peaks. The new formula is

[6.59] $$f7(x) = \frac{x(a + bx^2)}{1 + cx^2 + dx^4}.$$

I now have four coefficients to solve, so I'll need four equations, and I'll define four values of x, instead of three, for zero crossings. I won't bore you with the details here, but the technique is almost identical to that used previously; the matrix is simply one order larger. The optimum results are

[6.60]
$$a = 0.99995354$$
$$b = 0.42283387$$
$$c = 0.75531298$$
$$d = 0.05624676.$$

Compare these to the nonoptimized values:

$$a = 1.0$$
$$b = 0.52380952$$
$$c = 0.85714286$$
$$d = 0.08571429.$$

You can see that the minimax process changes the coefficients significantly, at least for this extreme range of 0 to 1. The smaller the range, the closer the coefficients come to their theoretical (i.e., nonminimaxed) values.

The error curve for this last optimization is shown in Figure 6.18. The maximum error is now down to 4.6×10^{-6}, which represents a reduction from the original error curve by a factor of more than 250. Now you can see how my empirical seventh-order approximation served my needs so well.

Figure 6.18 Relaxing the slope.

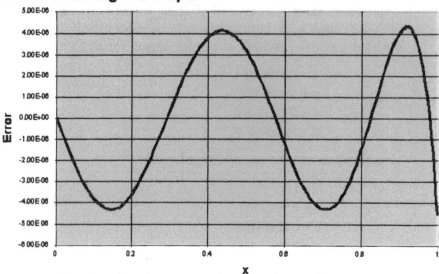

Combinations and Permutations

Amazingly enough, I have only begun to scratch the surface. Previously, you saw that I can reduce the errors dramatically by limiting the range of the input parameter. You just saw that I can also reduce the errors dramatically by applying minimax techniques to the coefficients. What remains is to combine the two techniques. So far, I've considered only the input range of 0 to 1, even ignoring the obvious offset by 22.5°.

Applying the minimax process to every possible combination of segment numbers and orders is tedious in the extreme. There are a lot of factors to consider.

- Do you care if the errors are discontinuous? If so, you must require the error to be zero at the boundaries; otherwise, you can relax this requirement and gain a little accuracy.

- Did you decide to force the error to zero at the boundary above? If so, you can use an even number of segments (equally sized in angle) because the error will be zero at $x = 0$; otherwise, you're better off using an odd number of segments.

Table 6.10   Minimaxed error, odd segments.

Order	Segments					
	1	3	5	7	9	11
1	0.048	1.52e–3	3.25e–4	1.18e–4	5.55e–5	3.04e–5
3	2.27e–3	7.08e–6	5.11e–7	1.00e–7	2.97e–8	1.04e–8
5	9.75e–5	3.11e–8	8.19e–10	3.26e–11	< 2.38e–12	—
7	4.58e–6	1.54e–10	3.42e–12	—	—	—
9	2.21e–7	—	—	—	—	—
11	7.71e–9	—	—	—	—	—

Although it's taken a lot of time, I've lovingly and laboriously performed the minimax optimizations on all the practical combinations of segment sizes and orders of approximation. Space does not permit giving the coefficients, but Table 6.10 gives the resulting maximum errors for odd segment numbers (as in Table 6.8). Table 6.11 gives similar results for even numbers of segments (assuming zero error at the boundaries). Compare these three tables, and you'll see how truly far the process of error minimization has come. Pick one of the combinations that fits your needs and enjoy the richness of its performance.

Table 6.11   Minimaxed error, even segments.

Order	Segments					
	2	4	6	8	10	12
1	5.24e–3	6.79e–4	1.87e–4	7.94e–5	4.021e–5	2.327e–5
3	5.13e–5	1.80e–6	2.19e–7	5.11e–8	1.774e–8	6.58e–9
5	5.56e–7	4.76e–9	2.64e–10	3.15e–11	7.55e–12	1.987e–12
7	5.31e–9	1.85e–11	9.33e–12	—	—	—
9	1.079e–10	—	—	—	—	—

A Look Backward

At this point, it's hard to remember how difficult this problem seemed to be in the beginning. Therefore, it might be worthwhile to review the process from start to finish.

I began with the power series of Equation [6.2], which converges so slowly that more than 7,000 terms would be needed to get even single-precision floating-point accuracy. At this point, things looked extremely bleak. A major breakthrough was deriving the continued fraction in Equation [6.18]. Truncating that continued fraction and simplifying, I got the various nth-order approximations given in Table 6.5. These formulas make the calculation of the arctangent feasible, but I still was left with large errors or, alternatively, high orders of approximation to keep the errors under control. I got the errors under control first by reducing the range and second by minimaxing the coefficients. The reduction in range showed that I could get dramatic reductions in the errors simply by the use of a few if tests to divide the range. Just when it seemed that reducing the range was more effective than going to a higher order formula, I tried minimaxing the formulas and found yet another dramatic increase in accuracy. You haven't seen this part yet, but you'll find that the higher the order, the greater the improvement wrought by minimaxing. In fact, the improvement is approximately equal to a factor of $2n + 1$, where n is the order of the approximation.

At this point, I almost have an embarrassment of riches. From a problem that seemed almost intractable, I've reached the point where I can not only achieve any accuracy I desire, but I can do so with relatively simple and fast code.

I'll end this chapter with a code example (Listing 6.7) implementing a specific version of the arctangent function. This example underscores better than all the words I could write how far I've come since the 7,000-term power series, because it provides measurable proof that I can achieve acceptable accuracy with very simple formulas and rapidly executing code.

I've chosen the case of three half-segments with a fifth-order formula optimized for zero error at the extremes. As discussed earlier, forcing the error to zero at the extremes costs a bit of accuracy, but it pays off with a seamless error function over the full range. The error is still more than acceptable for 32-bit accuracy ($\leq 3.7e\text{-}8$, worst case). Note that you can achieve this performance with only a fifth-order approximation and a single if test for range, even though the error at the extremes is forced to zero. This kind of performance is even better than the seventh-order formula I worked out empirically because the coefficients have been optimized more carefully.

Listing 6.7 incorporates all the ideas I've discussed so far, but I've folded them together into one function for the sake of efficiency. The code gives an excellent balance between accuracy, speed, and reasonable behavior. However, from the background information I've given in this chapter, you should have more than enough insight to roll your own algorithm to suit your specific needs. Use Tables 6.10 and 6.11 to guide your decisions. Have fun.

Listing 6.7   Arctangent with optimized coefficients.

```
/* Rational approximation for atan(x)
 * This polynomial uses a minimaxed 5th-order
 * ratio of polynomials.  Input angles are restricted
 * to values no greater than 15 degrees.  Accuracy is
 * better than 4e-8 over the full range of inputs.
 */
double arctan(double x){

  // constants for segmentation
  static const double b = pi/6;
  static const double k = tan(b);
  static const double b0 = pi/12;
  static const double k0 = tan(b0);

  // constants for rational polynomial
  static const double A = 0.999999020228907;
  static const double B = 0.257977658811405;
  static const double C = 0.59120450521312;
  double ang;
  double x2;
  int comp = FALSE;
  int hi_seg = FALSE;

  // make argument positive
  int sign = (x < 0);
  x = abs(x);

  // limit argument to 0..1
  if(x > 1){
    comp = TRUE;
    x = 1/x;
  }
```

Listing 6.7   Arctangent with optimized coefficients. (continued)

```
// determine segmentation
if(x > k0){
   hi_seg = TRUE;
   x = (x - k)/(1 + k*x);
}

/* argument is now < tan(15 degrees)
 * approximate the function
 */
x2 = x * x;
ang = x*(A + B*x2)/(1 + C*x2);

// now restore offset if needed
if(hi_seg)
   ang += b;

// restore complement if needed
if(comp)
   ang = halfpi - ang;

// restore sign if needed
if(sign)
   return -ang;
else
   return ang;
}
```

Chapter 7

Logging In the Answers

Last of the Big-Time Functions

In the past few chapters, I've shown algorithms for computing almost all the fundamental functions: square root, sine, cosine, and arctangent. I gave you optimized, thoroughly tested implementations useful in embedded-system applications. I've also shown how to compute the remaining trig functions, such as tangent, arcsine, and arccosine, from the functions given. Two fundamental functions remain to be discussed: the logarithm and its antifunction, also known as the exponential function.

Frankly, if I thought the logarithm (log) or the exponential (exp) were going to be as difficult as the arctangent, I might be tempted to give up and go on to something easy, like matrix differential equations. I know you'll be relieved to hear that they have much in common with the arctangent, sine, and cosine functions. This means that I'll be able to borrow from the material I've already presented and take a few shortcuts to the final forms.

The Dark Ages

I know I'm revealing my age when I tell you this, but when I was in high school, we didn't have pocket calculators. All arithmetic had to be done the hard way. Arithmetic, computed by hand, worked reasonably well for arithmetic and algebra classes, but it was another matter when it came to trigonometry. Multiplying five-digit trig functions by hand can take a while and is prone to error, so as a shortcut, we were taught logarithms. To multiply two numbers, you add their logs, and since adding is a lot easier than multiplying, this was a giant step forward. After a long series of such multiply-by-adding, we got the final answer by taking the antilog, that is, the exponential.

Those days are gone forever, thank goodness, and calculators and computers take care of the arithmetic for us. The need for logs and antilogs has diminished, but it's still there. Both logs and antilogs appear in other guises as the solutions to many math problems. Many natural processes, such as population and the cost of living, tend to increase exponentially with time. (I've noticed, on the other hand, that my salary does not.) The density of the air decreases roughly exponentially with altitude, and logs still provide the only way to raise a number to a nonintegral power.

A Minireview of Logarithms

If you're already comfortable with the concept of logs and antilogs, you can skip this section. If, like most folks, you haven't used logarithms since high school or college, this minireview might help. I'll begin at the beginning, which is the integral powers of 10 that are used in scientific notation.

Consider the square of 10:

$$10^2 = 10 * 10 = 100.$$

Similarly,

$$10^3 = 10 * (10^2) = 1,000,$$

$$10^4 = 10 * (10^3) = 10,000, \text{ and so on.}$$

If you examine the pattern of these numbers, you'll see that the number of zeros in each number is exactly the same as the exponent in the power-of-10 notation. Thus, 10 to any integral power n is just a one followed by n zeros.

Although the idea of scientific notation isn't really the topic of my current focus, I will be using it later, and I'm close enough to the idea here to give it a brief mention. Scientists often have to deal with numbers that are very large, like the number of stars in the universe, or very small, like the mass of an electron. Writing such large or small numbers gets to be tedious very fast, so scientists, being a lazy lot, have devised a notation that lets them express such numbers concisely. Simply take one power of 10, which gets the size of the number in the ballpark, and multiply it by a small number that defines the exact value. Thus, Avogadro's number (giving the number of molecules in a mole of gas) is

$$6.023 \times 10^{23},$$

which is equivalent to

$$602,300,000,000,000,000,000,000.$$

The first part of the number, 6.023, defines the first digits of the number; the power of 10 tells how many digits lie to the left of the decimal point. For the record, the number of stars in the universe is about 10^{22}, and the mass of an electron is 9.1×10^{-31} kg.

Suppose I have two numbers raised to integral powers,

$$x = 10^4 = 10,000$$
$$y = 10^3 =\ \ 1,000,$$

and I want to multiply them. Doing it the hard way, I find that

$$xy = 10,000,000.$$

Stated in the short form, this number is 10^7. But the exponent seven is also the sum of four and three, which are the exponents of x and y. Although a single example hardly proves anything, it certainly suggests that multiplying powers of 10 is equivalent to adding their exponents. Try it again with as many examples as you like, and you'll find that the rule always works.

To multiply powers of 10, add their exponents.

If you want to divide the same two numbers, you get

$$\frac{x}{y} = \frac{10,000}{1,000} = 10.$$

You can look at this result in two ways. On the one hand, you can observe that dividing two powers of 10 is equivalent to subtracting their exponents, as in

$$\frac{x}{y} = 10^{(4-3)} = 10^1 = 10.$$

On the other hand, you can think of this division as the product of two numbers: x and the reciprocal of y.

$$\frac{x}{y} = \frac{10,000}{1,000} = 10,000\left(\frac{1}{1,000}\right)$$

You already know that the first number is 10^4. You also know that you can multiply any two powers of 10 by adding their exponents. To make the rules come out right, you are left with the inescapable conclusion that the fraction, 1/1,000, is equal to a *negative* power of ten.

$$\frac{1}{1,000} = 10^{-3}$$

From this result, you can extend the notion concerning powers of 10 to negative powers, and thus to numbers less than one. The only thing is, you must change, a little bit, the way the rule is written. First, the exponent of powers of 10 tells you how many zeros to place after the 1. This rule doesn't really work properly if you try to extend to negative powers. A better way to state the rule is, the power of 10 tells you how many ways to move the decimal point, left or right. Thus, to write the number 10^4, first write the digit 1, followed by a decimal point:

1.0

Then move the decimal point four places to the right.

10,000.0

To get the number 10^{-4}, you still write the 1 but move the decimal point four places to the *left*:

1.0 → 0.0001

In this case, the number of zeros to the left of the one is not the value of the exponent, but one less. That's because the original decimal point was to the right of the one at the start. Nevertheless, if you think about the process as moving a decimal point, rather than adding zeros, you'll get the right value.

You need to consider one last facet of the rules. What if you divide two powers of 10 that are equal; for example,

$$\frac{x}{x} = \frac{10,000}{10,000} = 1.$$

Knowing that dividing is equivalent to subtracting the exponents, it's clear that subtracting two equal numbers gives us zero. Thus, the value 1 must be equivalent to 10^0. In fact, *anything* to the zero power is one. Table 7.1 shows the first few powers of 10 on either side of zero.

The concept of logarithms comes from a wonderful and ingenious flight of insight based on Table 7.1. If 10^0 is one, and 10^1 is 10, then there must be *some* power (obviously not an integer) chosen such that, when 10 is raised to that power, it will yield a number between 1 and 10.

Looking at it another way, you could plot the values in Table 7.1 on a graph (it works a lot better if you use a logarithmic scale). So far, you have only discrete points, not a continuous function. The flight of insight comes by recognizing that you can "smooth" the graph, filling in the gaps in the function to make it continuous, if you allow for nonintegral exponents. All the rules about adding and subtracting the powers should still apply, so you end up with the ability to multiply numbers by adding the corresponding exponents.

Table 7.1   Powers of 10.

n	10^n
−5	0.00001
−4	0.0001
−3	0.001
−2	0.01
−1	0.1
0	1.0
1	10.0
2	100.0
3	1,000.0
4	10,000.0
5	100,000.0

The logarithm (log), then, of a number is defined so that

[7.1] $10^{\log x} = x$.

A graph of the function is shown in Figure 7.1. As you can see, log(1) = 0, and log(10) = 1, as advertised.

Figure 7.1   The log function.

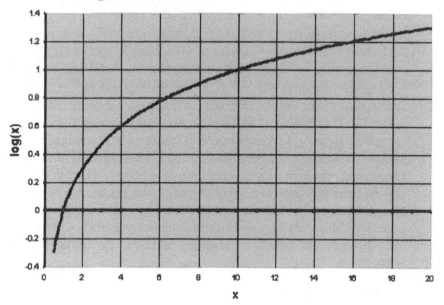

Take a Number

At this point, you might wonder how I obtained this curve. As it turns out, there is a power series for it, which you'll see in a moment, as well as much better representations, which you'll also see. Like most power series representations, the derivation of this one comes via calculus. However, if you know how to raise 10, or any other number, to a nonintegral power, you can conceivably find the log function even without such a power series by using simple iterative methods.

To find the logarithm of, say, three, pick a number, any number. I'll guess 1.0. 10^1 is 10, and that's way too high, so I'll cut my guess in half to 0.5. Now raise 10 to the power 0.5. What's that? You don't know how to raise a number to a fractional power? Sure you do, in this case, at least. Any number to the power 0.5 is the same as the square root of that number. The square root of 10 is 3.16, so I'm not too far off. I'll try 0.4. Ten to the power 0.4 is 2.51, so that's too low. I can keep trying numbers between 0.4 and 0.5 until I find $10^n = 3$.

But how did I know that $10^{0.4} = 2.51$? I cheated: I used my calculator. But my calculator used the logarithm to compute the value. I seem to be stuck in a twisty little maze of circular logic. To find the logarithm of a number, I need to raise 10 to a nonintegral power. But to do that, I seem to need the logarithm function that I don't know yet.

So how do I calculate the value of 10 to an arbitrary nonintegral power? One solution will always get me arbitrarily close to the correct answer. Because my guess of 0.5 was too high, I'll try half of that, 0.25. I'm now seeking $10^{1/4}$, which is simply the square root of the square root of 10. Its value is 1.77, which is way too low, but at least it gives me a starting point for the next guess.

By taking the average of the too-high and too-low values, I can get another value that's hopefully closer. If I continue this process, I will always get an exponent (logarithm) of the form

[7.2] $$\frac{P}{2^q}.$$

I know I can always raise 10 to this power, by a combination of multiplications and square roots. For example, the guess for the next step is

[7.3] $$\frac{1}{2}\left(\frac{1}{2}+\frac{1}{4}\right) = \frac{3}{8}.$$

Ten raised to this power is:

[7.4] $$10^{\frac{3}{8}} = (10^3)^{\frac{1}{8}} = \sqrt{\sqrt{\sqrt{1000}}}$$
$$= 2.37137$$

Using this method on my calculator, I get the sequence in Table 7.2.

Table 7.2   Logs by trial and error.

Guess	x	10^x
1	1	10.000000
2	0.5	3.162277660
3	0.25	1.778279410
4	0.375	2.371373706
5	0.4375	2.738419634
6	0.46875	2.942727176
7	0.484375	3.050527890
8	0.4765625	2.996142741
9	0.48046875	3.023213025
10	0.478515625	3.009647448
11	0.477539063	3.002887503

You can see that the process is converging, however slowly.

I have a confession to make: I used my calculator again, for convenience only, to generate Table 7.2. I didn't have to, however. You've already seen that you can raise 10 to any number of the form in Equation [7.2] by successive multiplications and square roots. The 11th guess for the logarithm in Table 7.2 is

$$0.47753906 = \frac{489}{1024} .$$

The denominator is 2^{10}, so you must take a square root 10 times. I must admit that I don't particularly relish the idea of taking the square root, 10 times, of a number that starts out to be one followed by 489 zeros. Fortunately, I don't have to. Remember, you form each guess by taking the average of the two previous numbers.

[7.5] $x = \frac{1}{2}(y + z)$

What's more, you already have the values of both y and z and the values of 10 raised to those powers. Raising 10 to the power x is equivalent to

$$10^x = 10^{\frac{1}{2}(y+z)}$$

[7.6]
$$= [10^{(y+z)}]^{\frac{1}{2}}$$

$$= \sqrt{10^{(y+z)}}$$

$$= \sqrt{10^y 10^z}.$$

Stated in words, all you need to do is to multiply the previous two powers and take the square root of the product. (This process, by the way, is called taking the *geometric mean*.) For example, for the fifth guess, I averaged the second and the fourth values.

[7.7]
$$x_2 = \frac{1}{2} \qquad\qquad 10^{x_2} = 3.162277660$$

$$x_4 = \frac{3}{8} \qquad\qquad 10^{x_4} = 2.3713737$$

$$x_5 = \frac{1}{2}\left(\frac{1}{2} + \frac{3}{8}\right) = \frac{7}{16} \qquad 10^{x_5} = \sqrt{3.162277660 \times 2.3713737}$$

$$= 2.7384196$$

That's exactly the same value I got in Table 7.2. As you can see, you could easily build Table 7.2 armed with nothing but a five-function calculator — the fifth function being the square root. If you lived in the Renaissance period, you could do it armed with only pen, ink, and a lot of paper.

I don't propose that you try to build log tables using the kind of successive approximation process I've used here. It was tedious enough to do it just for a single number, and even then I'm still a long way from an accurate solution. Still, I find it comforting to know that I can break the cycle of circular logic. Even if I didn't know calculus, and therefore didn't have the power series form, I could conceivably build a table of logarithms using only the sweat of my brow. It's somehow nice to know that this stuff can be done by actual humans, because ultimately, you may be the human that has to do it. Or at least program it.

Where's the Point?

As I've noted, in Figure 7.1, when the input arguments only range from one to 10, the output logarithm varies between zero and one. To deal with numbers outside this range, simply remember that raising 10 to an integral power moves the decimal point — to the right for positive integers and to the left for negative integers. If, for example,

$$\log(3) = 0.4771,$$

then

$$\log(30) = 1.4771$$
$$\log(300) = 2.4771$$
$$\log(3,000) = 3.4771$$
$$\log(30,000) = 4.4771.$$

and so on. More generally, because multiplying is equivalent to adding exponents, you can always split any number x into two parts, as in

$$x = n + f,$$

where n is the integer part (the part to the left of the decimal point) and f is the fractional part; therefore,

[7.8] $$10^x = 10^{(n+f)} = (10^n)(10^f).$$

The fractional part gives a number between one and 10, and the integral part serves to place the decimal point.

Other Bases and the Exponential Function

It should be clear that there's nothing magic about the number 10, other than the fact that we have 10 fingers. I could have just as easily used any other base number. The other two most likely choices are

- 2 (works well for binary computers) and
- The Euler constant e (= 2.718281828...).

Logarithms to the base e are called "natural logs," although in reality they only seem natural if you're a mathematician or happen to have 2.7183 fingers. At first glance, the use of base e may seem to be yet another of those unnecessarily confusing artifacts of math, like measuring angles in radians. It sometimes seems that mathematicians go out of their way to introduce new concepts or new measures, just to add to the fog factor. But as in the case of the use of radians for measuring angles, there really is a good reason for choosing base e. It's because the form e^x often shows up in the solutions of differential equations; in particular, the solutions to a whole class of problems called linear differential equations, which make up the great bulk of all the calculus problems that can be readily solved. Such problems tend to pop up in all kinds of unexpected places, from radioactive decay, to electronic circuits, to the population growth of jackrabbits.

To keep straight which base is associated with a logarithm, you should really include the base as part of the notation, as in $\log_2$, $\log_{10}$, $\log_e$, and so on.

However, as usual, mathematicians, engineers, and physicists are a lazy lot and tend to avoid writing any more symbols than are absolutely necessary. To avoid having to write the base explicitly, they've adopted the convention that the term "log" with no subscript is assumed to refer to base 10 and the term "ln" is the natural log, or base e. (No such convention exists for base 2, so you must use the subscript notation.)

Be careful when programming because some computer languages, notably FORTRAN and C, use log when they really mean ln. That is, to compute the natural log of x in most computer languages, you use log(x), not ln(x). Why? I have no idea. It's another one of those eccentricities that are tolerated because it's too late to change the convention.

The antilog in base e, of x, is the exponential function

$$\exp(x) = e^x.$$

This function is crucial to the whole process because, like the sine function, there's an infinite series definition that computes e^x. This makes it a kind of Rosetta stone that defines the antilogs for other bases. In computerese, exp() *always* implies base e, but you can use it to get the other antilogs because

[7.9] $10 = e^{\ln 10}$ and $2 = e^{\ln 2}$.

From this, you can write

[7.10] $10^x = (e^{\ln 10})^x = e^{x \ln 10}$

and

[7.11] $2^x = e^{x \ln 2}$

and, conversely,

[7.12] $\log x = \dfrac{\ln x}{\ln 10}$

and

[7.13] $\log_2 x = \dfrac{\ln x}{\ln 2}$.

The factors ln 2 and ln 10 are constants. These relations come in handy later on.

Can You Log It?

Now that you have the fundamentals down pat, I'll show you how you can compute these functions. Not surprisingly (after the lessons learned in previous chapters), both functions can be defined in terms of the infinite series

[7.14] $$e^x = x + \frac{x^2}{2!} + \frac{x^3}{3!} + \frac{x^4}{4!} + \frac{x^5}{5!} + \frac{x^6}{6!} + \dots$$

and

[7.15] $$\ln x = 2\left(z + \frac{z^3}{3} + \frac{z^5}{5} + \frac{z^7}{7} + \frac{z^9}{9} + \frac{z^{11}}{11} + \dots\right),$$

where

[7.16] $$z = \frac{x-1}{x+1}.$$

A graph of the function e^x is shown in Figure 7.2. Although it may not look like it, it is simply Figure 7.1 turned sideways. Only the scales are different because now it's the natural log instead of $\log_{10}$.

I'll look at Equation [7.14], the antilog function, later. This function should not give much trouble. Noting the factorial in the denominator of each term, you should expect this function to converge rapidly.

The series in Equation [7.15], on the other hand, is quite another matter. Like the series for the arctangent function, its terms have no factorials in the denominator, so you can expect very slow convergence. In fact, this time you don't even have the luxury of alternating signs. The magnitude of each term is almost completely determined by the power of z, so before I go any further, I'll examine the range of z.

Figure 7.2   The exponential function.

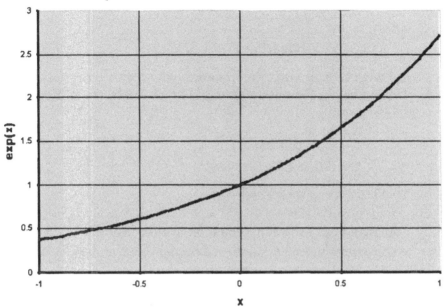

Figure 7.3 shows the relationship between x and z. As x varies from one to two, z goes from zero to $^1/_3$. This is encouraging because you don't have to deal with values of z near unity. Although true mathematicians would cringe at the idea of judging the convergence of a series by looking at one term, non-mathematicians can use a single term to at least get a feel for the number of terms needed in the equation by requiring the last term used in the truncated series to be less than some small value. The general form for the nth term is

[7.17] $$T_n = \frac{z^{2n+1}}{2n+1},$$

and, by setting this equal to the small value ε, gives

[7.18] $$\frac{z^{2n+1}}{2n+1} = \varepsilon$$

$$z^{2n+1} = (2n+1)\varepsilon.$$

If $n \gg 1$, I can ignore the 1 and write

[7.19]
$$z^{2n} = 2n\varepsilon$$
$$z = (2n\varepsilon)^{\frac{1}{2n}}.$$

If I were solving for z, this equation would give z as a function of n and ε. Unfortunately, I'm not doing that; I need to solve for n. Rewriting Equation [7.19] gives

$$z^{2n} = 2n\varepsilon$$
$$2n \ln z = \ln(2n) + \ln \varepsilon$$

[7.20]
$$2n = \frac{\ln(2n) + \ln\varepsilon}{\ln z}.$$

Although I can't solve this equation in closed form, I can do so by iteration. As an example, let $z = {}^1/_3$, the maximum value, and let $\varepsilon = 1.0 \times 10^{-8}$, to get

[7.21]
$$2n = \frac{\ln(2n) - 18.42}{-1.0986}.$$

A few iterations yields

$$2n = 14$$
[7.22]
$$n = 7.$$

Thus, I'll need at least eight terms (n starts at zero), or powers through z^{15}, to get reasonable accuracy. This actually isn't too bad — the fact that z never gets close to unity helps a lot. Still, I can do much better.

Figure 7.3   x versus z.

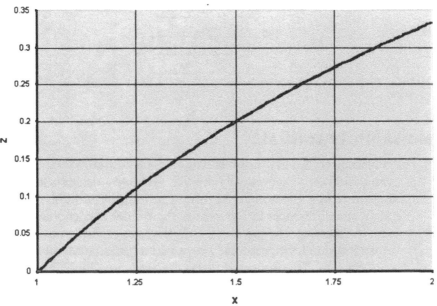

Back to Continued Fractions

You've already seen in Chapter 6 how a slowly converging series can be turned into a rapidly converging continued fraction. The process of computing the coefficients for the continued fraction is tedious but possible.

After all the work you did with the arctangent, you may not feel up to doing it all over again for the logarithm. Rejoice: you don't have to. Look again at the form of Equation [7.15]. Except for the 2 in the numerator and the signs of the terms, it's the same as the equation for the arctangent (see Equation [6.2]). The terms in the arctangent series alternate, whereas those in Equation [7.15] do not. You can get the equations to the same form if you replace every factor of z^2 in Equation [7.15] by $-z^2$. This suggests that the continued fraction will also have the same form.

If you prefer mathematical rigor, it may not be immediately obvious that this replacement is legal, but it is. Accordingly, you can write the continued fraction for the logarithm by inspection. Equation [7.23] is the result, where z remains as defined in Equation [7.16].

$$[7.23] \qquad \ln x = \cfrac{2z}{1 - \cfrac{z^2}{3 - \cfrac{4z^2}{5 - \cfrac{9z^2}{7 - \cfrac{16z^2}{9 - \cfrac{25z^2}{11 + \ldots}}}}}}$$

Rational Polynomials

As before, note that you must truncate the continued fraction at some level even though it is theoretically infinite. Once you decide where to truncate it, you can reduce the now-finite form to a ratio of two polynomials. Unfortunately, the differences in signs cause the rational polynomials to be different from the arctangent equivalent. Still, they are easily computed from the continued fraction form, especially if you let a tool like Mathcad do all the hard work for you. The truncated forms are shown in Table 7.3.

You can compute the maximum error in the formulas of Table 7.3 by setting $z = \frac{1}{3}$, its maximum value, and comparing the fractions with the exact value of the logarithm. The results are shown in Table 7.4.

The error behavior of these fractions is quite nice; note that the error in the 15th-order case is five orders of magnitude better than the value I predicted for the power series. Continued fractions really are a far superior approach. However, it will probably not surprise you to hear that you can do much better yet.

Table 7.3   Rational forms for the logarithm.

$(y = z^2)$

$2z$ — Order 1

$\dfrac{6z}{3 - y}$ — Order 3

$\dfrac{2z(15 - 4y)}{15 - 9y}$ — Order 5

$\dfrac{2z(105 - 55y)}{105 - 90y + 9y^2}$ — Order 7

$(y = z^2)$

$$\frac{2z(945 - 735y + 64y^2)}{945 - 1050y + 225y^2}$$ Order 9

$$\frac{2z(1155 - 1190y + 231y^2)}{1155 - 1575y + 525y^2 - 25y^3}$$ Order 11

$$\frac{2z(15015 - 19250y + 5943y^2 - 256y^3)}{15015 - 24255y + 11025y^2 - 1225y^3}$$ Order 13

$$\frac{2z(225225 - 345345y + 147455y^2 - 15159y^3)}{225225 - 420420y + 242550y^2 - 44100y^3 + 1225y^4}$$ Order 15

Table 7.4   Errors in rational approximation.

Order	Absolute Error	Relative Error
1	0.026	0.038
3	8.395e–4	1.211e–3
5	2.549e–5	3.677e–5
7	7.631e–7	1.101e–6
9	2.271e–8	3.276e–8
11	6.734e–10	9.715e–10
13	1.993e–11	2.876e–11
15	5.893e–13	8.502e–13

What's Your Range?

As usual, to get good accuracy from a truncated series or fraction, you need to restrict the range of the input variable. In this case, it's rather easy to do so. In fact, if you really want to be crude, you don't have to restrict it at all.

So far, I have not looked closely at the relationship between x, the original argument, and z, the argument of the power series or continued fraction. This relation is

[7.24] $z = \frac{x-1}{x+1}.$

Reverting this gives

[7.25] $x = \dfrac{1-z}{1+z}$.

As z traverses the range from -1 to $+1$, x takes on values from infinity to zero — the entire range of positive numbers. (The logarithm is undefined for negative arguments.) The relationship between x and z was shown in Figure 7.3, and the (reverted) relationship for a wider range of values of z is shown in Figure 7.4. (I've limited the range at the lower end of z in order to hold the value of x within reason.) If I really had to, I could use the continued fraction over the full range of z. The series from which the fraction is defined is valid for all values of z such that

$|z| \leq 1,$

which takes in all possible legal input arguments.

The problem with using such a wide range for z is the same one you saw in the arctangent algorithm: slower convergence. Stated another way, you need a higher order continued fraction to get reasonable accuracy. A more subtle problem is the loss of resolution that would occur at the larger values of x. Consider the values of z as x changes a small amount.

$z(100) = 0.98019801980198$

$z(101) = 0.980392156862745$

Although the value of x has changed by 1 percent, the value of z has changed by only 0.02 percent. Thus, I lose considerable resolution as x gets large. You have seen that this loss of resolution is totally unnecessary. The nature of the logarithm (base 10) is such that integral factors of 10 only change the decimal point, not the digits of the logarithm. Although this relationship is a little less obvious when discussing natural logs or logs of base 2, the same principle holds. You must assume that no matter what base you use, you have some simple mechanism for reducing the argument of the log function to a reasonable value.

Figure 7.4   *x* versus *z*, redux.

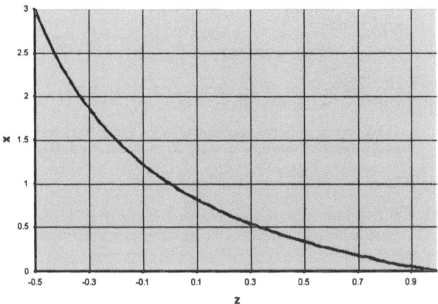

As a matter of fact, if the number is stored in floating-point format, this reduced range is built into the storage format for that number. A floating-point number typically has an exponent that is a power of two (not 10), followed by a mantissa that can range from 0.5 to just under 1.0. Multiplying the mantissa by two then gets me right into the range I've been using in my discussion so far. Alternatively, I could shift my focus to the natural range of the floating-point number, which is probably the easier task.

Familiar Ground

At this point, you should be getting a distinct feeling of déjà vu. I began with a power series that looks very much like the one for the arctangent, I generated a continued fraction based on the series, I truncated the continued fraction to get ratios of polynomials, and now I'm talking about limiting the range of the input argument. That's precisely what I did with the formulas for the arctangent, so it shouldn't surprise you to hear that most of rest of the process of finding a practical formula involves steps very similar to those I've already taken for the arctangent: dividing the argument range into segments to further limit the magnitude of z and performing minimax optimization of the coefficients.

In the case of the logarithm, the process of segmenting the range is easier than for the arctangent because it involves only a simple multiplication. The key formula is

[7.26] $\ln(kx) = \ln x + \ln k$:

conversely,

[7.27] $\ln x = \ln(kx) - \ln k$.

Suppose your initial argument is in the range from one to two, as used earlier. Simply divide by a constant, k, to change the range to

[7.28] $\frac{1}{k}$ to $\frac{2}{k}$.

It should be fairly apparent that the best choice for k is one that will give equal and opposite values of the logarithm at the two extremes; that is,

[7.29] $\ln\left(\frac{1}{k}\right) = -\ln\left(\frac{2}{k}\right)$

or

$\ln(1) - \ln k = (\ln 2 - \ln k)$.

But because $\ln(1) = 0$, this reduces to

$2 \ln k = \ln 2$

$\ln k = \frac{1}{2}\ln 2$,

or

[7.30] $k = \sqrt{2}$.

The range is now 0.707 to 1.414, which has a logarithmic range of −0.347 to 0.347. The corresponding value of z is 0.1716. This is just a little more than half the original maximum z, which was $1/3$, or 0.3333. At first glance, you might expect a reduction in error at the extremes by a factor of two, raised to the power of the approximation. In fact, you get considerably better than this. The accuracy of the various orders is shown in Table 7.5. The magnitude of the error goes roughly as $2^{(n + 2)}$, where n is the order of the approximation.

Table 7.5   **Effects of range and reduction on accuracy.**

Order	Full Range	Half Range
1	0.026	3.428e–3
3	8.395e–4	2.738e–5
5	2.549e–5	2.105e–7
7	7.631e–7	1.598e–9
9	2.271e–8	1.206e–11
11	6.734e–10	9.065e–14
13	1.993e–11	0
15	5.893e–13	0

Can I restrict x even further? Yes. Just as in the case of the arctangent, I can break the range into segments and basically use different values of k, depending on the value of x. In so doing, I can divide the range into any number N of half-segments. The first step (using the full range for z) corresponds to $N = 1$. The second step (using an offset k) corresponds to $N = 2$. The same argument I used in the case of the arctangent also applies: odd numbers of N are preferred over even ones, because they result in zero error when z is near zero (x near 1.0). As with the arctangent, using odd numbers seems to waste one half-segment, by using only the positive side of the one nearest $z = 0$. However, this is generally considered to be a small price to pay for linear behavior with zero error as z approaches zero.

For any number N, the first constant k is given by

[7.31] $k = 2^{\frac{1}{N}}$.

This is also the maximum range of the effective argument. In other segments, the value of k will be two half-segments larger. The maximum error, as a function of the number of half-segments, is shown in Table 7.6.

Table 7.6   **Error versus order and segmentation.**

	Segments				
Order	1	3	5	7	9
1	0.026	1.022e–3	2.216e–4	8.083e–5	3.805e–5
3	8.395e–4	3.635e–6	2.838e–7	5.283e–8	1.504e–8

Order	**Segments**				
	1	3	5	7	9
5	2.549e–5	1.245e–8	3.504e–10	3.329e–11	5.735e–12
7	7.631e–7	4.211e–11	4.272e–13	2.071e–14	2.151e–15
9	2.271e–8	1.416e–13	0	0	0
11	6.734e–10	0	0	0	0
13	1.993e–11	0	0	0	0
15	5.892e–13	0	0	0	0

Minimax

One step remains to improve accuracy, and that's to adjust the coefficients using a minimax approach as I did with the arctangent. To refresh your memory, the error without minimax ordinarily is very low except at the extreme of the range, at which it reaches a maximum. The idea of minimax is to let the errors oscillate between bounds and accept larger errors in the smaller values of z in order to reduce the worst case error. As I found with the arctangent, the improvement in accuracy can be quite dramatic, easily approaching or exceeding two orders of magnitude. As I perform the minimax process, I must decide whether or not to constrain the endpoint error. Clearly, I can get a smaller worst case error by allowing it to occur at the maximum range. However, I then must deal with the problem of discontinuous error curves where the segments meet. To avoid this, I can constrain the error at the maximum, z, to be zero. This costs a bit in the worst case error, but the trade is usually worth it in terms of having a well-behaved algorithm.

As an example, I can write the general form of the fifth-order approximation as

[7.32] $$f(z) = \frac{2z(a + bz^2)}{1 + dz^2}.$$

For the nonoptimized case, Table 7.3 shows that

$a = 1,$
$b = -4/15 = -0.2666666667,$
and
[7.33] $c = -9/15 = -0.6.$

The same coefficients, optimized for the case $N = 5$, are

$a = 1,$
$b = -0.2672435,$
and
[7.34] $c = 0.600576.$

The error curves before and after optimization are shown in Figures 7.5 and 7.6. In Figure 7.5, the two are shown to the same scale. Clearly, I get a dramatic improvement in accuracy using the minimax process. Figure 7.6 shows how the optimized error tends to oscillate about zero.

The errors for the minimaxed case for various segments are shown in Table 7.7. These errors were computed under the assumption of zero error at the extremes. Errors with the end value unconstrained would be lower yet. Note that for high-order approximations and large numbers of segments, the errors are too small for any of my methods to optimize. I've shown them as zero, but I know they're really not. However, for practical purposes, a zero indicates that these cases are so close to zero as not to matter and means that optimization is not necessary.

Figure 7.5   Effect of minimax; fifth order, five segments.

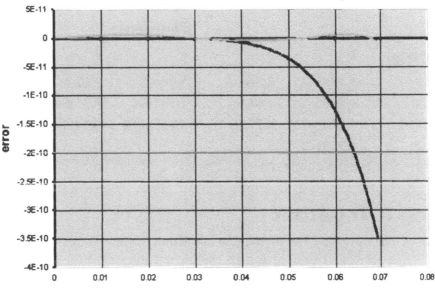

Figure 7.6 **Minimaxed error.**

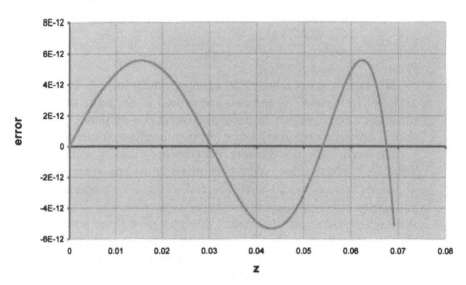

Table 7.7 **Error with coefficients minimaxed.**

		Segments			
Order	**1**	**3**	**5**	**7**	**9**
1	6.736e–3	2.56e–4	5.54e–5	2.023e–5	9.515e–6
3	5.646e–5	2.361e–7	1.781e–8	3.394e–9	9.447e–10
5	4.044e–7	1.978e–10	5.387e–12	5.148e–13	0
7	2.953e–9	0	0	0	0
9	0	0	0	0	0

Putting It Together

You now have the theoretical algorithm for any number of approximations for the logarithm, giving acceptable accuracy over the limited range. What remains is to turn the theory into usable code. For this purpose, I'll begin with the assumption that the input argument is stored in C double format; you'll recall from the square root algorithm that this format is simpler than the higher accuracy floating-point formats for Intel processors because of the way the exponent is defined. I can use the same software hack that I used in the square root calculation to give separate access to the exponent and mantissa of the floating-point word. The resulting code is shown in Listing 7.1. For the purposes of this exercise I've assumed a minimaxed fifth-order formula, with five half-segments. As can be seen from Table 7.7 and Figure 7.6, this form provides more than ample accuracy for most applications and may be considered overkill. Note that hacking into the floating-point format allows me to force the argument into the range I desire, neatly sidestepping the need to generate positive and negative return values. This step is taken care of by the segmentation.

Listing 7.1   The ln algorithm.

```c
/* Union to let us hack the exponent
 * of a floating point number
 * (also used in square root)
 */
typedef union{
   double fp;
   struct{
      unsigned long lo;
      unsigned long hi;
   }n;
} hack_structure;


/* Find the natural log for a floating point
 * number in the range 0..1
 *
 * Method: Limit the range to z = 0.069
 * Use minimaxed 5th-order rational
 * polynomial to calculate the log.
 */
double __ln(double x){
   const double a = 1;
   const double b = -0.2672435;
   const double c = -0.600576;
   double z = (x-1)/(x+1);
```

```
      double z2 = z*z;
      return 2*z*(a + b* z2)/(1 + c*z2);
}

double _ln(double x){
   const double limit1 = 0.879559563; // 0.5^(1/5)
   const double limit2 = 0.659753955; // 0.5^(3/5)
   const double k1 = 0.757858283; // 0.5^(2/5)
   const double k2 = 0.574349177; // 0.5^(4/5)
   const double ln_k = -0.138629436;  // ln(0.5^(1/5)

   if(x >= limit1)
      return(__ln(x));
   else if(x >= limit2)
      return(__ln(x/k1) + 2*ln_k);
   else
      return(__ln(x/k2) + 4*ln_k);
}

/* Main part of log function
 * Hacks the f.p. argument
 * Separates integer and fractional parts
 * Recombines after generating
 * kernel function.
 */
double ln(double X){
   const double ln2 = log(2);
   short expo;
   double retval;
   hack_structure x;
   if(X <= 0)
      return -BIG_FLOAT;
   x.fp = X;

   // grab the exponent
   expo = (short)(x.n.hi >> 20);
   expo -= 0x3fe;
```

```
    // force the exponent to zero
    x.n.hi &= 0x000fffff;
    x.n.hi += 0x3fe00000;

    // compute the log of the mantissa only
    retval = _ln(x.fp);

    // rebuild the result
    retval = ln2*(expo + retval/ln2);
    return retval;
}
```

Before leaving the natural log function, I should say something further about bases. You'll recall that I've talked about three useful bases:

- 10,
- *e*, and
- 2.

Unless stated otherwise, I've adopted the notation that log() refers to the logarithm, base 10, and ln() refers to the same logarithm, base *e*. In my implementation for the function, I have something of a dilemma. The power series of Equation [7.15] refers strictly to the natural logarithm, ln(). That's because the series is derived from calculus formulas, and in calculus, the use of the natural log is, well, natural. On the other hand, the exponent field of the floating-point number I'm processing is given as powers of two, so clearly base 2 is more "natural" for that portion of the programming. So which base should I use?

The decision is something of a toss-up, and for that reason, some computer systems and programming languages used to offer the logarithm, base 2, as well as base *e* and sometimes 10. However, remember that conversion between the bases is trivially easy, as noted in Equations [7.12] and [7.13]:

$$\log x = \frac{\ln x}{\ln 10}$$

$$\log_2 x = \frac{\ln x}{\ln 2}.$$

Because of the ease of conversion, most modern languages only support base *e*.

Remember that, when we do are doing logarithms (to the base 10) by hand, as we learned to do in high school, we divide the logarithm up into an integer part plus a fractional part. When taking the antilog, the fractional part gives the value of the digits of the number, while the integral part fixes the position of the decimal point:

[7.35] $10^x = 10^{(n+f)} = (10^n)(10^f)$

In floating-point notation, the number is stored in the format

[7.36] $x = 2^n f,$

where f is the mantissa constrained between 0.5 and 1. In taking the logarithm, you are really finding

$$\ln x = \ln(2^n f),$$

which is

[7.37] $\ln x = \ln(2^n) + \ln f$
$$= n \ln 2 + \ln f.$$

You can see that it really doesn't matter what the base is for the floating-point exponent (good thing, too, because some computers use base 4 or base 16). You need only remember to multiply n by the natural log of whatever the base is. It's a simple point, but one that's easy to get tripped up on.

On to the Exponential

Now that you can compute logarithms, it's time to focus your attention to the other end: the antilog, or the exponential function. The power series is given in Equation [7.14]. From experience with the sine/cosine series, you should recognize that the presence of the factorial in the denominator assures fast convergence of the series for any value of x, positive or negative. Even if x is very large, the powers of x will eventually be overwhelmed by the factorial in the denominator, and the series will eventually converge. However, that doesn't necessarily mean that it will converge *rapidly*. To make sure that happens, you need to take the usual step: restrict the range of the input variable and, indirectly, the output variable.

You'll notice that the terms in the series of Equation [7.14] all have the same sign. This is in contrast to, say, the series for the sine function, where the terms alternate. You might think that one could get better accuracy by restricting the argument to negative values, where the terms will alternate in sign. That logic seems sound enough, but it's wrong. This is partly because the value of the argument can approach unity, which would make it cancel the leading term of unity, and move all the significance to higher-order terms. For this and other reasons, it turns out to be better to restrict the range to positive values. When x is negative, the properties of exponents give

[7.38] $$e^x = \frac{1}{e^{-x}}.$$

The "wrapper" of Listing 7.2 takes care of this. In a moment, I'll show you how to restrict the range much more tightly. There's just one small problem. Although I've eliminated the negative half of the argument's range, I haven't limited the positive half, which can still go to infinity.

Listing 7.2   Wrapper for exponential function.

```
/* Wrapper to limit exponent range
 * The wrapper ensures that the
 * argument passed along is positive
 */
double exponential(double x){
   if(x < 0)
      return(1/expl(-x));
   else
      return(expl(x));
```

The series as shown in Equation [7.14] looks nice, but the form shown is terrible for evaluating the series because it involves both factorials and successively higher powers of x. Instead, you should use Horner's method, as shown in Equations [7.39] and [7.40]. Here, I've shown the method in two forms. The first is the more common and is easier to write; the second is preferred because it uses less stack space — an important consideration if your computer is using a math coprocessor.

[7.39] $$e^x = 1 + x\left(1 + \frac{x}{2}\left(1 + \frac{x}{3}\left(1 + \frac{x}{4}\left(1 + \frac{x}{5}\left(1 + \frac{x}{6}(\ldots)\right)\right)\right)\right)\right)$$

[7.40] $$e^x = \left(\left(\left(\left(\left((\ldots)\frac{x}{6} + 1\right)\frac{x}{5} + 1\right)\frac{x}{4} + 1\right)\frac{x}{3} + 1\right)\frac{x}{2} + 1\right)x + 1$$

Range Limits Again

The next level of range reduction is done easily in much the same way as for the log function. As before, I can always write the input argument as

[7.41] $x = n + f,$

where n is the integral part of x, and f is a fraction with absolute value less than one. Then again, as before, I can write

[7.42] $e^x = e^{(n+f)} = (e^n)(e^f).$

The first part of this equation is merely e raised to an integral power. I can handle this part separately. I need to evaluate only the second part in series, using as an argument f, which always has amplitude less than one. Now I have a practical algorithm as shown in Listing 7.3. The number of terms in the series, up to Order 11, gives eight-digit accuracy over the range.

Listing 7.3   A practical exponential.

```
/* Generate the exponential from its Hornerized
 * power series.  11th order formula gives 1e-8
 * accuracy over range 0..1
 */

double exp2(double x){
    return
(((((((((((x/11+1)*x/10+1)*x/9+1)*x/8+1)*x/7+1)*x/6+1)
      *x/5+1)*x/4+1)*x/3+1)*x/2+1)*x+1;
}


/* Separate integer and fractional parts of argument
 * Only fractional part goes to exp2.
 *Integer part computed using pow( )
 */

double exp1(double x){
    double retval;
    double intpart;
```

```
int n;
x = -x;
n = (int)x;
x -= n;
retval = exp2(-x);
intpart = pow(e, -(double)n);
return intpart*retval;
}
```

Though workable, this series still seems too long, so it's natural to try to reduce the range still further. In principle, that's easy. All I need to do is play with n and f a bit. Instead of requiring n to define the integral part of the input argument, I can let it be the number of any fractional segments in x,

[7.43] $$x = \frac{n}{m} + y,$$

where y is now restricted to the range 0 to $1/_m$. Clearly, I can reduce the argument to the exponential function to as small a range as I like simply by making m larger and larger. The only price I pay is the need to perform more multiplications to handle the "integral" portion of the function as embodied in n. The larger m is, the larger n must be to handle the integral part of the function, so I can expect to perform more multiplications.

Using the new definition of x, as embodied in Equation (45), the new formula is

[7.44] $$e^{\left(\frac{n}{m} + y\right)} = \left(e^{\frac{n}{m}}\right)(e^y) = \left(e^{\frac{1}{m}}\right)^n e^y.$$

Note that the constant factor raised to a power is now the mth root of e.

Many programming languages support exponentiation. If I were programming in FORTRAN, I'd use the built-in operator **. The C math library includes pow(). You must be careful, however, how these functions are used. Both are designed for the more general case, where a real number is raised to a real power. Unfortunately, they do this by using ln() and exp(), which I can hardly use here because I have to *supply* both functions. In any case, these functions can be very inefficient when the second argument is an integer. Some compilers (notably FORTRAN compilers) test for this condition and use different methods when the argument is an integer. For small values of the argument, compilers will even convert the operator to inline multiplications. However in general, you can't count on such favors. The function power shown in Listing 7.4 gives a reasonably efficient method for raising a number to an integer power. It basically treats the power as a binary number and uses successive squares to generate the various product terms. It is not the most efficient method one could possibly think of — that method would factor n, where possible, and use the largest factors to reduce the number of multiplications — however, the function represents a reasonable compromise between speed and simplicity.

Listing 7.4   An integer power function.

```
/* An efficient pow function for integer powers
 * Use only for positive integers
 */
double power(double x, int n){
    double retval = 1;
    double factor = x;
    while(n != 0){
        if((n&1) != 0)
            retval *= factor;
        factor *= factor;
        n >>= 1;
    }
    return retval;
}
```

The function in Listing 7.4 is a general-purpose function, which I highly recommend when you must raise a number to an integral power. It needs only as many multiplications as the number of bits in n.

However, in the present case, we can devise an even more attractive alternative because we know what the first argument is: it's the mth root of e. By precomputing, say, the first m such roots, we can use a case statement to break the process up into powers of m, rather than powers of two. This speeds the process considerably. A code fragment to do this is shown in Listing 7.5. There is no need to use a separate function for this, since the algorithm has little or no use outside the exp() function.

I am still left with two decisions to make: the range to use for y (determined by the choice of m) and the order of approximation. Table 7.8 shows the errors (relative errors this time) for various choices. Pick one that seems to meet your accuracy requirements. Listing 7.5 shows the complete exponential function using $m = 4$ and a sixth-order power series. This combination gives a maximum error of $\approx 1.0e{-}8$, which should be sufficient for most practical work. This routine is fast and accurate, and the code is simple.

Table 7.8   **Errors in exponential series.**

Order	Segments				
	1	**2**	**4**	**8**	**16**
1	0.718282	0.148721	0.034025	8.14845e-3	1.99446e-3
2	0.218282	0.023721	2.77542e-3	3.35953e-4	4.13339e-5
3	0.051615	2.83771e-4	1.7125e-4	1.04322e-5	6.43814e-7
4	9.9485e-3	2.3354e-5	8.4896e-6	2.59707e-7	8.03081e-9
5	1.6152e-3	2.3354e-5	3.51584e-7	5.3943e-9	8.3529e-11
6	2.2627e-4	1.65264e-6	1.24994e-8	9.61098e-11	7.44738e-13
7	2.786e-5	1.02545e-7	3.89223e-10	1.49925e-12	5.55112e-15
8	3.0586e-6	5.66417e-9	1.07809e-11	2.08722e-14	0
9	3.0289e-7	2.81877e-10	2.68674e-13	0	0
10	2.7313e-8	1.27627e-11	5.77316e-15	0	0
11	2.2606e-9	5.30243e-13	0	0	0

Listing 7.5   **A complete** exp **function.**

```
/* Generate the exponential
 *
 * Function limits the argument range
 * to 0..1/4
 */
double exp1(double x){
  /* The following constant is shown as computed.  In the
   * real world, it should be precomputed and
   * stored as a constant
   */
  const double root_e = sqrt(sqrt(exp(1)));
```

```
        double retval;
        double intpart;
        int n;
        x *= 4;
        n = (int)x;
        x -= n;
        x /= 4;
        retval = (((((x/6+1)*x/5+1)*x/4+1)*x/3+1)*x/2+1)*x+1;
        intpart = power(root_e,(double)n);
        return intpart*retval;
}
```

Listing 7.6   The optimized version.

```
/* Generate the exponential
 *
 * Function is optimized for the range
 * 0..1/4
 */
double exp1(double x){
    const double root_e = sqrt(sqrt(exp(1)));
    const double a= 1.000000034750736;
    const double b= 0.499997989964957;
    const double c= 0.166704077809886;
    const double d= 0.041365419657829;
    const double e= 9.419273457583088E-3;
    double retval;
    double intpart;
    int n;
    x *= 4;
    n = (int)x;
    x -= n;
    x /= 4;
    retval = (((((x*e+d)*x+c)*x+b)*x+a)*x+1;
    intpart = power(root_e,(double)n);
    return intpart*retval;
}
```

Minimaxed Version

As in the other series, I can optimize the coefficients to give a minimax behavior over the chosen range. For a power series such as Equation [7.14], I should be able to use Chebyshev polynomials for this. However, for this function it seems more reasonable to minimax the relative, rather than absolute, error. That takes me back to doing the minimax process, laboriously, by hand. Listing 7.6 shows the same function as Listing 7.5, but with the order reduced to fifth order, and the coefficients optimized. Despite using a lower order approximation, the error is now less than $2e{-}10$, which ought to be enough for most any embedded application.

The Bitlog Function

A discussion of logs and antilogs would not be complete without mention of a remarkable function that I ran across a couple of years ago. This function is basically an integer version of the logarithm, giving an integer output for an integer input. I've found it to be extremely useful when I need a very fast algorithm approximating the logarithm while computing in integer arithmetic. The best, or at least the most fun, way I can describe this function is to tell the story chronologically. That approach requires me to tell you something about the work I did at a "day job."

The impatient among you are welcome to skip the next couple of sections and cut right to the algorithm itself. The more curious are invited to follow along. If nothing else, the story of how I ran across the Bitlog function should disabuse you of the notion that I only do theory — that I am somehow immune to dirt under the fingernails — because the job, though technically challenging in more ways than one, was about as down and dirty as they come: converting existing, uncommented code from Z80 assembly language to C.

The Assignment

The assignment was for a company that already had existing products based on the Zilog Z80 processor. They were in the process of converting everything to a new product based on the Intel 80486. My job was to convert a major piece of the software from Z80 assembly language to C, then make it work in our new hardware using our PC-based cross-development system. It was important that the C-486 version exhibit exactly the same functionality as the Z80 version — no fair starting over from scratch. As is usually the case in such situations, the Z80 code was largely uncommented, and the original author was no longer with the company.

That author was Tom Lehman. I am going to go on record, right here and now, by asserting that, as far as I'm concerned, Tom Lehman is a genuine, bona fide, certified genius. However, commenting code is not his strong suit. Fortunately, I've been working with 8008s, 8080s, and Z80s since they first came out. I can read their assembly languages like a second tongue, and I can program either CPU without referring to reference books or cards. I know all the tricky idioms like putting passed, constant subroutine arguments inline and diddling the stack pointer to skip over them or swapping data to the stack and back. I thought I was a good Z80 programmer, but Tom Lehman put me to shame. I had never seen a Z80 work so hard or do so much in my life: 32Kb of code, stuffed full to overflowing, much of it executing 60 times a second, including many floating-point operations that used a library also written by Tom.

I wouldn't have believed it possible if I hadn't seen it with my own eyes. I stand in honest amazement and admiration at what Tom was able to accomplish. It almost goes without saying, though, that to do so he had to resort to every trick in the book, and then some, which didn't make the translation any easier. Numbers and flags were left in CPU registers or pushed onto the stack in whatever way made them easiest to access again later. Sometimes they were positive, sometimes they were negative, and sometimes they bounced from one to the other, all to save a byte or two. Great gobs of variables were accessed via Z80 index registers, which sometimes pointed one place, sometimes another, and sometimes were swapped.

The Approach

When translating code from one CPU to another, there are a number of approaches one can take. The fastest and least efficient method is to write a CPU emulator and leave the application code alone. I could have written a Z80 emulator for the 486, and simply executed Tom's Z80 code on it, but this would have been unacceptable for a number of reasons, not least of which would be speed. Another serious problem would have been the difficulty of disconnecting the Z80 application code from the "operating system" buried inside it and reconnecting to the new multitasking OS.

A method closely related to emulation is to write an automated assembly-to-assembly translator. I've written such translators in the past, and they do work more or less well depending on how similar the two processors are. An 8080-to-6800 translator produced code that was about three times larger and ran about three times slower than the original. The difference was primarily a result of the radically different behavior of the processor flags. Before and after many of the 8080 instructions, I had to insert instructions to make the 6800 flags behave like the 8080's. For processors as similar as the 8080 and 8086, the performance hit would not be nearly so drastic. For history buffs, this is the way Microsoft got their first versions of Microsoft BASIC and other apps running on the IBM PC. They used an 8080–8086 translator provided by Intel to speed conversion from the 8-bit to the 16-bit world.

Using an emulator or translator had certain appeal, not the least of which was the fun and challenge of building the translator and/or emulator. However, neither approach was considered acceptable in this case. First of all, I'd still have Z80 code or its equivalent in 8086 assembly language to maintain. The cost of maintaining C code compared to assembly code swings the decision strongly away from any kind of automated translation.

Most importantly, the product operated in medical equipment, where people and their continued good health are involved. I couldn't afford not to know, down to the last detail, exactly what was going on inside the system. I not only had to know the software inside and out, I had to understand what it was doing and why; therefore the translation involved as much psychoanalysis as code conversion. To effect a thorough and robust translation, I had to get inside Tom Lehman's head and understand not only what he was doing in a code section, but what he had been thinking when he wrote it.

I had to pull out every trick I could find in my toolbox, from flowcharts and pseudocode to structure diagrams, call trees, branch trees, data dictionaries, and some new angles I invented on the fly. One of the more powerful techniques was using a good macro editor to perform semi-automatic translations.

The Function

Translating software in this way is much like disassembling. Usually, beginning at memory location zero and working forward is a Bad Idea. Instead, you work in a bottom-up process. You look for low-level functions that are easily broken out from the main body of the software and figure out what they do. Once you've figured those out, you assign meaningful names to the functions and their related data items, then you look for the places they're called, which often gives you a hint as to the functionality of the next level up. The only difference, really, between translating and disassembling is that in the former case, you don't have to invent all the names on your own; with any kind of luck, the names in the source code have at least some mnemonic significance, although one must remain wary of misnamed items.

As I was analyzing Tom's Z80 code, I came across a subroutine named "Bitlog." The name of the routine should have given me a pretty strong suspicion as to its function, but neither the implementation or the usage gave me much of a clue. To add to the suspense, and impart some of the feeling of discovery I had when I was analyzing it, I'll show you the algorithm just the way I found it in Tom's code.

- Given a 16-bit integer, locate the leftmost nonzero bit.

- Define an integer, b, as the bit number of this bit, with bit 0 being the low-order bit.

- Now take the next three bits *after* the highest one (ignoring the leading 1). Call this three-bit number n.

- Define the Bitlog as

[7.45] $B(x) = 8*(b-1) + n$.

Because the input number was 16 bits, the largest value b can have is 15, and the largest n is seven. The range of the output number is 0 through 119, which fits comfortably in a single byte.

What was going on? Despite the seemingly obvious name of the function, it looked like some kind of floating-point format. I have worked with software floating-point routines and had, in fact, just finished translating Tom's short (24-bit) software floating-point package, which in turn was very similar to the one I'd written 20 years earlier and presented in a 3-part magazine column at the end of 1995 (Crenshaw 1995-96). The operations in Bitlog looked very much the same, with b playing the role of an exponent and n the lower three bits of a four-bit mantissa. This concept was also familiar to me. In floating-point numbers, any number except zero always has the high bit of the mantissa set. This being the case, there's no point in storing it. Instead, you sometimes gain an extra bit of resolution by storing only the bits after the most significant bit and assuming the highest bit is always a 1. This is called the "phantom bit" method. Could Tom possibly have invented not only a 24-bit floating-point format, but an eight-bit one as well? For what purpose?

Try as I might, I couldn't find any place where arithmetic was done using this format. In fact, in the only place I found the output of Bitlog used, it seemed to be used as a single integer.

What could this mean? Why take a number apart, store different parts of it in different ways, and then treat the whole thing as though it was a single integer?

Fortunately, I didn't have to worry about the deeper meaning. As I got further into the software translation, I realized that Bitlog was used only for output, and I was replacing the output portions of the software with totally different software, reflecting the differences in display technology. So for a time, at least, I set Bitlog aside as a curious oddity and went about the business at hand. Later, however, when I had more time to dig into it, I resolved to revisit the function and figure out what the heck Tom was doing there.

The Light Dawns

When I returned to Bitlog, it didn't take much intelligence to realize that the function is, in fact, generally logarithmic. I could see immediately that the top half of the function — the bit position b — changes by one when the input argument is doubled, so the output value must change by eight times the logarithm, base 2. But what about those low-order bits? Can I get reasonable behavior out of what seems to be a pseudo-floating-point operation? Things became clear when I wrote down a few numbers and calculated the output, beginning at the largest possible number (see Table 7.9).

Table 7.9   Bitlog largest arguments.

Input	Binary	Output
61440	1111xxxxxxxxxxxx	119
57344	1110xxxxxxxxxxxx	118
53248	1101xxxxxxxxxxxx	117
49152	1100xxxxxxxxxxxx	116
45056	1011xxxxxxxxxxxx	115
40960	1010xxxxxxxxxxxx	114
36864	1001xxxxxxxxxxxx	113
32768	1000xxxxxxxxxxxx	112
30720	0111xxxxxxxxxxxx	111
28672	01110xxxxxxxxxxx	110
26624	01101xxxxxxxxxxx	109

Input	Binary	Output
24576	01100xxxxxxxxxxx	108
22528	01011xxxxxxxxxxx	107
20480	01010xxxxxxxxxxx	106
18432	01001xxxxxxxxxxx	105
16384	01000xxxxxxxxxxx	104

Not only does the high half of the word change like the logarithm, according to the position of the highest nonzero bit, but the low part increases by one for every significant change in the next three bits. This can't possibly be equivalent to the logarithm, since the change is linear over the range of the lower bits. How far off is it? To see, I plotted the data shown in Table 7.9. The results are shown in Figure 7.7, where I've omitted the values for the smallest values of the argument.

For better comparison, I've added a line corresponding to the logarithm, base 2, of the input. As you can see, the two lines seem to have the same slope when plotted logarithmically. The offset, for the record, is eight, so I can make the correspondence even more dramatic by plotting the Bitlog along with the function.

[7.46] $g(x) = 8[\log_2(x) - 1]$

Figure 7.7   The Bitlog function.

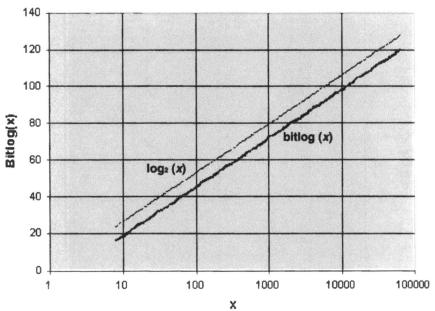

This graph is shown in Figure 7.8. The results are stunning. To the scale of Figure 7.8, the two functions are virtually indistinguishable. To put the final nail in the coffin, I looked at the error curve, shown in Figure 7.9. As you can see, the error is always less than one in the lowest bit, which is about all you can expect from an integer algorithm.

Figure 7.8   Bitlog and logarithm functions.

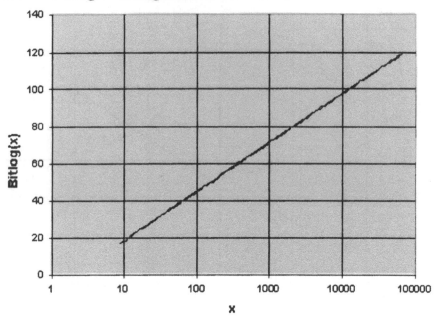

Figure 7.9   Error behavior.

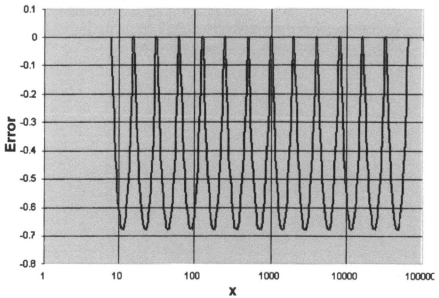

There still remains the behavior of the function near zero. The floating-point version of the log function has the embarrassing and sometimes frustrating tendency to go to negative infinity when the input argument is zero. With a little care, I can define a function behavior that is not so nasty, but to do so, I need to get a bit creative for the smaller values. The values of $B(x)$ for arguments of 32 and less are shown in Table 7.10.

Table 7.10   Bitlog smallest arguments.

Input	Binary	b	n	B(x)
32	0000000000100000	5	0	32
31	0000000000011111	4	7	31
30	0000000000011110	4	7	31
29	0000000000011101	4	6	30
28	0000000000011100	4	6	30
27	0000000000011011	4	5	29
26	0000000000011010	4	5	29
25	0000000000011001	4	4	28
24	0000000000011000	4	4	28
23	0000000000010111	4	3	27
22	0000000000010110	4	3	27
21	0000000000010101	4	2	26
20	0000000000010100	4	2	26
19	0000000000010011	4	1	25
18	0000000000010010	4	1	25
17	0000000000010001	4	0	24
16	0000000000010000	4	0	24
15	0000000000001111	3	7	23
14	0000000000001110	3	6	22
13	0000000000001101	3	5	21
12	0000000000001100	3	4	20
11	0000000000001011	3	3	19

Input	Binary	b	n	B(x)
10	0000000000001010	3	2	18
9	0000000000001001	3	1	17
8	0000000000001000	3	0	16
7	0000000000000111	2	3?	14
6	0000000000000110	2	2?	12
5	0000000000000101	2	1?	10
4	0000000000000100	2	0	8
3	0000000000000011	1	1?	6
2	0000000000000010	1	0?	4
1	0000000000000001	0	0?	2
0	0000000000000000	?	?	0

The values from $x = 15$ to $x = 8$ are straightforward enough; they follow the same rules as before. Note that because of the way the function is designed, only the highest four bits are considered significant. That's why the function for values of x, such as 24 and 25, give the same result. However, look at the case $x = 7$. Here, $b = 2$, and there are two ones following the high bit. It's tempting to say that $n = 3$. Therefore,

[7.47] $B(7) = 8*(2 - 1) + 3 = 8 + 3 = 11.$

But this is incorrect. It would imply a 0 between the leading 1 and the next two bits — a 0 that isn't really there. To follow the algorithm to the letter, I must take the value of n to be 110 binary, or six decimal, to give

[7.48] $B(7) = 8*(2 - 1) + 6 = 8 + 6 = 14.$

In other words, I must assume a "phantom" trailing 0 to round out the three low-order bits. Because of this phantom bit, the values of the Bitlog change by two, instead of by one for succeeding values of $x < 8$. This rule should continue all the way to $x = 0$.

When I consider the implementation, I can imagine that, after I've located the highest order bit, I shift the input argument right until the next three bits are in the three low-order positions. I then mask off the high 1 bit. What the result for values less than eight shows us is that, at this point (i.e., when $b \leq 2$, I must stop shifting. Note that I also don't mask the high bit at this point; the phantom bit in the "mantissa" takes care of itself.

In this way, the function extends all the way down to $x = 0$ and gives a reasonable result: zero. Although this result may hardly seem reasonable from a logarithm function and may not make sense from a strictly mathematical point of view, it's certainly a lot more comfortable than negative infinity, and the end result is a function that's well behaved while also close to logarithmic over its full range, as shown in Figure 7.10 of the complete function. As expected, the function perfectly parallels the logarithm down to $x = 8$, at which point the Bitlog function flattens out to arrive at zero for zero input, while the true logarithm continues its inexorable slide to minus infinity.

Figure 7.10 Bitlog, full range.

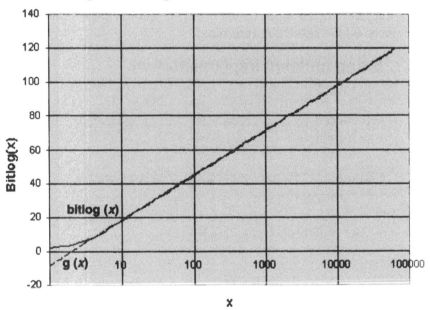

For those of you who like analogies, consider this one. It's a well-known fact that human hearing is logarithmic in nature; a sound that contains double the power doesn't sound twice as loud, only a fixed amount (3dB) louder. Accordingly, potentiometers that are used as volume controls are designed to generate a logarithmic response; turn the knob through x degrees, and you get y more decibels. If you're ordering a potentiometer from a vendor, he'll offer your choice of linear or logarithmic response curves. All audio amplifiers that still use analog components use logarithmic pots for the volume controls. But this rule clearly can't go on forever, because the knob has only a finite range of rotation. Regardless of the scale used, in terms of rotation angle per decibel, it would take a lot of turns to get from —∞dB to anything useful. Therefore, the response curve of a "logarithmic" pot is not truly logarithmic; it's tweaked to give a true zero output when set fully counterclockwise, but a finite volume after the first few degrees of rotation. This behavior is very much analogous to the Bitlog function.

The Code

Now that you know the rules, the implementation (Listing 7.7) is fairly straightforward — almost trivial. To make the function more universal, I decided to use an unsigned long integer as the input argument. The output value, 247 for an input of 0xffffffff, still fits comfortably in a single byte, much less the short I'm returning.

Listing 7.7   Bitlog function implementation.

```
/* Bitlog function
 *
 * Invented by Tom Lehman at Invivo Research, Inc.,
 * ca. 1990
 *
 * Gives an integer analog of the log function
 * For large x,
 *
 * B(x) = 8*(log(base 2)(x) - 1)
 */

short bitlog(unsigned long n)
{
  short b;

  // shorten computation for small numbers
  if(n <= 8)
    return (short)(2 * n);

  // find the highest non-zero bit
  b=31;
```

Listing 7.7   Bitlog function implementation.

```
while((b > 2) && ((long)n > 0))
{
  --b;
  n <<= 1;
}
n &= 0x70000000;
n >>= 28;
cout << b << ' ' << n << endl;
return (short)n + 8 * (b - 1);
}
```

Because it's easier to test a bit for nonzero value if you know where it is, I chose to shift left instead of right. This way, I can always test the sign bit for a 1. Once it's been found, I strip it off, move the three bits back down where they belong, and build the result.

The only tricky part is to remember to stop shifting when I reach what is ultimately to become bit 3. I tried a few tricky ways to do this by monitoring the value of *b*, but then it dawned on me that for input arguments less than or equal to eight the output is simply twice the input. So I short-circuit the entire calculation for these cases, which simplifies the code and is faster for small inputs.

Why Does It Work?

At this point, I had an understanding of how the Bitlog function worked. However, I was still left with a couple of questions, primary of which was why the algorithm worked so well. How did Tom Lehman know that removing the highest nonzero bit would lead to a function that works like the logarithm? If you look at the error curve in Figure 7.9, it's easy to convince yourself that something profound is going on here. This is not the error curve of a process chosen by random trial. It's telling me that there's something fundamentally *right* about the approach.

To see the reason, you must look to the power series expansion of the natural log function. One such function (not that used in [7.15]) is

[7.49] $\ln(1 + x) = x - \frac{x^2}{2} + \frac{x^3}{3} - \frac{x^4}{4} + \dots$.

Letting $z = 1 + x$, this becomes

[7.50] $\ln(z) = z - 1 - \dfrac{(z-1)^2}{2} + \dfrac{(z-1)^3}{3} - \dfrac{(z-1)^4}{4} + \dots$

Naturally, the more terms in the series, the more accurate it's likely to be. On the other hand, this power series is notoriously slow to converge. The denominator of each term doesn't contain a factorial or any other mechanism that makes the fraction get small very rapidly. It increases only as n, so you can't expect much difference in size between the 1,000th and the 1,001st term. I do not show the graphs here, but trust me, you don't get much improvement by going from the first-order approximation to the fourth. Hence you might just as well use the first-order approximation, and write

[7.51] $\ln(x) = x - 1.$

This curve is shown in Figure 7.11 over the range 1 to 2 along with the natural log function I'm trying to fit. The results are pretty terrible. The curves start out together and agree closely for small x, but the approximation curve, being linear, continues in a straight line while the logarithm function curves downward. At $x = 2$, the error is considerable.

Figure 7.11   Linear approximation.

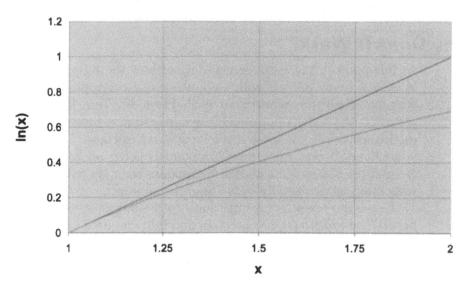

However, I can improve the fit dramatically by using a scale factor in the approximation. Instead of requiring that the two curves be parallel for small z, I'll change the function to

[7.52] $f(z) = C(z-1)$

and choose C so that the two functions agree at $z = 2$. Therefore,

$$f(2) = \ln(2)$$
$$C(2-1) = \ln(2)$$

and

[7.53] $C = \ln(2)$

The function now becomes

[7.54] $f(z) = \ln(2)(z-1)$.

This curve is plotted along with $\ln(z)$ in Figure 7.12. As you can see, the fit is much better. More dramatic is the error curve in Figure 7.13. It should look familiar because it's the same shape as the error curve of the Bitlog function (Figure 7.9).

Figure 7.12   Scaled linear approximation.

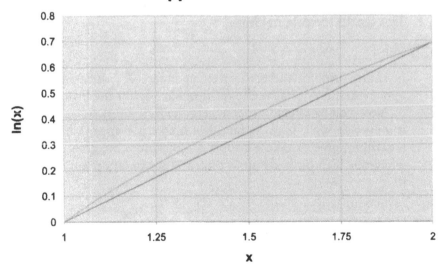

Figure 7.13 Error in scaled approximation.

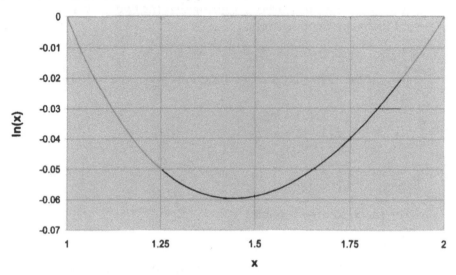

One more step needs to be taken. Recall that the Bitlog function is related to the logarithm, base 2, not the natural log. The two are related by the formula

[7.55] $\log_2(x) = \frac{\ln(x)}{\ln(2)}$.

This recalls the original form,

[7.56] $\log_2(x) = x - 1$.

In short, most of the error seen in Figure 7.13 goes away if you use the approximation not as the logarithm base *e*, but the logarithm base 2.

Now you can begin to see where the Bitlog function comes from and why it works so well. It is, in fact, the first-order approximation of the base 2 logarithm, give or take a couple of constants. Imagine a four-bit number in which the high bit is always a one, and the binary point immediately follows it:

     1.zzz.

The lower three bits can vary between 0 and 7, giving a fractional part between zero and $^7/_8$. To get *z* for this approximation, I must subtract one, which is equivalent to stripping the high bit. That's exactly how the Bitlog function works. From my high school days, I know that I can deal with integral parts of the log simply by shifting the number left. This is the function of the *b* component of the Bitlog. The factor of eight is simply to get over the three fractional bits.

Only one question is unresolved: Why subtract 1 from *b*? Why is the function

$$g(x) = 8[\log_2(x) - 1]$$

and not simply

$$g(x) = 8\log_2(x)?$$

The short answer is, if I didn't subtract 1, I would not get a function that passes near zero at $x = 0$. More to the point, however, note that the smallest value I can operate on is the binary number that still has a 1 before the decimal point:

1.000.

In order to take the number all the way down to zero, I must assume an offset of one. I must admit, this additive factor is a bother. If I want to do arithmetic using the logarithm, I need to be able to revert Equation [7.46] to read

[7.57] $$\log_2(x) = \frac{1}{8}B(x) + 1.$$

Adding 1 is an extra step I'd rather not have to do. I can always redefine the Bitlog to come closer to the true logarithm, base 2, but to do so I'd have to give up the nice behavior near zero. I'm not ready to do that, so the Bitlog function stands as Tom Lehman wrote it.

What Did He Know and When Did He Know It?

Once I had satisfied myself that the Bitlog function is indeed a logarithm of sorts and has a very nice behavior, I still found myself wondering where it came from.

Did it spring full-blown from the fertile brain of Tom Lehman, or did it come from some dry, dusty handbook that I didn't know about? Was there a paper in an ACM or IEEE journal that Tom borrowed it from? Had he done his own analysis, paralleling the process I just went through here?

There's nothing like the direct approach, so I called Tom up and asked him where he got the algorithm. Turns out, he didn't find it in a book or journal. He didn't derive it from some analysis. In fact, he didn't think it was a very significant accomplishment and had all but forgotten it. It did, indeed, spring full-blown from his fertile brain, which is why I think he's a genius.

The Integer Exponential

Not long ago, a fellow asked if I knew of a satisfactory integer exponential function. I told him no, other than the series expansion, I did not. I wish I could take back that answer, because I see now what the integer exponential ought to be. It ought to be the inverse of the Bitlog function. I'll call this function

[7.58] $\quad E(z) = B^{-1}(z)$

so that

[7.59] $\quad E(B(x)) = B(E(x)) = x$,

at least within the limits of integer arithmetic. Now consider the following.

$$B(x) \approx 8[\log_2(x) - 1]$$

$$2^{B(x)} \approx 2^{8[\log_2(x) - 1]} = \left(\frac{1}{2^8}\right)2^{8\,\log_2(x)}$$

Then

[7.60] $\quad 2^{B(x)} \approx \left(\frac{x}{2}\right)^8$.

Now let $B(x) = z$, which implies that

[7.61] $\quad x = B^{-1}(z) = E(z)$.

Then Equation [7.60] becomes

$$2^2 \approx \left[\frac{E(z)}{2}\right]^8$$

$$2^{z+8} \approx [E(z)]^8$$

$$E(z) \approx 2^{\frac{z+8}{8}}$$

[7.62] $\quad E(z) \approx 2^{\frac{z}{8}+1}$.

This, then, is the relationship between the inverse Bitlog function (the Bitexp), and the exponential function. Calling the integer exponential is

equivalent to raising two to a fractional power. For verification, you can
seek any of the values for which the Bitlog function is exact. For example,
consider $x = 1,024$, whose Bitlog is 72. Dividing this function by eight gives
nine. Adding one gives 10, and raising two to the 10th power gives 1,024.
The functions do indeed invert each other, at least in the cases where division by eight is an integer.

I should stress, though, that you should not generate $E(z)$ from Equation
[7.62], but rather design it so that it is the inverse of $B(x)$ over the entire
range. To do so, you mainly need to reverse the process used to generate
$B(x)$. The code to do so (Listing 7.8) is a straightforward inversion of Listing 7.7. It gives the true inverse of $B(x)$, and the two functions are one-way
invertible, as required by Equation [7.59]; that is,

[7.63] $B(E(x)) = x$.

The opposite is not quite true because $B(x)$ generates the same values for
different inputs. However, the two functions do revert within the limitations
of integer arithmetic, which is the best one can hope for.

Listing 7.8   Integer exponential.

```
/* Bitexp function
 *
 * returns an integer value equivalent to
 * the exponential.  For numbers > 16,
 *
 * bitexp(x) approx = 2^(x/8 + 1)
 */

unsigned long bitexp(unsigned short n)
{
    unsigned short b = n/8;
    unsigned long retval;

    // make sure no overflow
    if(n > 247)
        return 0xf0000000;

    // shorten computation for small numbers
    if(n <= 16)
```

Listing 7.8   Integer exponential.

```
            return (unsigned long)(n / 2);


    retval = n & 7;
    retval |= 8;
    cout << dec << b << ' ' << hex << retval << endl;

    return (retval << (b - 2));
}
```

Wrap-Up

I hope you've enjoyed this excursion into the Bitlog function and its origins. As with the audio volume control mentioned earlier, it's useful whenever you need a response that's roughly logarithmic in nature. Many such situations exist. A typical case is when you need a function similar to an automatic gain control (AGC). That was Tom Lehman's original use. I'm currently using the function in a different context, in a production program, and it performs flawlessly. It's blazingly fast and more than accurate enough for the application.

Postscript

Since I first published the Bitlog function and its inverse in my column in *Embedded Systems Programming* magazine, I received several e-mail messages from folks who were using similar algorithms. All were very pleased with the algorithm, and most thought they had invented it (and in a sense, they did). Apparently, there is more than one genius out there.

References

Crenshaw, J. W. "Programmer's Toolbox: Floating-Point Math," *Embedded Systems Programming*, Vol. 8, #11, November 1995, pp. 25–36.

Crenshaw, J. W. "Programmer's Toolbox: Floating-Point Math, Part 2," *Embedded Systems Programming*, Vol. 8, #12, December 1995, pp. 29–40.

Crenshaw, J. W. "Programmer's Toolbox: Floating-Point Math, Part 3," *Embedded Systems Programming*, Vol. 9, #1, January 1996, pp. 19–34.

Numerical Calculus

Chapter 8

I Don't Do Calculus

When I began this book, I wrestled with the problem of how far I needed to go into calculus. That I was going to have to get into it, I had no doubt. The reason is simple: embedded systems are the bridge between two worlds — the discrete world of digital computers, with their loops, tests, and branches, and the analog world that we live in. The digital computer obeys programmed commands, predefined by its builders to perform certain mathematical or logical operations, in sequence as written by the programmer and scheduled by the CPU clock circuitry. The real world, on the other hand, is an analog one. Events change with time, and the behavior of such events is described by calculus, specifically, according to the laws of physics. Those laws, like it or not, are almost exclusively described by differential equations.

Any embedded system more complex than a television has sensors that go out and measure some of these real-world events. It often controls actuators whose job it is to alter these events, so again, even though one may wish it to be otherwise, the behavior of real-world events is described by calculus. Every feedback loop, every PID controller, every low-pass or high-pass filter, every simulation (even your favorite action game), and every fast Fourier transform (FFT) is a case of calculus in action.

But how many people, even trained engineers, really feel comfortable with calculus?   Let's face it: we're a nation of mathophobes. How often have you heard someone say, "Well, I've never been very good at math. I just don't have the gift for it."? At your next cocktail party, mention a phrase like "time rate of change" or "the integral of ..." and watch everyone's eyes glaze over. Even graduate, practicing engineers or computer scientists may turn ashen and suddenly remember that they have a pressing date for a root canal when shown a formula like

$$\int_{-\pi}^{\pi} f(x)\sin(nx)dx.$$

Face it: because of past bad experiences, most of us feel uncomfortable with math in general and calculus in particular. I used to teach remedial math at a local junior college, so I got to see the walking wounded — people who were taught algebra by the football coach and came away thinking that their brains were somehow miswired. Balderdash! To paraphrase an old saw, there are no bad math students, there are only bad math teachers.

You may have gathered that my heart is in clearing away the fog and the mystery associated with math and making the tools of math accessible to everyone. Nowhere is this more true than in the case of calculus, because even though you may have taken calculus in college, few are really comfortable with it, and you need to be. I generally try to avoid making dogmatic statements, but I'm going to make one now.

If you are working with embedded systems and don't feel comfortable with calculus, you are limiting and damaging your career.

My next tough decision was how far to go in explaining calculus. Many of my readers are engineers and quite familiar with it already. They might feel insulted if I "talked down" to them by explaining what seems to be obvious. On the other hand, other readers haven't been exposed to much math or, like the folks described above, learned calculus in college but never really "got it." It was a tough to decide how to be sure most readers understood the things I was going to say while not insulting the intelligence of those who already knew the subject.

What finally swayed my decision was to look at the way I learn. Personally, I am never insulted when someone tells me something I already know. On the other hand, if they fail to tell me something I *need* to know, that affects my ability to understand everything that follows. Many of my future chapters will rely heavily on both analytical and numerical calculus, and I don't want to lose you the first time I mention the word "derivative."

What follows is a no-bull, fog-free review of calculus, free of fancy words or theorems. If you already know calculus, consider it a review — a review that most people can use. If you don't know calculus, this is where you learn it, in a stress-free, painless environment. If this level is beneath you, please don't feel insulted, just skip this chapter and move on to the parts that interest you.

Clearing the Fog

The first step is to recognize that you deal with calculus every day without knowing it. For example, a speedometer measures a rate. To you, it's called speed but to a mathematician it is the time rate of change of position. Within the speedometer, there's an odometer. To you, it measures your car's mileage. To a mathematician, it's integrating the speed. Under your foot is a pedal called the accelerator. It's the most direct application of calculus you're likely to find, because it controls, more or less directly, your car's acceleration, which is the first derivative of speed, and the second derivative of position. See? You've been using applied calculus for years, and you didn't know it.

As with so many topics, the gulf between the layman and the expert is not one of intelligence or understanding but of terminology. You don't need a course in calculus or a brain transplant; you only need the magic decoder ring that lets you understand and speak the language.

Galileo Did It

Long before Newton got beaned by that apple, people wondered what was going on when things fell. The first person to get it right was not Newton, but Galileo Galilei in 1638. He did something that had been considered bad form by the Greek culture before him: he performed an experiment. He wanted to time falling bodies to see how far they fell in each time interval. To slow them down enough to observe, he rolled balls down a grooved, slightly sloping board and timed their motion. He came up with a remarkable (for his time) conclusion: the speed of a falling body increases at a constant rate, independently of its mass. He proved it to the skeptics by that famous experiment, which may or may not have happened, at the Leaning

Tower of Pisa. From his conclusion, Galileo was able to derive all the equations describing accelerating bodies without using a lick of calculus.

Look at Figure 8.1, which shows the speed of a falling body as a function of time. As you can see, the curve describing this behavior is a straight line. Galileo, being an excellent mathematician, would have known immediately that such a curve is described by the equation

[8.1] $v = gt.$

The constant g, which is called the *acceleration of gravity*, has a value in the customary U.S. system of units of about 32 feet/second/second, or 32 ft/s^2. This means that every second, the speed increases by 32 ft/s over what it was the previous second. From a mathematician's point of view, g is merely the slope of the straight line obtained by reverting Equation [8.1].

Figure 8.1   Speed of a falling body.

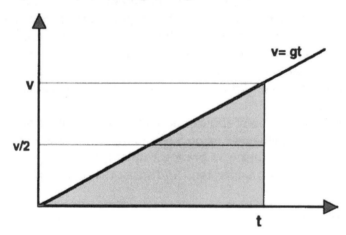

[8.2] $g = \dfrac{v}{t}$

Unfortunately, Galileo didn't have Figure 8.1. Because radar guns hadn't been invented yet, he couldn't measure speed directly — only position and time. What he had, then, was a graph like that in Figure 8.2. However, he could deduce the relationship between speed and time through a simple but ingenious line of reasoning, as follows. Look at Figure 8.1 again. Choosing some arbitrary time t, indicated by the vertical line, how far can I expect the object to move during that time? If the speed is *constant*, that would be easy: it's simply that speed multiplied by the time. But because the speed is not constant, I have to use the average speed. Because the speed is changing linearly with time, that average speed is halfway between zero and the current speed. In other words, it's $v/2$, so the distance traveled, x, must by the product of this average speed and t:

$$x = (v/2)t = \frac{1}{2}(gt)t$$

or

[8.3] $x = \frac{1}{2}gt^2.$

Figure 8.2   A falling body.

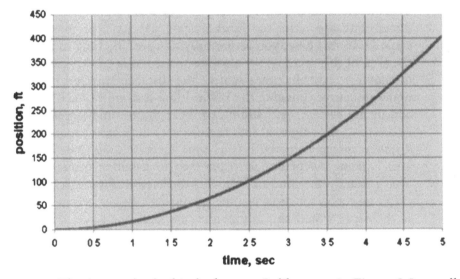

That's exactly the kind of curve Galileo saw in Figure 8.2, so all he had to do was fit the data to find the value of g.

One important aspect of all this is that the parameters in Equations [8.1] to [8.3] are not dimensionless; they have units. The acceleration g has units ft/s^2. Multiplying by t, as in Equation [8.1], gives the units $(ft/s^2)s = ft/s$, the units for speed. Similarly, the units in Equation [8.3] are $(ft/s^2)s^2 = ft$, the units for position. Notice that the units cancel, just as if they were algebraic variables. You may not have seen units bandied about this way, but it works, and in fact, you manipulate units the same way whenever you compute an average speed or a trip time.

Seeing the Calculus of It

As you've seen, Galileo's analysis was done without the benefit of calculus, which wouldn't be invented for another 28 years by Isaac Newton. (Newton, by the way, was born on Christmas day, the year Galileo died. He would invent calculus, his three laws of motion, and his law of gravitation and explain the motion of the planets all before the age of 24. Sort of makes you feel inadequate, doesn't it?) Nevertheless, lurking in this simple relationship between speed, position, and time lie practically all the key concepts of calculus, which I will now coax into the light.

Look at Figure 8.1 yet again. I got the distance traveled in time t by multiplying this time, which represents the base of a rectangle, by its height $v/2$. When you multiply the base of a rectangle by its height, you get its area, so the distance traveled is simply the area of the rectangle. (Note again that the units work out. If you multiply length and width of, say, a rug, whose dimensions are units of distance, you get its area in units of distance squared. In this case, you're multiplying speed by time to get distance.) But the area of this rectangle is also the shaded area of the triangle bounded by the curve $v = gt$. In other words, the distance traveled is simply the area under the curve. Imagine the vertical line starting at $t = 0$ (the origin of the coordinate system) and sliding to the right as time increases. Because v is also zero at the origin, at first there is very little area being swept out, so the distance changes slowly. But as time increases, the speed is greater, more area is being swept out for each step in t, and x increases more rapidly. That's exactly the kind of behavior depicted in Figure 8.2. I didn't show a graph of g as a function of time because it's rather boring. g is a constant, so the curve would be a horizontal line. Still, the same idea applies. At any time t, the area under this line would be gt. In other words, the area under the g curve is exactly the same as the speed given by Equation [8.1] and plotted in Figure 8.1.

If you think of g, v, and x as functions of time, which I'll indicate by the notation $g(t)$, $v(t)$, and $x(t)$, then

$v(t)$ = area under the curve $g(t)$ and

$x(t)$ = area under the curve $v(t)$.

(Of course, g is really a constant in this case, but it doesn't hurt to think of it as a function of time that just happens not to change.)

The area under the curve is something special: it's the *integral*. There's a math notation for it, but I won't show you that right now because I really need to do a little more work before the notation makes much sense. For now, just remember the following.

The area under the curve $f(x)$ is called the *integral* of $f(x)$.

Before going any further, I need to clarify the term, "area under the curve," because I recall that it left me feeling less than confident the first time I heard it. I mean, when people say "under the curve," just how far "under" do they mean? And because the curve can extend from zero to infinity (or sometimes to negative infinity), isn't that area also infinite? Questions like this used to bother me in college, and left me with the nagging feeling that I was missing something, and I was.

Two points help clear up any ambiguities. First, saying "under the curve" is somewhat sloppy terminology. What is really meant is the area between that curve and the x-axis [i.e., the axis for which $f(x) = 0$]. What's more, "area" is defined in a way that's not very intuitive to the layman. In this terminology, any time the curve is above the x-axis, the area is positive. If it goes below the axis, that area is considered negative (because it's a negative value multiplied by the width of the section). A curve like the sine wave can range equally above and below the x-axis, and end up with a total area of zero. That concept takes some getting used to, but it is critical in understanding the terminology.

Second, not all the area under the curve is included, only the area from one specified point (often $x = 0$) to another specified point. In other words, the area under the curve is a thoroughly bounded area delimited above by the curve, below by the x-axis, to the left by the starting value of the independent variable, and to the right by the ending value of the independent variable.

Generalizing the Result

The results in the preceding section are for the special case of a falling body, but you can easily extend the idea to any sort of motion. It's obviously not limited just to gravity — you'd get the same results for any case where the acceleration is constant — just substitute the general acceleration a for g in the equations above.

In the case where speed doesn't begin at zero, you'd have a trapezoid instead of the triangle of Figure 8.1. Still, the average speed would be halfway between the beginning and ending speeds. In this case, the area under the curve is

[8.4] $x(t) = v_0 t + \frac{1}{2} a t^2,$

where v_0 is the initial speed. You can do similar things if the position or the time doesn't start at zero. The most general forms for the equations are

[8.5] $v(t) = v_0 + a(t - t_0)$

and

[8.6] $x(t) = x_0 + v_0(t - t_0) + \frac{1}{2} a(t - t_0)^2.$

   This is a good place to discuss the difference between *definite* and *indefinite* integrals. It's a distinction that often causes confusion, and I wish someone had been around to clarify the difference for me.

   There are two ways to look at Figure 8.1. First, imagine the time t to be a general parameter, starting at zero and increasing as far as you like. As t increases, imagine the vertical line sweeping out a larger and larger area, and if you plot that area against time, you get a new function $x(t)$, as shown in Figure 8.2. In other words, t is a dynamic parameter that takes on a range of values, generating $x(t)$ as it does so. This is the indefinite integral, so called because it holds for all values of t when you haven't specified a particular value.

   On the other hand, you can think in more static terms. Imagine two specific values, t_1 and t_2, which delimit areas in Figure 8.1. In this case, the area is not a function, because nothing changes. It's simply a number to be found, and you can do so by subtracting the area for t_1 from that for t_2. Alternatively, since Figure 8.2 already represents these areas for all t, you can simply use it to calculate the difference $x(t_2) - x(t_1)$. This is the definite integral. Therefore, the essence of the relationship between the two kinds of integrals is that

- the indefinite integral is a *function* and
- the definite integral is a *number*.

To get the definite integral of any function, evaluate the indefinite integral at the desired two points and take the difference. Simple.

A Leap of Faith

Getting back to Galileo's problem, recall that the speed and position are related to acceleration as areas under curves. But now I've identified this "area under the curve" as the integral of a function. At this point, I bid farewell to Galileo, who only considered the case of constant acceleration. I'm going to make a giant leap of faith here and assert without proof that the same kind of relationship works for any kind of acceleration. In other words,

$v(t)$ = integral of $a(t)$, and

$x(t)$ = integral of $v(t)$.

Figure 8.3 shows an example for a more complicated case. For the moment, don't worry about how I calculated these integrals. Simply notice that, although it's not obvious that the three curves are related through integrals, they do seem to hang together. When acceleration is high, the speed increases most rapidly. When it's constant, the speed increases linearly, just as in Galileo's problem. When the acceleration is zero, the speed is constant. Also note that when the speed is zero, the curve for position is flat; that is, the body is standing still.

Everything seems to boil down to this idea of measuring the area under the curve. In the case where the curve is a straight line, it's trivially easy. It gets harder as the function to be integrated gets more complex. It's a lot easier if you can write down an analytical expression describing the curve; the problem of finding an analytical integral, given such an expression, is the essence of integral calculus. I'll touch on this later on, but I'll warn you of a depressing fact in advance: the set of functions that can't be integrated analytically is infinitely larger that the set of those that can. Often, you have to resort to numerical or other techniques to estimate the area under the curve.

Figure 8.3 Position, velocity, and acceleration.

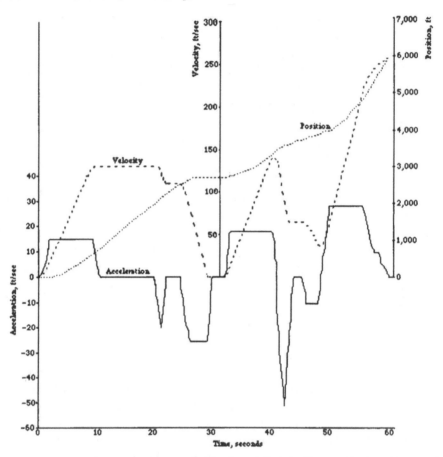

Lots of ingenious methods have been devised. As you might expect, computing the area enclosed by a curve is important in the science of surveying, where it seems that almost no plot of land is cooperatively rectangular. I suppose surveyors now use a computer for this, but they used to use a marvelous mechanical gadget called a planimeter, which looks something like a pantograph with rolling wheels. You simply trace a map of the area in question, and the resulting area can be computed by reading some dials connected to the wheels. The planimeter is, in fact, a mechanical analog computer.

There are other, cruder ways. I used to do it by plotting the curve on graph paper and simply counting all the little squares enclosed by the drawing. It takes a while and it's not super-accurate, but it's often sufficient. My favorite trick, developed by chemists who have lab-quality balances, is to plot the curve on graph paper, cut out the area with scissors, and weigh it. By comparing the weight of the cutout with that of a rectangular piece from the same paper, you get an accurate measurement of the area. The makers of graph paper carefully control the density of their paper for precisely this reason.

Down the Slope

I'll come back to the integral later. For now, I'll turn the telescope around and look through the other end. I've already discussed the idea of a slope, and I've shown that the slope of the curve $v = gt$ is simply g. The concept of slope can be understood intuitively. It shouldn't surprise you that a horizontal line has a slope of zero.

Nevertheless, the term "slope" is just one of many terms, like speed, acceleration, power, that have both everyday, intuitive meanings and more rigorous, mathematical meanings. You can get in trouble if you mistake one usage for the other.

From a mathematical point of view the slope of any straight line is defined by choosing two points on the line, such as the points P1 and P2 in Figure 8.4, which in turn are defined by the ordered pairs (x_1, y_1) and (x_2, y_2). Then the slope is defined as

[8.7] $$m = \frac{y_2 - y_1}{x_2 - x_1}.$$

Mathematicians and physicists are a lazy lot, so whenever they see the same form popping up over and over, they tend to create a shorthand notation for it. In this case, the notation for the difference between two "adjacent" numbers is

[8.8] $$\Delta x = x_2 - x_1$$

and the slope is

[8.9] $$\text{slope} = m = \frac{\Delta y}{\Delta x}.$$

Get used to this delta notation because you'll see a lot more of it.

According to this definition, as well as intuition, the slope of a horizontal line is zero. The slope of a line at a 45° angle is one because Δx and Δy are equal. The slope of a vertical line is infinite because the denominator, Δx, is zero. Lines sloping up to the right have positive slopes because both Δx and Δy have the same sign. Lines sloping down to the right have negative slopes.

This method of defining slopes is all well and good for straight lines, but how does it apply to curves? Does a curve have a slope? Absolutely. It just has a slope that changes as you move along the curve, as shown in Figure 8.4. I can choose any point on the curve and draw a line that's tangent to the curve. By definition, the slope of the curve is the same as that of the tangent line. Graphically, you can find this slope by picking two points on the tangent line, measuring x and y at the two points, and computing the slope from Equation [8.7].

Figure 8.4   The slope of a curve.

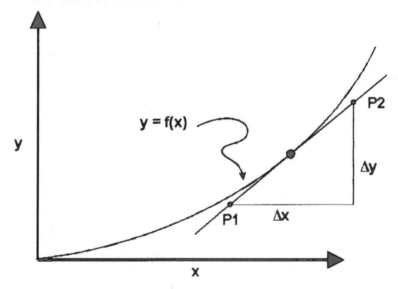

If I hadn't told you the value of g, you could have easily gotten it by plotting a graph as in Figure 8.1 (from Galileo's data, perhaps). At the end of one second, you'd find that the speed is 32 ft/s, so by definition the slope is 32 ft/s^2. But that's exactly the value of acceleration. If you were to plot Figure 8.2 with a scale and measure the slope by using the tangent lines, you'd find that it's always exactly the same as the speed at that instant. This is just like the integral relationship, except backwards. Not surprisingly, mathematicians have a name for this slope, too.

Figure 8.5   Approximating the slope.

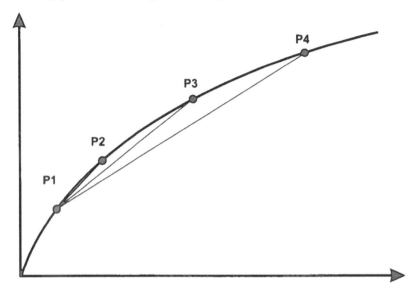

The slope of the curve of $f(x)$ is called the *derivative* of $f(x)$.

I can now assert that

$v(t)$ = derivative of $x(t)$ and

$a(t)$ = derivative of $v(t)$.

In other words, the integral and the derivative have a reciprocal relationship with each other. The derivative of the integral of $f(t)$ is simply $f(t)$ again. You can't quite say that the opposite is true, but almost: the integral of the derivative will differ from $f(t)$ only, at most, by an additive constant.

Symbolism

So far, I've studiously avoided showing you calculus notation. I thought it more important to start with the concepts. But it's time now to look at the symbolic forms used to indicate concepts like "integral" and "derivative."

The first form is the easy one because I've already alluded to it: the delta notation of Equation [8.8].

$$\Delta x = x_2 - x_1$$

Over time, the uppercase Greek delta (Δ) has become universally recognized by mathematicians to imply a difference of two numbers. In the discussion of slopes, I've been deliberately vague about just what points should be used for (x_1, y_1) and (x_2, y_2), partly because, for straight lines, it doesn't really matter. Similarly, if you're measuring the slope of a curve graphically, it doesn't matter; once you've drawn the tangent line, as in Figure 8.4, you can pick any two points on the line to do the deltas.

Imagine that someone gives you a mathematical function, $f(x)$, and asks you to plot its slope over some range in x. How would you do it? The obvious approach would be to first plot the graph, to as high an accuracy as you can, then draw tangent lines at various places. You could determine the slope of each of those lines using the method I've discussed and plot those slopes against x.

Of course, drawing the tangent line to a curve is not the easiest thing in the world to do. I've done it many times, and one can get a reasonable result. But after all, the accuracy of that result is only as good as the accuracy with which you can draw. (I feel compelled to add here that I'm old enough to remember when an engineer's skill with drafting instruments was at least as highly valued as the quality of his Table of Integrals. Before Quattro Pro and Excel, engineers used a drafting table, a T-square, French curves, and, among the most useful gadgets of all, an adjustable triangle, which not only allowed you to draw a line at any angle, but also to measure one to a fraction of a degree. If this strikes you as primitive and crude, let me remind you that it's this kind of thing that got us to the Moon. Those were the good old days.)

Although your results using this approach might be crude, you can certainly imagine, in your mind's eye, improving the accuracy to any desired degree of precision. All you need to do is simply change the scale of your graph. You could pick a tiny section of the curve of $f(x)$, plot it on as large a table as you need, draw the tangent line, and measure its slope as carefully as possible. It's as though you took a magnifying glass and expanded each section of the curve, plotting it to whatever accuracy you need.

Conceptually, you could continue this process indefinitely until you get the desired accuracy. It's a tedious process, but if, for example, the survival of the free world were at stake (as it was during World War II) or the success of the Apollo program were at stake (as it was shortly afterward), you'd find a way to get those numbers.

In any case, regardless of how difficult it really is to do something, mathematicians have no problem envisioning the process going on forever. They can imagine infinitely wide graph paper and infinitely accurate plotting of the function and its tangent line. They can envision a magic magnifying glass that lets them blow up any section of the graph without losing resolution. Eventually, if you continue this process, the curve $f(x)$ becomes indistinguishable from a straight line and the task becomes trivial.

Now, I'll change the experiment a little bit. If you are using a computer, you do not have a graph; all you have is an equation for the function $f(x)$. Because there's no computer instruction called "find a tangent to a curve," you have to resort to an approximation, as in Figure 8.5. If I want to know the slope at point P1, I could (rather foolishly) choose the right-hand point, P4, do the delta thing, and find the slope using Equation [8.9]. This gives the slope of the straight line connecting the two points, and that slope is indeed *approximately* the same as the slope at P1. (This method, by the way, is called the method of divided differences and is often used for computer work when no other data is available.) As you can see, it's only a rough approximation. Because I chose a point rather far from P1 and the curve has considerable curvature over that region, the line is obviously not really tangent to the curve. Choosing P3 gets me closer and P2 closer yet. At this point, the eyeball says that the two slopes are just about the same, but numerically I still have only a few digits of accuracy.

Using a computer, I'd have a problem getting a much better result. I can continue to move the second point closer to P1, and, for a while at least, I will continue to get better results. Unfortunately, I will also begin to lose accuracy because, eventually, the numbers get so nearly equal that their difference comes close to zero. The floating-point numbers used by computers are not infinitely accurate, and computers don't take kindly to being asked to compute the ratio 0/0.

Nevertheless, even though any computer, as well as any graphical method, eventually runs out of gas in this problem, mathematicians have no such limitations. Their magic magnifying glass works forever, and they have no trouble at all imagining the second point getting closer and closer to P1 until it's infinitesimally close, at which point the computation of the slope would also be infinitely accurate. As my major professor used to say, "Conceptually, there's no problem." To indicate this downhill slide to infinitesimal sizes and infinite resolution, I'll change the notation to read

[8.10]   slope $= m = \dfrac{dy}{dx}$.

Mathematicians describe d in rigorous terms that involve things like taking limits as distances go to zero. I prefer to think of d as a "little bitty Δ." In other words, imagine moving the two points in Figure 8.5 so close together that the error in computing the slope is exactly zero.

At this point, you may wonder how I'm going to produce a real-world version of the mathematician's magic magnifying glass or of a computer that can compute the smallest of differences with zero error. The thing that makes these theoretical mathematical concepts work is that you don't really have to rely on just graphical or computer methods — you can get an analytical expression for the slope, and analysis can give you all the accuracy there is.

To see how it works, consider the function

[8.11] $y = f(x) = x^3$.

I want to find the slope (derivative) of y at some arbitrary value x_1. To do that, I need another point, which I'll let be

[8.12] $x_2 = x_1 + \delta$,

where δ is Δx and assumed to be small. Expanding the cube in Equation [8.11], I get

[8.13] $y_2 = x_1^3 + 3x_1^2\delta + 3x_1\delta^2 + \delta^3$.

Then

$$\Delta y = y_2 - y_1 = x_1^3 + 3x_1^2\delta + 3x_1\delta^2 + \delta^3 - x_1^3,$$
$$= 3x_1^2\delta + 3x_1\delta^2 + \delta^3,$$

and

$$m = \frac{\Delta y}{\Delta x} = \frac{3x_1^2\delta + 3x_1\delta^2 + \delta^3}{\delta}$$

or

[8.14] $\dfrac{\Delta y}{\Delta x} = 3x_1^2 + 3x_1\delta + \delta^2$.

To complete the process, you need one last leap of faith, but at this point it's not a very large one. As you can see, the first term in Equation [8.14] doesn't contain δ, but the others do. It should be apparent that, as I shrink δ to get closer and closer to the desired slope, these terms are going to contribute less and less and will eventually vanish, sort of like the Cheshire cat. In fact, now that I have Equation [8.14], I can simply set δ to 0 and declare that

[8.15] $$\frac{dy}{dx} = 3x^2.$$

This is the essence of differential calculus, and you've just seen a derivative calculated. Remember, I didn't say anything at all about which value of x I used to find the derivative, so this result is perfectly general — it applies everywhere along the function $f(x) = x^3$. Finding the derivative of a function is called *differentiation*.

Big-Time Operators

So far, I've always used the symbols Δ and d as part of a larger term like Δx. But it's also possible to think of them as operators, just like the unary minus. Applying Δ to x, for example, always means taking the difference of two adjacent values of x. Similarly, taking a derivative of any function with respect to x means computing the slope of the function, as I did to get Equation [8.15], so you might see the operators standing alone with nothing to operate on; for example, d/dx or d/dt. As long as everyone agrees what is meant by these operators, that's fine. In fact, to some extent the terms in this fraction can be manipulated as though they were algebraic variables. For example, I can talk about the "naked" differential dy and write

[8.16] $$dy = \frac{dy}{dx}dx.$$

It's as though the two dx terms cancel each other.

Some Rules

I can use this concept of a derivative and compute the derivative for any given function in exactly the same manner that I did above for $f(x) = x^3$. Recall that I got the result for that case by expanding a power of $x_1 + \delta$, using a binomial expansion. For the polynomial term, the leading term in the expansion was canceled, and the second term was $3x^2$ multiplied by δ, which eventually got canceled. It should come as no surprise that the same process works for any power of x. This leads to the very important rule

$$\frac{d}{dx}x^n = nx^{n-1}.$$

In other words, to differentiate a power of x, simply multiply by that power and subtract 1 from it. This general rule also works for negative powers of x; for example,

$$\frac{d}{dx}\left(\frac{1}{x}\right) = \frac{d}{dx}(x^{-1}) = -x^{-2} = -\frac{1}{x^2}.$$

The rule even works when $n = 0$. There are whole books that give the derivatives of various functions. I obviously can't show all the derivatives here, but fortunately only a few will cover most of the cases you're likely to encounter. The key ones are shown in Table 8.1. Also shown are a few rules, similar to the associative/commutative rules of algebra, that define what to do with combinations of various forms. Armed with these formulas and a little practice, you'll find that there are not many expressions you can't differentiate.

Table 8.1   Derivatives.

Derivatives of Functions

$$\frac{d}{dx}x^n = nx^{n-1}$$

$$\frac{d}{dx}k = 0 \qquad\qquad\qquad\text{(k any constant)}$$

$$\frac{d}{dx}\sin x = \cos x$$

$$\frac{d}{dx}\cos x = -\sin x$$

$$\frac{d}{dx}e^x = e^x \qquad\qquad\qquad\text{(yes, that's right!)}$$

$$\frac{d}{dx}\ln x = \frac{1}{x}$$

Combination Rules

$$\frac{d}{dx}kf = k\frac{df}{dx}$$

$$\frac{d}{dx}(f + g) = \frac{df}{dx} + \frac{dg}{dx}$$

$$\frac{d}{dx}(fg) = g\frac{df}{dx} + f\frac{dg}{dx}$$

$$\frac{d}{dx}\left(\frac{f}{g}\right) = \frac{1}{g^2}\left(g\frac{df}{dx} - f\frac{dg}{dx}\right)$$

It's important to realize that none of the rules are black magic, and you don't need to be Einstein to derive them. They can all be derived using the same method I used for finding the derivative of $f(x) = x^3$. However, you don't have to derive them, because someone else already has. A table of integrals (or derivatives) is not something that was handed down from the sky on stone tablets; it's simply the tabulation of work done by others.

Getting Integrated

So far, I've been pretty successful at explaining the concepts of calculus by simple examples and with a modicum of common sense. Now I'll attempt the same thing with integrals.

I'll go back to the idea of the integral as the area under a curve. Just as the definition of a derivative came from trying to find the slope of a curve, you can understand integrals by trying to find the area under it. Look at Figure 8.6.

Suppose you want to know the integral of $f(x)$, the area under the curve, from zero to some arbitrary value x. To approximate this area, divide the distance x into N intervals, each having length

[8.17] $\Delta x = \frac{x}{N}.$

For each such interval Δx, you can construct a rectangle whose height touches the curve at the top left corner. To approximate the area, all you have to do is to add up the areas of the rectangles. In other words,

[8.18]   integral $(f(x))$ = area = $A = A_0 + A_1 + A_2 + A_3 + ... + A_{N-1}$

The height of each rectangle is the same as the value of $f(x)$ at that point, and the base is Δx, so each area has the form $f(x)\Delta x$. Thus, the total area is

$$f(x_0)\Delta x + f(x_1)\Delta x + f(x_2)\Delta x + \ldots + f(x_{N-1})\Delta x .$$

Figure 8.6 Approximating the integral.

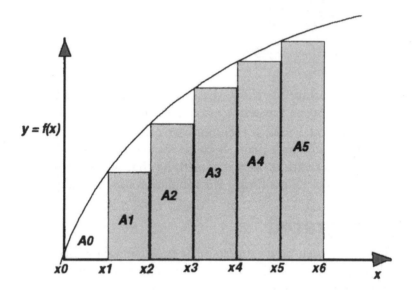

I can simplify all this by using the notation for summing terms:

[8.19] $\displaystyle \text{integral} = \sum_{i=0}^{N-1} f(x_i)\Delta x .$

Of course, Δx is the same in all terms, so by all rights, you could factor it outside of the summation. For where I'm going, though, it's better to leave it where it is.

As you can see from Figure 8.6, this kind of estimate is pretty awful when N is small; that is, when Δx is large. But I can fix that easily enough the same way I did when estimating the slope in Figure 8.5: by reducing the size of Δx. As I do so, the number (N) of terms inside the summation gets larger and larger. At the same rate, Δx is getting smaller, so the sum, which contains Δx as a factor, remains bounded. Eventually, I get to the point where Δx has earned the right to be called the "little bitty" value dx.

At about this point, students of calculus begin to have trouble believing that all those missing little triangles of area don't add up to something bigger. I've always wished for an animated film that would show the process of a sum converging to some value. Lacking that, the best I can do is Figure 8.7, which shows how the approximation changes as I increase the number of divisions. Here, the function to be integrated is $f(x) = x^2$ over the range 0 to 4. To produce Figure 8.7, I started with the truly terrible first approximation, using only one division, then doubled the number of divisions for the next step (hence the logarithmic scale). That first step gave terrible results because the only rectangle has zero height. As you can see, the fit gets better and better as N increases, and it is converging to the exact value, which happens to be $64/3 = 21^{1}/_{3}$. I took the calculation out past a million terms, at which point the result is 21.33330+. That's convincing enough for me.

Figure 8.7   Converging on the exact value.

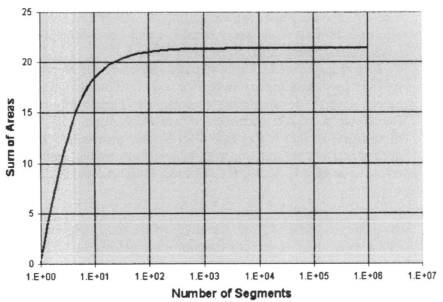

Nobody wants to compute an integral using a million terms, but you don't have to. Remember, that little exercise was just to convince you of the concept. At this point, I'm ready to assert that, conceptually at least, I can let N go all the way to infinity and replace Δx by dx. When I do, I no longer have an approximation, but an exact result. To emphasize that fact, I'll replace the summation sign by a new symbol, and write

[8.20] $integral = \int f(x)dx$.

This is an indefinite integral, which means that it is a function. Remember that, as in Figure 8.1, I'm talking about the area under the curve from 0 to some **arbitrary** value of x. A definite integral looks much the same, except that in this case I'm talking about a specific number, which is the area enclosed by two bounding values of x. To indicate the definite integral, I can annotate the integral sign with the range information, like this:

[8.21] $I = \int\limits_{0}^{4} x^2 dx$.

More Rules

I now have a notation for the integral that makes a certain sense when you see how it was developed from a summation of areas. What's still missing is a set of analytical rules for computing the integral. After all, you can't just add up a million terms every time. If at all possible, it would be better to have a set of rules similar to those in Table 8.1.

As you might recall, I was actually able to derive one of those rules analytically by expanding a power of x + δ in a binomial expansion. Unfortunately, I can't do similarly for integrals. Fortunately, I don't have to. Remember the duality between derivatives and integrals. The derivative of the integral of $f(x)$ is $f(x)$ again. So to determine what the integrals are, I need only ask what function will, when differentiated, yield the function I'm trying to integrate. Recall, for example, Equation [8.15].

$$\frac{d}{dx}x^3 = 3x^2$$

The answer, then, is that the function that yields x^2 as a derivative is simply one-third of x^3; in other words,

[8.22] $\int x^2 dx = \frac{1}{3}x^3$

and, more generally,

[8.23] $\int x^n dx = \frac{1}{n+1}x^{n+1}$.

This and some other useful rules are shown in Table 8.2.

Table 8.2   Integrals.

Integrals of Functions

$$\int dx = x$$

$$\int x^n dx = \frac{1}{n+1}x^{n+1} \qquad\qquad (n \neq -1)$$

$$\int \frac{dx}{x} = \ln x$$

$$\int e^x dx = e^x$$

$$\int \sin x\ dx = -\cos x$$

$$\int \cos x\ dx = \sin x$$

Combination Rules

$$\int k f(x)dx = k\int f(x)dx$$

$$\int (u+v)dx = \int u\ dx + \int v\ dx$$

$$\int u\ dv = uv - \int v\ du$$

Now you can see how I knew what the correct value should have been in Figure 8.7. I simply evaluated Equation [8.21] as a definite integral. You've heard me say that to evaluate a definite integral, you evaluate the function at the two extreme points and subtract them. You may never have seen it done or seen the notation used to do it, though, so I'll walk through it here.

$$\int_0^4 x^2 dx = \left|\frac{1}{3}x^3\right|_0^4 = \left[\frac{1}{3}(4)^3\right] - \left[\frac{1}{3}(0)^3\right] = \frac{64}{3}$$

The funny parallel bars that look like absolute value enclosures mean, in this context, to evaluate what's inside for the two limits shown; that is, substitute both limits into the expression and subtract the value at the lower limit from the value at the higher limit. It's easier to do it than to describe it.

Some Gotchas

I'd like to be able to tell you, as with Table 8.1, that most of the integrals you'll need can be found in Table 8.2. I can't. There are lots of forms that can be integrated, and each form is rather special. That's why whole books of tables of integrals are published. If you don't find what you need in Table 8.2, it's time to go to the nearest library.

You'll also notice that there are not nearly as many helpful rules, so if you don't find the exact form you need, you won't get much help from them. Face it: integration is tougher than differentiation.

If possible, it's always best to get an indefinite integral because it's an analytical function and so is valid for all x. To evaluate a definite integral, you then need only plug the range values into the equation. Failing that, you are often forced to resort to such tricks as counting squares or weighing graph paper. But you don't need such tricks anymore because everyone has computers now, right?

Because this subject is so important, a lot of methods have been devised for computing numerical solutions, and they don't require using a million terms. I'll be talking about such numerical methods in Chapter 9.

One other tricky little bit to cover is the strange case of the additive constant. In Table 8.1, you'll see that the derivative of a constant is zero. The significance of that comes in when you consider what function will, when differentiated, give the function you're trying to integrate. I've given the integration formula as though I could answer the question, and all those books of tables, including Table 8.2, reflect that. But really, you can't answer the question that precisely. If some constant was added to the function, you wouldn't know based on the derivative because that information gets lost in the process of differentiating. To be perfectly precise, the best you can say is that the integral is thus-and-so plus a possible additive constant. The constant is not shown in the tables, but you should always assume one. For definite integrals, this is not a problem. Even if there were such a constant, it would cancel during the subtraction process. But for indefinite integrals, always include an additive constant as part of the answer. What value should it have? Often it's decided by the boundary conditions for the problem. For the falling body problem, for example, the constant shows up in Equations [8.5] and [8.6] as v_0 and x_0. It's especially important to note how v_0 appears in the formula for x. Because of the second integration, it's no longer just a constant — there's a time-dependent term that you would have missed without it.

A last gotcha is hidden in Table 8.2 in the relation

[8.24] $\int u\,dv = uv - \int v\,du\,.$

This is the first time I've mentioned an integral over any variable but x. Don't let it throw you. After all, you can call the variables by any name you choose, but there's an even more important, and much more subtle, implication to Equation [8.24]. In it, assume that u and v are *themselves* functions of x. But remember that you can treat these d operators as though they are algebraic, so you can write the *chain rule*

$$du = \frac{du}{dx}dx$$

(and similarly for dv). What I'm really talking about here is a change of variables. If you can't find the integral that you need in Table 8.2 or in a handbook, you still often can solve the problem by substitution of variables. Sometimes, when faced with a particularly nasty integral, it's possible to separate it into the two parts: u and dv. If dv can be integrated to get v, you can use Equation [8.24] to get an equivalent, and hopefully nicer, integral. The only thing to remember is that you must substitute not only for the function, but also for the differential. I know this part sounds confusing, and it is. I won't lie to you: solving integrals analytically is tough, even on the best of days. That's why people tend to resort to numerical methods so often. Of all the ways to get the values of messy integrals, by far the most effective is a change of variables. Unfortunately, there's no easy way to guess what sort of change will get the answer. It involves trial and error, a lot of insight, more than a little luck, and a lot of practice.

Just remember my major Professor's words: "Conceptually, there's no problem."

Chapter 9

Calculus by the Numbers

Calculus is a funny sort of discipline. Volumes and volumes have been written about it, and college courses seem to go on *ad infinitum*; yet, the basic concepts can be stated in two fundamental principles. Suppose some function $y = f(x)$. The graph of the curve described by this function is depicted in Figure 9.1. The foundation of calculus is based on the two fundamental definitions:

- The *integral* of $f(x)$ is defined to be the area under the curve.
- The *derivative* of $f(x)$ is defined to be the slope of the line that is tangent to the curve at some point x. The slope, in turn, is defined to be the change in y divided by the change in x.

From these two definitions, everything else in calculus can be derived. The concepts are simple — the devil, as they say, is in the details. In particular, the integral of some given analytical function depends, not surprisingly, on the nature of that function, and one can conceive of an infinity of such functions. Reams of books have been written giving tables of integrals (and derivatives) for functions mathematicians have encountered in the past.

If you are good with calculus, are armed with a good table of integrals, and can solve a problem analytically, that's great. Getting a neat, closed-form solution to a tough problem is always satisfying and makes a mathematician or physicist feel like a real hero. From a more practical perspective, it's also a more useful result, because the analytical integral of a function $f(x)$ is another function that also holds over all x, or at least the part of it for which the function and its derivatives are defined. This means you don't have to perform the integration or differentiation again if one constant in the equation changes. You have an analytical solution, so you need only substitute the new value of the constant to adapt to the change.

Figure 9.1   Slope and area.

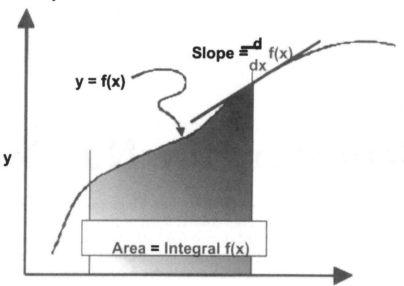

Unfortunately, we live in the real world, and in the real world, Murphy's Law still holds. The sad but true fact of life is that there are many more calculus problems that *can't* be solved than there are problems that *can* be — an infinite number, in fact. No matter how thick your book of integrals, it always seems that the problem you have to solve at the moment isn't in there. When an integral or derivative can't be found, when that elusive closed-form solution doesn't exist, what do you do?

Simple: let the computer do it. In other words, do the calculus numerically. A number of useful methods are available to find integrals and derivatives (mostly the former) when the analysis gets too hairy, and these methods are used every day to solve most of the important math problems that are solved by computer. The methods are useful in three areas: numerical integration (two types), numerical differentiation (usually for some kind of search procedure), and numerical interpolation.

Aside from missing out on the sheer self-satisfaction of having found an analytic integral, the downside of numerical methods is that the constants of the problem are all tangled up in the integral to be found. The numerical result is just what the name implies: A number, or at best, a set of numbers. Change any parameter in the equation, and you must redo the calculation to get a new set of numbers.

The upside is that you can throw away your books of tables, and you don't have to be a calculus wizard. The two basic principles and some ingenuity are all you need to solve the problem. Computing things such as areas and slopes should be duck soup for a computer. Unfortunately, Murphy isn't finished with you yet. Getting a numerical solution is easy enough; getting an accurate one with limited computing power takes a little longer.

The discipline of performing calculus numerically is called, for reasons that are lost in the mists of time, Numerical Analysis. That's what I will talk about in this chapter. I'll begin with the basic concepts and give you techniques that will work for both the quick and dirty cases and those more sophisticated cases in which high accuracy is needed. I'll begin by looking at numerical integration of the first type, called *quadrature*. You'll see the general approaches and ideas that underlie all of the numerical methods. In the process, I'll also show you one of my favorite "speed secrets," a neat method for generating integration and interpolation formulas to any desired order.

Return to Figure 9.1. If the integral (the area under the curve) is taken between two specific values of x, the result is a single number, the area, and is called a *definite* integral. An *indefinite* integral is another function. To get it, imagine that the first limit is $x = 0$ and the second slides to the right from zero to infinity. As the right border moves, the two limits enclose more area, and the resulting area can be plotted against x and is considered a new function of x.

It can be shown that the integral and the derivative are inverses of each other. That is, the derivative of the indefinite integral of $f(x)$ is simply $f(x)$ again. And vice versa, except for a possible additive constant.

You have to admit, these concepts are pretty darned straightforward. If you have any problems with them, just stare at Figure 9.1 until it soaks in. Generally, I find that it's not the basic concepts that stop people in their tracks, but the notation, which tends to look like so much chicken-scratching when you first see it.

I'm afraid there's not much I can do about the notation: It's one that's been held over for centuries. After a while, though, you'll find that it begins to make a certain perverse sense. For example, compare the approximate and exact expressions for the integral of $f(x)$, as depicted in Figure 8.6.

[9.1]
$$A = \sum_{i=0}^{N-1} f(x_i)\Delta x$$

[9.2]
$$A = \int_{x_0}^{x_6} f(x)dx$$

As you can see, the notation is similar enough to look familiar, but just different enough so the two forms can't be confused.

In a similar way, you can determine the slope of a line by picking two points on the line, say (x_1, y_1) and (x_2, y_2). The slope is defined to be

[9.3]
$$\text{slope} = \frac{\Delta y}{\Delta x}.$$

(It doesn't really matter which two points you choose to find this slope because the ratio of the two differences is constant for any two distinct points on the line.)

Again, the notation used for the derivative is a minor modification of Equation [9.3].

[9.4]
$$\text{slope} = \frac{dy}{dx} = \frac{df(x)}{dx}$$

First Approximations

Equation [9.1] is a first, and crude, method to approximate the integral. The integral was based on the notion of splitting the area into rectangular segments, as in Figure 8.6. The math was given in Equation [8.19] ?? and is repeated here.

[9.5]
$$\int f(x)dx \approx h \sum_{i=0}^{N-1} f_i$$

where, for simplicity of notation, I've let f_i represent $f(x_i)$, and replaced Δx by h. (Why h? I don't know. Like so many symbols in math, this one is traditional.) For obvious reasons, the method of Figure 8.6 and Equation [9.1] is called *rectangular integration*.

Similarly, Equation [9.3] gives the first approximation to the derivative of $f(x)$.

[9.6] $$\frac{dy}{dx} \approx \frac{\Delta y}{\Delta x} = \frac{\Delta f}{\Delta x}.$$

In practice, you need only choose two arbitrary points in the neighborhood of x_i, compute the differences Δx and Δy, and divide them. Not surprisingly, this is called the method of *divided differences*. In a pinch, you can use x_i as one of the two points; in fact, this method is often practiced. But I think you can see that you're likely to get a somewhat better approximation if you choose two points that straddle x_i.

If I had unlimited computing power, I could wrap this chapter up right now. Both of the methods just shown would work fine if I had unlimited word lengths and plenty of computing time. I have no such luxury in the real world, so I end up having to compromise accuracy against computing time.

There are two main sources of error. First, the rectangular approximation is awfully crude. Figure 8.7 shows that the estimate of the integral didn't really stabilize until I'd divided the function into about 500 segments. Before that, I was underestimating the integral, and one look at Figure 8.6 shows why. As you can see, every segment in the figure leaves out a small triangular area that should be included, so the estimate is always less than the actual area under the curve (if the curve sloped downward, the function would overestimate it). For reasons that will become obvious later, this error is called *truncation error*. You can always reduce truncation error by increasing the number of steps, but that costs computer time.

Even if you could afford the computer time needed to add several hundred, or even a million, numbers, you're not completely out of the woods because another source of error creeps in at the other end of the spectrum. This error relates to the finite word length of floating-point arithmetic. Remember, a floating-point number can only approximate the actual number to a precision that depends on the number of bits in its mantissa. Each time you add a small number to a larger sum, you lose some of the lower order bits in the result. The result is rounded to the finite length of the floating-point word. As I add more and more terms, I accumulate these tiny rounding errors in a random way, and eventually the results become meaningless. This is called *round-off error*. Round-off error can be controlled by *reducing* the number of steps — but that makes the truncation error worse again. In practice, you always end up walking a tightrope, balancing the two effects by a judicious choice of the step size h. The optimal step size is the one in which the two errors are equal in magnitude. In general you don't know what value of h this is, so you typically end up trying smaller and smaller steps until the result seems "good enough" (how's that for mathematical rigor?).

Of course, there's no law that says you have to use the areas shown in Figure 8.6 for the approximation. You could just as easily let the height of each rectangle be the value of $f(x)$ at the *end* of each interval, in which case you'd end up with

[9.7] $$\int f(x)dx \approx h \sum_{i=1}^{N} f_i .$$

The only difference between Equations [9.5] and [9.7] is subtle: the range of summation. The first sum leaves out the last data point; the second sum leaves out the first data point. In this second case, every subarea will tend to *overestimate* the actual area. Not surprisingly, this second approximation is going to be larger than the first and larger than the exact value, but you can feel pretty confident that the two results will converge if you make N large enough.

To find out how large N needs to be, I'll try these two methods on a real problem. I'll use the same integral that I used before.

[9.8] $$I = \int_0^4 x^2 dx,$$

The results are shown in Figure 9.2. As expected, the original approximation underestimates the integral for small numbers of intervals, whereas the second approximation overestimates it. Eventually, they converge at around 300 to 500 steps (intervals).

Two things become immediately apparent. First, the method *does* work. Although the first few approximations differ quite a bit, both formulas seem to be converging on an answer around 21.333, or 64/3, as expected. The second thing that hits you is the large number of steps required. Although it's not immediately apparent at the scale of Figure 9.2, even after approximately 1,000 steps you still have barely three good digits of accuracy. Extrapolating a bit, it appears that it will take something like 1,000,000,000 steps to get nine-digit accuracy (except that round-off error would destroy the results long before that). So, although the concept of rectangular areas works just dandy for analytical work, where I can happily "take the limit as N approaches infinity," in the real world the method stinks.

Figure 9.2   Two approximations.

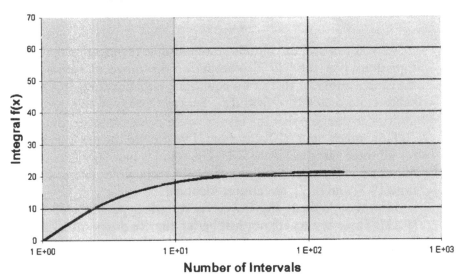

Improving the Approximation

One improvement comes immediately to mind. If the estimate given by Equation [9.5] is always low (for functions sloping upward), and that given by Equation [9.7] is always high, try averaging the two estimates. If

[9.9] $$A_{low} \approx h \sum_{i=0}^{N-1} f_i$$

and

[9.10] $$A_{high} \approx h \sum_{i=1}^{N} f_i,$$

then the average is

[9.11] $$A = \frac{1}{2}(A_{high} + A_{low}).$$

In summing the two terms, note that every f_i except the first and last are added twice, thus canceling the factor of $1/2$ in the front. Only those two terms are treated differently. The result can be written

[9.12] $$\int f(x)dx \approx h\left[\frac{f_0}{2} + \sum_{i=1}^{N-1} f_i + \frac{f_N}{2}\right].$$

As you can see, the middle summation is identical to the terms found in Equations [9.5] and [9.7], except for the range of summation. You can transform either of these two equations into Equation [9.12] by adding or subtracting one-half of f_0 and f_1. Because these formulas are so similar, it's difficult to imagine that the results could be much better, but they are.

How much better? To find out, I've plotted the error as a function of N for all three cases in Figure 9.3. The trend is more obvious if I use logarithmic scales. As you can see, thr erros for rectangular integration, as in Equations [9.5] and [9.7], are almost identical.

However, look at the lower curve, which is obtained from Equation [9.12]. These results are not just better, they're *dramatically* better. Not only does the error start out four times smaller than for the simpler approximations, but it has a slope that's twice as steep. That's very important, because it means you can obtain practical accuracies: around one part in 10^8 in a reasonable number of steps — about 32,000.

Figure 9.3   Error behavior.

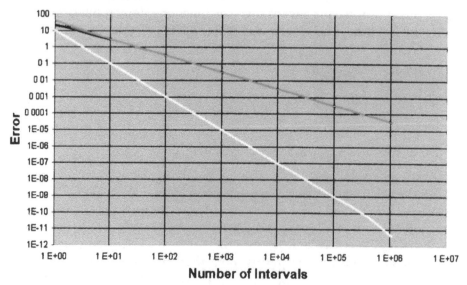

A Point of Order

The behavior of the error curves in Figure 9.3 certainly indicates there is something fundamentally different between the first two and the third methods. The reason is that Equation [9.12] represents a higher order approximation than the other two. It's time to explain the term "order" and put it on a more rigorous foundation. To do that, I can derive Equation [9.12] from a different viewpoint.

Instead of approximating the area under the curve by rectangles as I did in deriving Equations [9.5] and [9.7], I'll use trapezoidal areas, as shown in Figure 9.4. The area of each trapezoid is

$$A_i = \frac{h}{2}(f_i + f_{i+1}),$$

so I can write the total area as

$$A = \frac{h}{2}[(f_0 + f_1) + (f_1 + f_2) + (f_2 + f_3) + \ldots + (f_{N-2} + f_{N-1}) + (f_{N-1} + f_N)].$$

Figure 9.4 **A better approximation.**

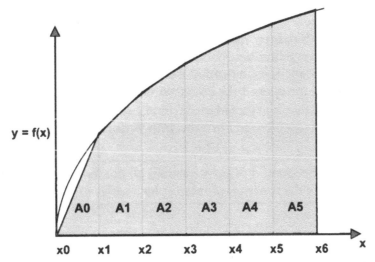

This is exactly the same form that led to Equation [9.12], and in fact the equations are identical. In other words, averaging the two methods for rectangular integration gives exactly the same results as using the trapezoidal integration of Figure 9.4.

This sheds some light on what mathematicians mean when they talk about the "order" of a polynomial. With rectangular integration, you're assuming that the curve $f(x)$ is approximated, for each interval, by a polynomial of Order 0; that is, a constant. Since you're integrating this polynomial to get area, the integral is a polynomial of Order 1. Similarly, with trapezoidal integration you're using a polynomial of Order 1 (a straight line), which has an integral of Order 2. In other words, the term "order" implies the order of an equivalent polynomial that approximates the integral. It's common to make this order explicit by writing the equations

[9.13] $$\int f(x)dx = h \sum_{i=0}^{N-1} f_i + O(h)$$

and

[9.14] $$\int f(x)dx = h\left[\frac{f_0}{2} + \sum_{i=1}^{N-1} f_i + \frac{f_N}{2}\right] + O(h^2).$$

The last term in each equation is read "order of ... " and implies that the error term is of Order 1 or 2. Knowledge of the order is important because it tells how you can expect the error to vary with changing values of the step size h. In the first-order method, halving the step size halves the error. In the second-order method, halving the step size reduces the error by a factor of four. That's exactly the behavior that Figure 9.3 displays.

Another point here is far more important, and I can't emphasize it enough. I got the first-order method by fitting a zeroth-order polynomial (the constant x_i) and the second-order method by fitting a first-order polynomial. This is a thread that runs through every one of the methods of numerical calculus, so burn this into your brain:

The fundamental principle behind all numerical methods is to approximate the function of interest by a series of polynomial segments.

The reason is very simple: we know how to integrate and differentiate polynomials. We also know how to fit a polynomial through a series of points. Once you get this concept fixed in your head, the rest, as my major professor used to say, is merely a matter of implementation.

Higher and Higher

If I get such dramatically better improvement using second-order integration, can I do even better with third order or higher? Yes, although the algorithms become a bit more complex for each step. For the third order, consider the three points at x_0, x_1, and x_2. I can attempt to fit a quadratic (i.e., a second-order polynomial) through them, yielding a third-order integral. Assume the approximation

[9.15] $f(x) = ax^2 + bx + c$.

I want to integrate this function over the range x_0 to x_2. But don't forget that

$$x_0 = x_1 - h$$

and

$$x_2 = x_1 + h.$$

Thus I can write the integral

$$I = \int_{x_1-h}^{x_1+h} f(x)dx$$

$$= \int_{x_1-h}^{x_1+h} (ax^2 + bx + c)dx$$

$$= \left| \frac{a}{3}x^3 + \frac{b}{2}x^2 + cx \right|_{x_1-h}^{x_1+h}$$

or

[9.16] $$I = \frac{a}{3}[(x_1+h)^3 - (x_1-h)^3] + \frac{b}{2}[(x_1+h)^2 - (x_1-h)^2] + c[(x_1+h) - (x_1-h)].$$

Expanding the polynomials gives

$$(x_1 + h)^3 - (x_1 - h)^3$$
$$= (x_1^3 + 3x_1^2 h + 3x_1 h^2 + h^3) - (x_1^3 - 3x_1^2 h + 3x_1 h^2 - h^3)$$
$$= 6x_1^2 h + 2h^3.$$

Similarly,

$$(x_1 + h)^2 - (x_1 - h)^2$$
$$= (x_1^2 + 2x_1 h + h^2) - (x_1^2 - 2x_1 h + h^2)$$
$$= 4x_1 h$$

and

$$(x_1 + h) - (x_1 - h) = 2h,$$

then

[9.17]
$$I = \frac{a}{3}[6x_1^2 h + 2h^3] + \frac{b}{2}[4x_1 h] + c[2h]$$
$$= \frac{h}{3}[a(6x_1^2 + 2h^2) + 6bx_1 + 6c].$$

It's certainly not obvious, but this is equal to

[9.18]
$$I = \frac{h}{3}[a(x_1 - h)^2 + b(x_1 - h) + c + 4ax_1^2 + 4bx_1$$
$$+ 4c + a(x_1 + h)^2 + b(x_1 + h) + c].$$

(The easiest way to prove this is to expand Equation [9.18]. You'll find that most of the terms cancel, leaving you with Equation [9.17].)

Look at the nine terms in Equation [9.18] and compare them to Equation [9.15]. You can see that the first three are simply $f(x_0)$, or f_0. Similarly, the next three are $4f_1$, and the last three f_2, so I end up with

[9.19] $$I = \frac{h}{3}(f_0 + 4f_1 + f_2).$$

Now it's finally starting to look usable. To complete the job, I need to add up all the areas for each set of three points. Remember, this area I spans two intervals: that from x_0 to x_1 and that from x_1 to x_2. I must repeat the process and add up all the areas for every pair of intervals in the desired range. Note that this requires that N be an even number. In the general case,

$$[9.20] \qquad \int f(x)dx = \frac{h}{3} \sum_{\substack{i=0 \\ n \text{ even}}}^{n-2} (f_i + 4f_{i+1} + f_{i+2}) + O(h^3).$$

This is Simpson's Rule. It's one of the most popular methods for this kind of integration because it represents a nice balance between accuracy and simplicity. Because I began with a second-order formula, it's reasonable to expect a third-order integral. Simpson's rule is, however, *fourth* order. Because it's a fourth-order method, I can expect the error to go down by a factor of sixteen each time I cut the step size in half.

I can continue in the same vein to get formulas of even higher order. How high should the order be? In general, the higher, the better. All other things being equal, a higher order method requires fewer steps to get the same accuracy. Because it requires fewer steps, the higher order method is less likely to cause round-off error. On the downside, the polynomial fits tend to become unstable as the order goes higher. Imagine that I have 10 points as in Figure 9.5. Chances are that the curve they represent is smooth, like the dashed curve. But if I try to fit a ninth-order polynomial through these points, there's a good chance I'll end up with the crooked curve shown, with a lot of extraneous wiggles in it. This effect is caused by numerical errors in the point values. Even though the upper and lower lobes of the wiggly polynomial ought to pretty much cancel each other out, you just have to believe that some error is going to creep in because of the wiggles, and indeed it does. Because of this, there's a practical limit to the orders you can use. I've seen integrations done with orders as high as 13, but most work gets done with orders of three through seven. I'll come back to this point and to higher order integration later. But first, I'd like to take a look at the other end of the spectrum.

Figure 9.5   Fitting data and instability.

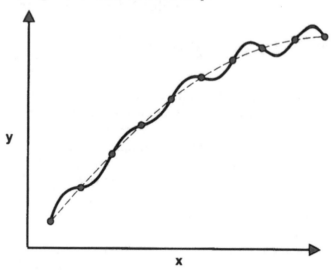

Differentiation

Finding the derivative of a function is called *differentiation*. (Why isn't it derivation? Again, who knows?) Numerical differentiation is not used nearly as often as numerical integration, mainly because it's not needed nearly as often. Differentiating a function analytically is far, far easier than integrating one, so you can usually find the derivative you need without having to resort to numerical methods. Furthermore, numerical differentiation is notoriously inaccurate and unstable, so programmers tend to use it only when there is no alternative. Another look at Figure 9.5 should convince you that if integration using higher orders can be unstable, differentiation, which involves the slope of the fitted polynomial, can be disastrous. Still, many problems exist for which numerical differentiation is the only practical approach, so it's important to have the methods in your bag of tricks.

If you followed my advice and burned the main principle into your brain, you already know the general approach to use: approximate the given function by segments of polynomials and then take the derivative of the polynomials. The first-order approximation is easy. I simply fit a straight line between two adjacent points (as in the trapezoidal integration of Figure 9.4).

[9.21] $\dfrac{dy}{dx} = \dfrac{\Delta y}{\Delta x} = \dfrac{1}{h}(f_{i+1} - f_i)$.

The next problem is which value of i to use. The obvious but somewhat use-less answer is to use the one that gives the best approximation to the actual derivative.

You can really have two kinds of problems, each of which requires a somewhat different approach. First, suppose that I have a computer func-tion that can evaluate $f(x)$ for any x. In this case, it makes sense to let f_i be represented by $f(x)$. I can replace f_{i+1} by $f(x + h)$, and since I'm not talking about tabular points at the moment, I can let h be any value that's conve-nient. Thus,

[9.22] $\dfrac{dy}{dx} \approx \dfrac{1}{h}[f(x + h) - f(x)]$.

Choosing a good value for h is even trickier for this problem than for integration, but I'll follow the same informal idea: start with a fairly large value of h and reduce it until the result seems to be "good enough." The same rules about errors apply. You can go too far in reducing h, because as h decreases, $f(x + h)$ becomes almost identical to $f(x)$, so you end up sub-tracting two nearly equal numbers. As anyone who's ever worked much with floating-point numbers can tell you, this is a sure way to lose accuracy. It's the same old trade-off: truncation error if h is too large, round-off error if it's too small. The difference is that, because the terms are subtracted in Equation [9.22], rather than added as in the integration formulas, the round-off error becomes much more of a serious problem, and that's why numerical differentiation has a deservedly bad reputation.

There's no law that says that I must choose the second point as the one to the right of x. As I did with the first-order integration, I could just as easily choose a different approximation for the derivative by using a point to the left of x. This leads to

[9.23] $\dfrac{dy}{dx} \approx \dfrac{1}{h}[f(x) - f(x - h)]$.

As I did in Figure 9.2, I can compute the derivatives using both methods. Comparing the two results gives me more confidence that I'm on the right track.

I'm sure you've already figured out what comes next. If either Equation [9.22] or [9.23] will work and will give me a first-order answer, I'll bet the average of the two will give me a second-order answer. Let

$$\text{slope}_{\text{high}} = \frac{1}{h}[f(x+h) - f(x)]$$

and

$$\text{slope}_{\text{low}} = \frac{1}{h}[f(x) - f(x-h)].$$

Then the average is

$$\frac{dy}{dx} \approx \frac{1}{2}(\text{slope}_{\text{low}} + \text{slope}_{\text{high}})$$

or

[9.24] $$\frac{dy}{dx} \approx \frac{1}{2h}[f(x+h) - f(x-h)].$$

This certainly doesn't look much like a second-order formula, so how can I be sure it is? The easiest way is to substitute the second-order polynomial and expand it as before. Let $f(x)$ be the polynomial given by Equation [9.15]. Then

$$f(x+h) = a(x+h)^2 + b(x+h) + c$$
$$= a(x^2 + 2hx + h^2) + b(x+h) + c.$$

Similarly,

$$f(x-h) = a(x^2 - 2hx + h^2) + b(x-h) + c.$$

When I subtract these two expressions, most of the terms cancel to give

$$4ahx + 2bh.$$

Dividing by $2h$ gives the derivative

[9.25] $$\frac{dy}{dx} = 2ax + b + O(h^2).$$

Differentiating Equation [9.15] analytically shows that the derivative is $2ax + b$ and the result is good to the accuracy of the second-order approximation.

More Points of Order

Sharp-eyed readers may have detected a seeming inconsistency with the way I've been defining orders here. When I used a first-order polynomial to approximate the function, I then integrated to get a *second*-order integral. When I used a first-order polynomial for the derivative, I still called it a first-order solution. By all rights, it would seem, I should really call this a *zeroth*-order formula, since differentiating should reduce the order by one.

But recall that the most meaningful definition of the order of an approximation is the error characteristic displayed as h is changed. To verify the error behavior, I used the equations above to approximate the derivative of

$$f(x) = e^x \qquad \text{for} \qquad x = \frac{1}{2}.$$

Sure enough, I got error characteristics identical to those in Figure 9.3. Equations [9.23] and [9.24] display first-order error behavior, and Equation [9.25] displays second-order behavior. Strange as it may seem, the terminology I've been using is consistent.

I could easily take the idea of numerical differentiation farther and obtain higher order formulas, but I won't pursue that here, because in a few moments I'm going to show you how to generate formulas of *any* arbitrary order.

Tabular Points

While finding the derivative of a continuous function is possible, as I've described it above, you rarely need to actually do it that way since, given a mathematical formula for a function, you can almost always find the derivative analytically. A much more likely situation is the case in which you're given not an analytical function, but a table of values at discrete points x_i. The task is to estimate a corresponding table of derivatives. For this case, I can go back to the form of Equation [9.21] and write approximations equivalent to Equations [9.22] through [9.24].

[9.26] $\dfrac{dy}{dx} = \dfrac{1}{h}(f_{i+1} - f_i) + O(h)$

[9.27] $\dfrac{dy}{dx} = \dfrac{1}{h}(f_i - f_{i-1}) + O(h)$

[9.28] $\dfrac{dy}{dx} = \dfrac{1}{2h}(f_{i+1} - f_{i-1}) + O(h^2)$

Each of these equations estimates the derivative at the point x_i. If I want the value of the derivative at a different, nontabular point, it's a matter of interpolation, which comes next.

Inter and Extra

Quick, what's the sine of 33.35678°? If you're the typical student of today, you'll simply key 33.35678 into your calculator and press the "sin" key. Back when I was in school, I had to do it the hard way (I also walked 10 miles, barefoot, in the snow). I had a book that included a table of sines of every angle from 0° through 90°, in 0.1° increments. To find the actual value, I had to *interpolate* between two adjacent angles — in this case, 33.3° and 33.4°. Extrapolation is exactly like interpolation, except that you don't necessarily have tabular points straddling the desired point. Consider Figure 9.6. The function $y = f(x)$ is defined by the straight line passing between points P_0 and P_1. What is the value of $f(x)$ at point P?

Figure 9.6   **Interpolation/extrapolation.**

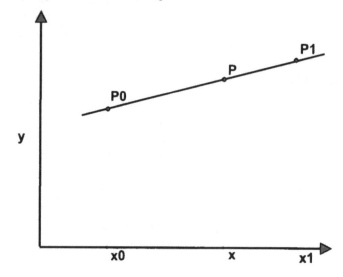

The slope of the line can be computed from any two pairs of points. I can write

$$\text{slope} = \frac{f_1 - f_0}{x_1 - x_0} = \frac{f(x) - f_0}{x - x_0}.$$

Solving for $f(x)$, I get

$$f(x) - f_0 = (x - x_0)\left(\frac{f_1 - f_0}{x_1 - x_0}\right)$$

or

[9.29] $\quad f(x) = f_0 + (x - x_0)\left(\frac{f_1 - f_0}{x_1 - x_0}\right).$

This formula is exact when the curve is a straight line, as in Figure 9.6. If the curve is not a straight line, it's still a good formula as a first-order (linear) approximation. This is the formula for linear interpolation, which is the kind most often used. Note that if the function is truly a straight line, it makes no difference whether x lies between x_0 and x_1 or not. If it does not, then you're really doing extrapolation rather than interpolation, but the principle, and even the formula, is the same. Of course, if the curve is not actually a straight line, then the formula is only approximate, and it stands to reason that the farther x is from the range between x_0 and x_1, the worse the approximation is going to be. That's life when it comes to extrapolations.

It is possible to derive higher order interpolation formulas. The method is the same as usual: fit polynomial segments to the tabular points then evaluate the polynomial at the point of interest. I won't do that now because I have other matters to discuss that are far more important. After you learn these additional techniques, the interpolation formulas will be much easier to understand and implement.

The Taylor Series

You might remember the Taylor series from college calculus. It's another one of those formulas that seem intimidating, but it's really quite simple and incredibly powerful. Take another look at Equation [9.29]. Because the slope of a curve is simply its first derivative, I can write this as

[9.30] $\quad f(x) = f_0 + (x - x_0)\left(\frac{df}{dx}\right)_0.$

The subscript zero on the derivative is used to indicate that the derivative should be evaluated at $x = x_0$. As long as I assume the curve is a straight line, it really doesn't matter where the slope is evaluated, but I'm not going to maintain that assumption any longer.

What if the curve is quadratic? It turns out that I can extend Equation [9.30] to read

$$f(x) = f_0 + (x - x_0)\left(\frac{df}{dx}\right)_0 + \frac{1}{2}(x - x_0)^2\left(\frac{d^2f}{dx^2}\right)_0$$

and, in the general case,

[9.31]
$$f(x) = f_0 + (x - x_0)\left(\frac{df}{dx}\right)_0 + \frac{1}{2!}(x - x_0)^2\left(\frac{d^2f}{dx^2}\right)_0 + \dots$$

$$+ \frac{1}{n!}(x - x_0)^n\left(\frac{d^nf}{dx^n}\right)_0 + \dots.$$

This is the Taylor series. It's an infinite series with which you can extrapolate any function from x_0 to any other point x, providing you know not only the value of $f(x)$ at $x = x_0$, but also every one of its derivatives, all evaluated at $x = x_0$. So much for interpolation and extrapolation; this formula covers them all. There is that small matter of knowing all those higher order derivatives, which might seem at first to be insurmountable. But what the Taylor series lets you do is to separate the problem into two separate parts: find the derivatives however you can, then use the series to interpolate or extrapolate. Often, you can find the derivatives by using the Taylor series "backwards." I'll show you how in a moment, but first I need to build a base to work from.

Deltas and Dels

Until now, I've loosely referred to the difference Δx as the difference between two adjacent values of x. That's true enough, but now I need to get a lot more rigorous in defining just what "adjacent" means. For this section, and most of what follows, I'll assume that I always have a set of tabular values. What's more, I'll assume that the interval is always equal, with a value h.

The forward difference, Δ, is now defined as

[9.32] $\Delta x_i = x_{i+1} - x_i.$

The backward difference, ∇ (pronounced *del*), is defined as

[9.33] $\nabla x_i = x_i - x_{i-1}.$

I've been very specific about the definitions by attaching an index to the difference as well as to the individual values. As the name implies, the forward difference is computed from the current value and the next (forward) value in the table. Similarly, the backward difference uses the *previous* value and the current value. These two differences are not independent, since

[9.34] $\Delta x_i = \nabla x_{i+1}$.

In other words, changing from one difference to the other is equivalent to moving a step forward or backward in the table.

Using these definitions, I can now rewrite Equations [9.26] and [9.27].

[9.35] $\dfrac{df_i}{dx} = \dfrac{1}{h}\Delta f_i + O(h)$

[9.36] $\dfrac{df_i}{dx} = \dfrac{1}{h}\nabla f_i + O(h)$

Remember that the second-order equation was obtained by averaging these two, so

$$\dfrac{df_i}{dx} = \dfrac{1}{2h}(\Delta f_i + \nabla f_i) + O(h^2).$$

The term in parentheses leads to the definition of a new difference, called the *central* difference, which is

[9.37] $\delta x_i = \dfrac{1}{2}(\Delta x_i + \nabla x_i)$,

so I can write the second-order formula as

[9.38] $\dfrac{df_i}{dx} = \dfrac{1}{h}\delta f_i + O(h^2).$

Comparison of Equations [9.28] and [9.38], or simply expansion of Equation [9.37], should convince you that

[9.39] $\delta x_i = \dfrac{1}{2}(x_{i+1} - x_{i-1})$.

Big-Time Operators

At this point, you may be wondering where I'm going with all this and wishing I weren't getting so bogged down in definitions. But, believe it or not, I'm right on the verge of some concepts that can crack this case wide open.

Take another look at Equation [9.37]. As you can see, every value of x in the equation is evaluated at the point x_i. In a way, I can simply think of the differences as multiplying x_i. More precisely, I must think of them as *operators* that operate upon something; in this case, x_i. Acting very naively, I could "factor" x_i out, leaving the relation

[9.40] $$\delta = \frac{1}{2}(\Delta + \nabla).$$

At first glance, an equation like this seems strange. I've got differences, but they're not differences *of* anything. The deltas are pure operators. In practice, they have to operate on something, either x_i, f_i, or whatever. Equation [9.40] really only makes sense if I apply these operators to a tabular value and perform the arithmetic that the definitions of the operators imply. But it turns out that, in many situations, I can just blithely write equations such as Equation [9.40], pretending that these are merely algebraic symbols, and manipulate them as such. Happily, I always get the right answer. This concept of operator notation is incredibly powerful and is the key to determining higher order formulas. I can perform complex algebraic manipulations on the operators, without worrying about what they're going to operate on, and everything always comes out okay in the end.

Using the same technique, I can define a relationship between the operators Δ and ∇. Consider that

$$x_1 - x_0 = \Delta x_0 = \nabla x_1.$$

Then I can write

$$x_1 = x_0 + \Delta x_0.$$

Again, if I assume that I can "factor out" the x_0, treating Δ as simply an algebraic quantity, I get

[9.41] $$x_1 = (1 + \Delta)x_0.$$

Alternatively, I could have used the second form to get

[9.42] $$x_1(1 - \nabla) = x_0.$$

For both of these relations to hold, it must be true that

$$x_1 = (1 + \Delta)x_0 = \frac{x_0}{1 - \nabla}.$$

"Dividing out" the x_0 gives

[9.43] $\qquad 1 + \Delta = \dfrac{1}{1 - \nabla}.$

From this somewhat remarkable relation, I can solve for either difference in terms of the other:

[9.44] $\qquad \Delta = \dfrac{1}{1 - \nabla} - 1 = \dfrac{\nabla}{1 - \nabla}$

or

[9.45] $\qquad \nabla = \dfrac{\Delta}{1 + \Delta}$

Now I have a neat way to convert between forward and backward differences. But there's much, much more to come. Recall that I started this little excursion into operators using Equation [9.40], which gave δ in terms of Δ and ∇. Now I can use it with Equations [9.44] and [9.45], to get

$$\delta = \frac{1}{2}(\Delta + \nabla)$$

$$= \frac{1}{2}\left(\Delta + \frac{\Delta}{1 + \Delta}\right)$$

$$= \frac{1}{2}\left[\frac{\Delta(1 + \Delta) + \Delta}{1 + \Delta}\right]$$

$$= \frac{1}{2}\left[\frac{\Delta(2 + \Delta)}{1 + \Delta}\right]$$

or

[9.46] $\qquad \delta = \dfrac{\Delta(2 + \Delta)}{2(1 + \Delta)}.$

Similarly,

[9.47] $\qquad \delta = \dfrac{\nabla(2 - \nabla)}{2(1 - \nabla)}.$

Before I leave the strange case of the central difference, notice that by comparing Equations [9.45] and [9.46] I can write

$$\delta = \frac{\Delta(2-\nabla)}{2}$$

or

[9.48] $\delta = \Delta - \dfrac{\nabla\Delta}{2}.$

Alternatively, from Equations [9.44] and [9.45]

[9.49] $\delta = \nabla + \dfrac{\nabla\Delta}{2}.$

Setting the two expressions for δ equal gives

$$\Delta - \frac{\nabla\Delta}{2} = \nabla = \frac{\nabla\Delta}{2}$$

or

[9.50] $\Delta - \nabla = \nabla\Delta.$

Can this be possible? Can the difference of the two operators possibly be the same thing as their *product*? This hardly seems reasonable, but amazingly enough, that's *exactly* what it means. If you don't believe it, you can easily prove it to yourself by performing each operation on x_i. You can also easily prove that

[9.51] $\Delta\nabla = \nabla\Delta.$

The *z*-Transform

The z-transform is the foundation of all digital signal processing, and learning how to use it is the subject of whole courses or curricula. You may not have noticed, however, that it fell out of my analysis. That it did so only serves to illustrate the power of operator algebra.

Take another look at Equation [9.41],

$$x_1 = (1 + \Delta)x_0,$$

and Equation [9.42], which can be written

$$x_1 = \left(\frac{1}{1 - \nabla}\right)x_0 \, .$$

Before, I used these two equations to get a relationship between the forward and backward difference operators. But I can look at them another way. In either equation, I can consider the term in **parentheses** as a new operator. Let

[9.52] $z = 1 + \Delta = \dfrac{1}{1 - \nabla} \, .$

Then I can write

[9.53] $x_1 = zx_0 \, .$

As you can see, z is a *shift* operator; it advances the index by one. Here it changes x_0 to x_1, but I'm sure you can see that it will work for any index. Again, if I pretend that z is really an algebraic quantity instead of an operator, I can blithely revert Equation [9.53] to

[9.54] $x_0 = z^{-1}x_1 \, .$

In other words, z^{-1} is a *delay* operator; it reduces the index by one. What's more, higher powers of z^{-1} simply reduce the index by more than one. This is the basis of all digital signal processing. Because z serves to advance the index by one, it has little value in practical applications. The data arrives sequentially, and unless you are clairvoyant, you have no way of knowing what the next data point, beyond the last one measured, will be. On the other hand, the inverse operator z^{-1} is extremely useful and practical because it's easy to generate delays electronically using an analog or digital bucket brigade shift register. In a computer, I can get the same result simply by saving past values of certain variables. If I can transform the equations describing the system of interest into a form that only depends on powers of z^{-1}, I'll have a solution that can be easily implemented in either hardware or software.

The Secret Formula

At this point, I've determined relationships between the forward and backward difference operators, the central difference operator, and the shift operator, z. What remains to be done, and what makes everything else work, is to relate these difference operators to the "smooth" world of calculus, derivatives, and functions. As I do that, you're going to get to see my all-time favorite formula — a remarkable and elegant relationship that connects differences with derivatives and, through them, discrete systems with continuous systems. I've used it to derive all kinds of numerical integration methods, but the thing that intrigues me the most about it is the incredible ease with which one can derive it, using the concepts of operator arithmetic. Once you've seen it, I think you'll like it as much as I do.

I'll begin with the Taylor series for the special case in which $x = x_1$. In this case,

$$x - x_0 = x_1 - x_0 = \Delta x_0 = h,$$

so the Taylor series becomes

$$f(x_1) = f_1 = f_0 + h\left(\frac{df}{dx}\right)_0 + \frac{1}{2!}h^2\left(\frac{d^2 f}{dx^2}\right)_0 + \dots + \frac{1}{n!}h^n\left(\frac{d^n f}{dx^n}\right)_0 + \dots .$$

Now define the *derivative* operator D such that

[9.55] $$Df(x) = \left(\frac{d}{dx}\right)f(x) = \frac{df}{dx}.$$

Then the Taylor series can be written

$$f_1 = f_0 + hDf_0 + \frac{1}{2!}h^2 D^2 f_0 + \dots + \frac{1}{n!}h^n D f_0 + \dots$$

or, again "factoring out" the f_0,

[9.56] $$f_1 = \left(1 + hD + \frac{h^2 D^2}{2!} + \dots + \frac{h^n D^n}{n!} + \dots\right)f_0 .$$

You should recognize the series in the parentheses as the exponential function I discussed in Chapter 7 "Logging In the Answers". As a matter of fact, it's identical in form to the power series given in Equation [7.14]. If you take it one step further to the notion that these operators can be manipulated as though they are simple algebraic variables, you can write

[9.57] $$x_1 = e^{hD} x_0 .$$

Surely, this time I've gone too far. Treating operators like algebraic variables is one thing, but how do I raise Euler's constant, *e*, to the power of an operator? Easy — I just expand the exponential back to its series again. You can think of Equation [9.57] as simply a shorthand form of Equation [9.56]. But incredible as it may seem, I can continue to treat this exponential as an operator. It always works.

There's more. Comparing Equations [9.53] and [9.57], you can see that

[9.58] $z = e^{hD}$.

This remarkable relationship is the Rosetta stone that connects the world of continuous time that we live in with discrete time in the world of computers and digital signal processing. Using this relationship, plus the others I've already derived, you can express all the other operators in terms of *D* and vice versa. A complete set of these relations is shown in Figure 9.7. Some of these relations are going to seem totally screwball to you, but trust me: they work. Remember, all of the "functions" shown have series expansions.

Figure 9.7   Operator relationships.

$$\Delta = \frac{\nabla}{1 - \nabla}$$

$$= z - 1$$

$$= e^{hD} - 1$$

$$= e^{\sinh^{-1}\delta} - 1$$

$$\nabla = \frac{\Delta}{1 + \Delta}$$

$$= 1 - z^{-1}$$

$$= 1 - e^{-hD}$$

$$= 1 - e^{\sinh^{-1}\delta}$$

(continued)

$$\delta = \frac{1}{2}(\Delta + \nabla)$$

$$= \frac{\Delta(2+\Delta)}{2(1+\Delta)}$$

$$= \frac{\nabla(2-\nabla)}{2(1-\nabla)}$$

$$= \frac{z^2-1}{2z}$$

$$z = 1 + \Delta$$

$$= \frac{1}{1-\nabla}$$

$$= e^{hD}$$

$$= e^{\sinh^{-1}\delta}$$

$$D = \frac{1}{h}\ln(1+\Delta)$$

$$= -\frac{1}{h}\ln(1-\nabla)$$

$$= \frac{1}{h}\ln z$$

$$= \frac{1}{h}\sinh^{-1}\delta$$

Using these relations, I am now in a position to derive formulas for any kind of integration, differentiation, interpolation, and so on to any desired order. This means that we can solve practical problems to any desired degree of accuracy.

At this point, I'm sure you must be asking, "But *how* do I use the formulas? Yes, the math may be impressive, and the shorthand notation might be slick, but how does it help me solve practical problems?"

I understand how you feel, but trust me: The answers to all your practical problems in numerical analysis are all right there, in Figure 9.7. Sometimes the solution is staring you right in the face, but you can't see it. That often seems to be the case with these formulas. All we need to do is to see how to tease the answers out into the open, so we can apply them. We'll be addressing those issues in the next chapter.

Chapter 10

Putting Numerical Calculus to Work

What's It All About?

In the last chapter, I began with the fundamental concepts of calculus and defined the terms, notation, and methods associated with the calculus operations called the integral and derivative of continuous functions. In looking at ways to explain these operations and in thinking about how to implement them in a digital computer, several other operators naturally popped up. These were the forward, backward, and central difference operators, as well as the shift operator, z.

[10.1] $\qquad \Delta x_i = x_{i+1} - x_i \qquad\qquad\qquad$ forward difference

[10.2] $\qquad \nabla x_i = x_i - x_{i-1} \qquad\qquad\qquad$ backward difference

[10.3] $\qquad \delta x_i = \frac{1}{2}(\Delta x_i + \nabla x_i) \qquad\qquad$ central difference

[10.4] $\qquad z x_i = x_{i+1} \qquad\qquad\qquad\qquad$ shift operator

Finally, I introduced the all-important relationship

[10.5] $z = e^{hD}$,

where h is the step size, or time interval between successive points, and D is the ordinary derivative operator with respect to time.

[10.6] $D = \dfrac{d}{dt}$

Relationships between various operators were given in Figure 9.7.

The problem most folks have with Figure 9.7 is a simple case of incredulity; they simply can't believe that it contains the answers to all their problems. The operator notation used is so concise and elegant that it's easy to miss the most fundamental point — that this single set of equations contains within it all the information needed to solve virtually any problem in numerical calculus. I'm not talking theory here; I'm talking down and dirty, computer programming solutions to real-world problems to an arbitrary degree of accuracy. Sometimes equations look so simple, it is difficult to imagine that they contain the answers to life, the universe, and everything. That was certainly the case, for example, with Einstein's famous equation $E = mc^2$. It's also true of Equation [10.5], which I consider to be among the most important relationships on the planet. I can't emphasize this point enough. Despite its seeming simplicity, it holds the answers to all of our problems.

Why is this particular formula so important? Because the world and the universe we live in is governed by the laws of physics. Quantum mechanics notwithstanding, this universe and its laws are based on the notion that time is continuous, flowing like a river, carrying us inexorably downstream. The laws of physics describe the state of the universe at a particular point in time through a set of differential equations. Those equations involve, among other things, derivatives with respect to time — derivatives related to the D operator of Equation [10.6].

The digital computer, on the other hand, doesn't see things that way at all. Imagine a simple embedded system that works by reading some input voltage V from some source — say, a thermometer. Over time, that voltage will in all probability vary in a continuous manner, although it may be corrupted by noise. However, the computer doesn't see the voltage as varying continuously. It *samples* the voltage, using some analog-to-digital (A/D) converter, and it samples that voltage at specific points in time. Thus, whereas the real world sees the voltage as the continuous function

[10.7] $V = V(t)$,

the computer sees it as a sequence of discrete values

[10.8] $V(n) = v_0, v_1, v_2, v_3, \ldots, v_{n-2}, v_{n-1}, v_n, \ldots$.

To operate on this voltage in an embedded system, you need a way to relate the discrete series of measured voltages (or more precisely, the set of integers delivered by the A/D converter that purport to encode those voltages) back to the real world of continuous time. Equation [10.5] provides that ability. It's the bridge between the world of continuous time, described by derivatives D, to the internal world of discrete time, described by the operator z and its cousins. That's why I call it the "Rosetta Stone" connecting the two worlds.

In the last chapter, you not only learned how calculus works, but you saw how to derive the formulas of Equations [10.1] through [10.5] — particularly the latter. Even so, if you're not familiar with the concise operator notation used in these equations, it may not be immediately apparent that these equations contain all the information you need to solve real problems.

That's what this chapter is all about. In it, I'll apply the equations to real problems, and you'll see, step by step, how they work. In the process, I think you will be surprised how easily theory can be turned into real-world practice.

Differentiation

The most straightforward use of Equation [10.5] is the most handy: compute the derivative of a function. Remember, the secret to using these operators is to forget that they're operators and naively manipulate them as though they were simple algebraic variables. If you do that and revert Equation [10.5], you get

[10.9] $hD = \ln z$.

But what does this really mean? How can you take the logarithm of an operator? Simple: as in Equation [10.5], the formula is meaningless until you expand the function back into a series. To do that, recall the relationship between z and Δ given in Figure 9.7:

[10.10] $z = 1 + \Delta$.

Equation [10.9] becomes

[10.11] $hD = \ln(1 + \Delta)$.

From a table of math functions,

$$\ln(1 + x) = x - \frac{x^2}{2} + \frac{x^3}{3} - \frac{x^4}{4} + \dots .$$

If you replace x by Δ, you finally get a formula you can use:

[10.12] $$hD = \Delta - \frac{\Delta^2}{2} + \frac{\Delta^3}{3} - \frac{\Delta^4}{4} + \dots .$$

At this point, you're probably thinking that's clear as mud. *How* do you use it? The equation above doesn't seem to be much help, but again, the answer is right there, almost shouting at you. An example should make everything clear. The first step is to use those operators to operate on something.

The Example

Using Equation [10.12] to operate on f_i gives

[10.13] $$Df_i = \frac{1}{h}\left(\Delta f_i - \frac{\Delta^2 f_i}{2} + \frac{\Delta^3 f_i}{3} - \frac{\Delta^4 f_i}{4} + \dots \right).$$

The first difference is defined in Equation [10.1]. The higher order differences are simply differences of differences; that is,

$$\Delta^2 f_i = \Delta(\Delta f_i),$$
$$\Delta^3 f_i = \Delta(\Delta^2 f_i) = \Delta(\Delta(\Delta f_i)_i),$$

and so on. You can compute all these differences from the table of the function itself. Try it with the function

[10.14] $$f(x) = x^2 - 3 \sin(x),$$

which is shown in Figure 10.1.

Figure 10.1 A sample function.

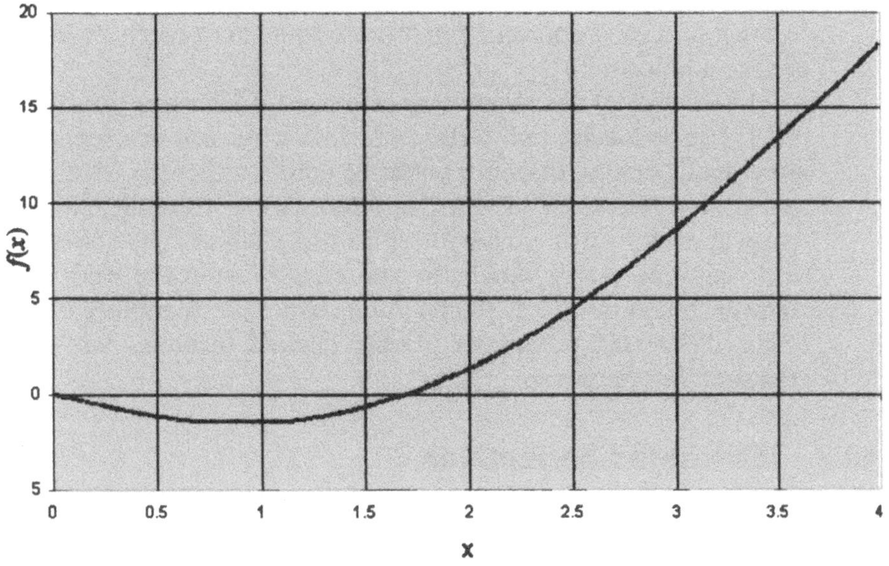

What is the derivative of $f(x)$ when $x = 2$? To find out, you can build a table of the values and their differences in that vicinity (Table 10.1).

Table 10.1 Difference table.

x	f(x)	Δf	Δ²f	Δ³f	Δ⁴f
2	5.27210772	0.133221591	0.001477525	4.7108e–06	–2.70233e–08
2.01	5.40532931	0.134699115	0.001482236	4.68378e–06	
2.02	5.540028425	0.136181351	0.001486919		
2.03	5.676209777	0.137668271			
2.04	5.813878047				

Start with just the values of x and $f(x)$. Since, from Equation [10.1],

$$\Delta f_i = f_{i+1} - f_i,$$

all you have to do to fill the next column is subtract the current value of $f(x)$ from the next one in the table. Put another way, each difference is given by simply subtracting two values from the column to its left: the entry directly to the left, and the one below it. If you want to generate a table of derivatives, do this for every row (except the last few, for which there are no entries). In this case, however, you only need enough differences to fill the row for $x = 2$. I've taken the differences out to Order 4, but there's no reason that you can't continue the process to higher orders, as long as numeric precision holds out.

In Table 10.2, I've shown the estimated derivative, truncating Equation [10.13] at one term, two terms, and so on. As you can see, the estimate is quite good, even with only two terms, and is exact (to 10 digits) with four terms. Not bad for a "crude" approximation. Actually, this is about the worst approximation you will see in this chapter. The series for the logarithm converges very slowly, as you can see from the nearly equal coefficients in Equation [10.13]. The saving grace is the rapidly decreasing sizes of the differences themselves. Other derived formulas will be at least as effective, if not more so.

Table 10.2   Estimated derivatives.

Order	Derivative
1	13.32215905
2	13.2482828
3	13.24843983
4	13.24844051
Exact	13.24844051

Now that you've seen this example, I hope you can see how the seemingly esoteric formulas of Figure 9.7 really do obtain practical results. What's more, the "translation" from the operator notation to a practical implementation is straightforward. Whatever the problem at hand, manipulate the equation to express the thing you're looking for (in this case, Df), expand the resulting formula into a power series of differences, then build a table of the differences to get the final result. It is simple and practical. The nice part about Equation [10.13] is that you can see immediately how to extend the solution to any desired order of approximation. You should also see that this extension doesn't take much extra work or extra CPU time.

Backward Differences

In Equation [10.13] and the above example, note that I chose to use forward differences. I could just as easily have used backward differences. Substituting for z in Equation [10.9], I could have written

$$[10.15] \qquad hD = \ln\left(\frac{1}{1-\nabla}\right) = -\ln(1-\nabla).$$

The logarithm series is exactly the same as before, except you replace Δ by $-\nabla$. The result is the backward equivalent of Equation [10.12]:

$$[10.16] \qquad hD = \nabla + \frac{\nabla^2}{2} + \frac{\nabla^3}{3} + \frac{\nabla^4}{4} + \dots.$$

Applying this formula to the function $f(x)$ gives

$$[10.17] \qquad Df_i = \frac{1}{h}\left(\nabla f_i + \frac{\nabla^2 f_i}{2} + \frac{\nabla^3 f_i}{3} + \frac{\nabla^4 f_i}{4} + \dots\right),$$

which, as you can see, is virtually identical to Equation [10.13] except for the signs and the backward differences.

Given a choice between Equations [10.13] and [10.17], you should probably use Equation [10.13], because theoretically, the alternating signs should result in faster convergence (i.e., fewer terms are required). However, remember that Equation [10.17] has the great advantage that it requires only backward differences, which makes it useful when the data is streaming in in real time. More to the point, in practice, I find no appreciable difference in accuracy. The results from applying the backward difference formula to the example problem match Table 10.2 so closely that I won't even bore you by repeating it. However, I do think it's important to show the difference table so you can see the structural differences.

As you can see in Table 10.3 and as the name implies, the backward difference table always looks to the past, not the future, so you need only those entries in the table that help to build the bottom row. Contrast this with Table 10.1, where the top row was used in the solution.

Table 10.3   Backward differences.

x	f(x)	∇f	∇²f	∇³f	∇⁴f
1.96	4.753901438				
1.97	4.881260581	0.128817553			
1.98	5.010078134	0.130280782	0.001468022		
1.99	5.140358916	0.131748804	0.001472787	4.73795e–06	
2	5.27210772	0.133221591	0.001477525	4.7108e–06	–2.70233e–08

Seeking Balance

You've probably noticed that whichever formula you choose, you end up with a one-sided result. Equation [10.13] uses only the current and later values of x, whereas Equation [10.17] uses only the current and earlier values.

It's reasonable to suppose that you'd get an even better approximation if you used values on both sides of the desired x. I suppose you could use a combination of forward and backward differences, but this is clearly the kind of problem that begs for the balanced arrangement that central differences offer.

The material in this chapter was first published in *Embedded Systems Programming* (Crenshaw 1993). In that article, I tried develop a formula using central differences in a very simple way using a technique I've found helpful in the past: average the results of two similar algorithms. I wrote a formula derived by averaging Equations [10.13] and [10.17]. The results were mixed for two reasons. First, I made a really dumb sign error in my expression for the central difference that led into a blind alley of ridiculous statements and wrong conclusions. Second, and more important, averaging the two equations turns out to give a result that, while not exactly wrong, is not optimally correct, either. I expected the averaged equations to simplify down to a set of terms involving only central differences, but that didn't happen.

Even though I didn't feel right about what I'd written, under the pressure of a tight deadline I didn't have the time to fix it. Later, I published an errata straightening out the definition for the central difference, but I never revisited the balanced equation for the derivative.

Fortunately, I've learned some things since 1993, from which you will benefit. Fixing the silly sign error changed the relationships in Figure 9.7 for the better, so this time around, I can give you a formula involving only central differences. The results are astonishing; I think you will be amazed at how well the formula works.

I also understand now why the averaging didn't work. Although it's mostly only of historical interest, I'll briefly discuss the differences between the two approaches.

As it turns out, I was making the problem much more difficult than it needed to be. The key to the central difference formula is found in Figure 9.7:

[10.18] $D = \frac{1}{h}\sinh^{-1}\delta$.

To get the central difference formula, I need only expand the power series. My trusty table of integrals (Pierce 1929), which also has series expansions, tells me that the series for $\sinh^{-1}(x)$ is

$$\sinh^{-1}(x) = x - \frac{1}{2}\frac{x^3}{3} + \frac{1\cdot 3}{2\cdot 4}\frac{x^5}{5} - \frac{1\cdot 3\cdot 5}{2\cdot 4\cdot 6}\frac{x^7}{7} + \frac{1\cdot 3\cdot 5\cdot 7}{2\cdot 4\cdot 6\cdot 8}\frac{x^9}{9} - \cdots.$$

Thus, the formula for the derivative is

[10.19] $D = \frac{1}{h}\left[\delta - \frac{1}{2}\left(\frac{\delta^3}{3}\right) + \frac{1\cdot 3}{2\cdot 4}\left(\frac{\delta^5}{5}\right) - \frac{1\cdot 3\cdot 5}{2\cdot 4\cdot 6}\left(\frac{\delta^7}{7}\right) + \frac{1\cdot 3\cdot 5\cdot 7}{2\cdot 4\cdot 6\cdot 8}\left(\frac{\delta^9}{9}\right) - \cdots\right].$

The beautiful aspect of this equation is that, as you can see, it involves only odd orders of the differences. This means that, to get the same order of accuracy, you'll need only about half as many terms in your approximation as you would with either the forward or backward difference formulas.

Note the pattern of the coefficients: the product of odd integers in the numerator and even integers in the denominator and a δ term divided by the power of the next odd integer. Once you see the pattern, you can extend this formula to any order desired. However, as you will see, if you need more than Order 9, something's seriously wrong.

Table 10.4 shows the differences for the sample problem.

Table 10.4   Central differences.

x	f(x)	δf	δ²f
1.95	4.627995855		
1.96	4.753901438	0.126632363	
1.97	4.881260581	0.128088348	0.001458402
1.98	5.010078134	0.129549167	0.001463222
1.99	5.140358916	0.131014793	0.001468015
2	5.27210772	0.132485197	0.00147278
2.01	5.40532931	0.133960353	0.001477518
2.02	5.540028425	0.135440233	0.001482229
2.03	5.676209777	0.136924811	0.001486913
2.04	5.813878047	0.138414059	
2.05	5.953037894		

x	δ³f	δ⁴f	δ⁵f
1.95			
1.96			
1.97			
1.98	4.80642e–06		
1.99	4.77896e–06	–2.74006e–08	
2	4.75162e–06	–2.72771e–08	1.24833e–10
2.01	4.72441e–06	–2.71509e–08	
2.02	4.69732e–06		
2.03			
2.04			
2.05			

Remember, each central difference entry is based on one point each side of the current point (ahead and behind). Note that I had to extend the difference table to one more order because the fourth-order difference isn't used. The results of the computation are shown below in Table 10.5.

Table 10.5   Results for central difference.

Order	Derivative
1	13.2485197
3	13.24844051
5	13.24844051
Exact	13.24844051

As you can see, I didn't need the fifth-order approximation (and there-fore the fourth difference); the estimation was already dead-on by the time I got to third order. This is a rather remarkable result because I get a high-accuracy approximation to the derivative with only two terms in the for-mula:

[10.20] $$D = \frac{1}{h}\left[\delta f_i - \frac{\delta^3 f_i}{6}\right].$$

For the record, in the process of developing this formula, I attempted to graph the error curve for the fifth-order case. I couldn't; it was too small. The errors ranged from zero up to about 1.0e–14, relative, which put them right at the limit of the numerical accuracy of the Intel coprocessor, and made the results too noisy to be usable. The *maximum* error using the third-order formula of Equation [10.20] is on the order of one part in 10^{10}.

Looking back, I can see why my idea of averaging forward and back-ward differences didn't work as well. Both of those formulas were based on the logarithm series

$$\ln(1 + x) = x - \frac{x^2}{2} + \frac{x^3}{3} - \frac{x^4}{4} + \dots .$$

Averaging the two formulas is equivalent to writing

$$\frac{1}{h}\ln(1 + \Delta) = -\frac{1}{h}\ln(1 - \nabla)$$

$$D = \frac{1}{2h}[\ln(1 + \Delta) - \ln(1 - \nabla)]$$

[10.21] $$D = \frac{1}{2h}\left[\ln\left(\frac{1 + \Delta}{1 - \nabla}\right)\right].$$

As it happens, another version of the logarithm series has, like the $\sinh^{-1}$ series, only odd terms.

[10.22] $\ln\left(\dfrac{1+x}{1-x}\right) = 2\left[x + \dfrac{x^3}{3} + \dfrac{x^5}{5} + \dots\right]$

The form of Equation [10.21] is so nearly the same as Equation [10.22] that it's tempting to try to force-fit it. However, although the thing I'm taking the log of in Equation [10.21] looks *almost* like the argument in Equation [10.22], it is not quite. The numerator has a Δ in it, whereas the denominator has a ∇. That seemingly small difference is enough to mess up everything. Because the two differences are almost equal, a formula based on Equation [10.22] would almost work. However, it would not be exact, and I won't settle for anything less here. This explains why averaging the two solutions didn't give me a formula involving only central differences, as I'd hoped; the correspondence wasn't exact. I could have tried to fix the problem by expressing one of the differences in Equation [10.21] in terms of the other, but I suspect that would have led me farther down the garden path. It is far better to go with the direct D-to-δ relationship, which, as you've seen, led to a formula that's not only accurate, but super-simple.

Getting Some z's

The main advantage of the series formulas derived above in terms of the differences is the very fact that they are extensible. You can see the pattern in the series and thus extend them as far as you like. This is especially useful in hand computations, because you can see the values of the differences and thereby tell when it's pointless to add more terms. But some people prefer to compute things directly from the values of f_i, without having to compute the differences. No storage is saved by doing this — it's either store past (or future) values of f_i or store the differences — but fewer additions or subtractions are required. The disadvantage is that you lose the extensibility; the results are completely different depending on where you truncate the series.

For those who prefer to work only with function values, you could begin with the equation directly relating D and z in Figure 9.7. However, this is no good because the power series for the logarithm involves the argument $(1 + x)$, not x. Thus, you're better off starting with Equation [10.16].

$$hD = \nabla + \frac{\nabla^2}{2} + \frac{\nabla^3}{3} + \frac{\nabla^4}{4} + \dots$$

This time, however, replace each instance of ∇ by its equivalent term in z:

$$\nabla = 1 - z^{-1}.$$

Note carefully that I began with Equation [10.16] instead of Equation [10.12], which uses Δ, because backward differences give me equations in terms of z^{-1}, not z. I can realize the former with the use of delay which isn't true for z. The formula for the derivative now becomes

[10.23]

$$D = \frac{1}{h}\left[(1 - z^{-1}) + \frac{(1-z^{-1})^2}{2} + \frac{(1-z^{-1})^3}{3} + \frac{(1-z^{-1})^4}{4} + \ldots\right]$$

This formula, like the others, is extensible to any order — with a difference: you can't get a formula that's explicitly in powers of z unless you expand each polynomial in the numerators. First, decide how far out to take the approximation, then expand the polynomials and collect terms. The process is tedious but perfectly straightforward. A symbolic algebra program like Maple, Mathcad, Mathematica, or Matlab will generate the formulas with ease.

In Figure 10.2, I've shown the resulting approximations for Orders 1 through 10, which ought to be enough for any problem. The coefficients for the higher order formulas look messy, but the algorithms themselves are quite simple. Computationally, if you're using the z-operator formulation, you don't need to compute columns of differences, so you only need to maintain enough storage to keep the necessary past values.

Figure 10.2   Derivative formulas using z operator.

Order 1

$$D = \frac{1}{h}(1 - z^{-1})$$

Order 2

$$D = \frac{1}{2h}(3 - 4z^{-1} + z^{-2})$$

Order 3

$$D = \frac{1}{6h}(11 - 18z^{-1} + 9z^{-2} - 2z^{-3})$$

Order 4

$$D = \frac{1}{12h}(25 - 48z^{-1} + 36z^{-2} - 16z^{-3} + 3z^{-4})$$

Order 5

$$D = \frac{1}{60h}(137 - 300z^{-1} + 300z^{-2} - 200z^{-3} + 75z^{-4} - 12z^{-5})$$

Order 6

$$D = \frac{1}{60h}(147 - 360z^{-1} + 450z^{-2} - 400z^{-3} + 225z^{-4} - 72z^{-5} + 10z^{-6})$$

Order 7

$$D = \frac{1}{420h}(1089 - 2940z^{-1} + 4410z^{-2} - 4900z^{-3} + 3675z^{-4} - 1764z^{-5} + 490z^{-6} - 60z^{-7})$$

Order 8

$$D = \frac{1}{840h}(2283 - 6720z^{-1} + 11760z^{-2} - 15680z^{-3} + 14700z^{-4} - 9408z^{-5} + 3920z^{-6} \\ - 960z^{-7} + 105z^{-8})$$

Order 9

$$D = \frac{1}{2520h}(7129 - 22680z^{-1} + 45360z^{-2} - 70560z^{-3} + 79380z^{-4} - 63504z^{-5} \\ + 35280z^{-6} - 12960z^{-7} + 2835z^{-8} - 280z^{-9})$$

Order 10

$$D = \frac{1}{2520h}(7381 - 25200z^{-1} + 56700z^{-2} - 100800z^{-3} + 132300z^{-4} - 127008z^{-5} \\ + 88200z^{-6} - 43200z^{-7} + 14175z^{-8} - 2800z^{-9} + 252z^{-10})$$

How well does the z operator version work? In Table 10.6, I show the results of the derivative of the sample function for each order.

Table 10.6   Derivative using the _z_ operator.

Order	Derivative
1	13.17488035
2	13.24828144
3	13.2484412
4	13.24844051
Exact	13.24844051

Although I don't show the table for the backward differences, you should know that the results obtained here for the _z_ operator are identical to those obtained using backward differences. This shouldn't surprise you because both the _z_ operator and the backward difference operator use exactly the same data, the tabular values of $f(x)$ from $x = 1.96$ through $x = 2.0$.

Numerical Integration

Quadrature

Now that you've seen the general method for applying the "Rosetta Stone" formula, you should also see that it can be used for virtually any problem involving tabular data. About the only limitation, other than imagination, is the restriction that the data be tabulated with a fixed interval, _h_.

There are two kinds of numerical integration. In the first and simplest kind, called quadrature, you are given a function $f(x)$ or a table of its values and are asked to find the area under the curve. I've already covered this case, so I'll limit the discussion here to a review and clarification of relationships between theoretical concepts, reality, and numerical approximations to it.

In Chapter 8, I defined the integral of a function $f(x)$ as the area under the curve. Then I discussed how to go about approximating the function by a series of rectangular areas (see Figure 8.6). I used that model as a springboard into the mathematical concept of an integral, where a summation approximating the area under the curve is converted to an integral exactly equal to that area as the number of rectangles increase without limit.

In Chapter 9, I figuratively turned the telescope around and looked into the other end. Instead of using the concept of adding rectangular areas as a springboard to the theory of integration, I considered its use to numerically approximate the integrals.

I found that the use of rectangular areas, which is equivalent to fitting the curve of $f(x)$ with a zero-order polynomial (a constant), yielded a first-order integral, meaning that the error in the approximation went as order(h) [often written $O(h)$], where h is the step size, or the difference between successive values of x. The results were mixed; the method works, but because of the low order, a large — perhaps impractically large — number of steps is required for a good approximation.

In Equation [9.12] and Figure 9.4, I arrived at a model using trapezoidal areas, corresponding to a first-order polynomial (straight line) fit of $f(x)$ over each interval. Not surprisingly, the first-order fit led to a second-order integral, $O(h^2)$.

Finally, I introduced the concept of a second-order-fit polynomial which, together with some creative summation, produces the quadrature formula known as Simpson's Rule (see Equation [9.20]). Because the fitting polynomial is of second order, you have every right to expect Simpson's Rule to lead to a third-order integration formula. However, you get a pleasant surprise. Thanks to some fortuitous cancellations of terms, Simpson's Rule turns out to give a fourth-order result. This means you can get results that are essentially perfect — results are exact to within the accuracy of the computer's floating-point word length — with a surprisingly small number of steps and a formula that is negligibly more complex than a simple summation.

An Example

To bend an old homily, an example (with pictures) is worth a thousand words. To see numerical quadrature in action, I'll consider the integral of the following function, which is similar to the function used in the discussion of slopes.

[10.24] $$f(x) = \sin(x) + \frac{x^3}{40}$$

A graph of this function is shown in Figure 10.3. If I chose a crude approximation, using only four intervals, I'd end up fitting the function with two quadratics, as shown in Figure 10.4. The corresponding integral is given by

$$\int f(x)dx = \frac{h}{3}[(f_0 + 4f_1 + f_2) + (f_2 + 4f_3 + f_4)]$$

[10.25]

$$= \frac{h}{3}(f_0 + 4f_1 + 2f_2 + 4f_3 + f_4).$$

Figure 10.3   A function to integrate.

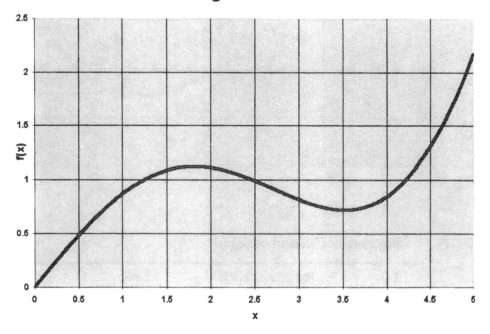

For the given function, the numerical value is 5.102053584793906. This is admittedly pretty far off the actual value of 4.622587814536773. However, you get the idea from Figure 10.4 that things are going to get much better as the number of intervals increases. Table 10.7 shows the values as N increases.

Figure 10.4 Four intervals.

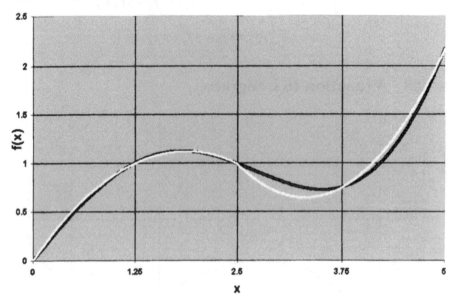

Table 10.7 Simpson's Rule integral.

N	Integral f(x)	Error
2	5.10205358479391	0.4794657
4	4.63446383999906	0.0118760
8	4.62322450680946	0.0006366
16	4.62262621328102	3.8398e–5
32	4.62259019350767	2.3789e–6
64	4.62258796289825	1.4836e–7
128	4.62258782380431	9.2675e–9
256	4.62258781511592	5.7914e–10
512	4.62258781457297	3.6193e–11
1,024	4.62258781453904	2.2622e–12
2,048	4.62258781453691	1.3944e–13
4,096	4.62258781453678	7.9936e–15

Note that I get acceptable accuracy with as few as 32 steps and more than enough accuracy for most purposes with 256 steps. Beyond 4,096 steps, I run into the limits of the Intel math coprocessor. A log–log graph of these errors, shown in Figure 10.5, displays the telltale slope of a fourth-order approximation: every crossing of a grid line on the x-axis corresponds to four grid lines on the y-axis.

Figure 10.5   Simpson's Rule error.

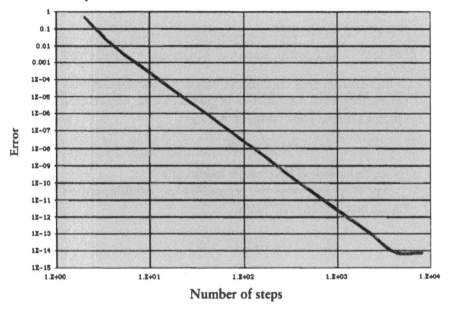

Number of steps

In the world of numerical integration, the fourth order is one that commands a high degree of respect. The most popular general-purpose integration formula is the fourth-order Runge–Kutta formula, which I'll discuss later. Except for things like interplanetary trajectory generators, people rarely need to go much higher in order. In fact, the fourth-order Runge–Kutta method reduces to Simpson's Rule when applied to a problem of quadrature. Therefore, I'll accept Simpson's Rule here as a perfectly adequate solution to that problem and won't seek further improvements.

Implementation Details

Before leaving the subject of quadrature, I should say something about both the structure of a practical integrator and the relationship between low- and high-order implementations. Consider Equation [9.5] again.

$$\int f(x)dx = h \sum_{i=0}^{N-1} f_i$$

This equation implements the integrator derived from the concept of rectangular areas.

Despite the elegance and brevity of summation signs, sometimes I find it helpful to write out the summation in its long form. To be specific and still keep the notation short enough, suppose I have only 10 steps in the function (and therefore 11 tabular values, including beginning and ending values). Equation 9.5 becomes

[10.26] $I1a = h(y_0 + y_1 + y_2 + y_3 + y_4 + y_5 + y_6 + y_7 + y_8 + y_9).$

Recall that this formula is based on the assumption that the heights of the rectangles are set at the beginning value of each interval. Alternatively, when I set it at the ending value, I get

[10.27] $I1b = h(y_1 + y_2 + y_3 + y_4 + y_5 + y_6 + y_7 + y_8 + y_9 + y_{10}).$

(Note the subtle difference in indexing.)

The next step was the second-order formula, which I got by averaging Equations [10.26] and [10.27].

$$I2 = \frac{1}{2}(I1a + I1b)$$

$$= \frac{h}{2}[(y_0 + y_1 + y_2 + y_3 + y_4 + y_5 + y_6 + y_7 + y_8 + y_9)$$

$$+ (y_1 + y_2 + y_3 + y_4 + y_5 + y_6 + y_7 + y_8 + y_9 + y_{10})]$$

[10.28] $= h\left(\frac{y_0}{2} + y_1 + y_2 + y_3 + y_4 + y_5 + y_6 + y_7 + y_8 + y_9 + \frac{y_{10}}{2}\right)$

The second-order formula looks almost exactly like either of the first, except that the first and last points require a (trivially) different treatment. This is an encouraging sign that increasing the order will not necessarily increase the workload appreciably, at least for quadrature.

Look again at Simpson's Rule, which was given in Equation [9.20].

$$\int f(x)dx = \frac{h}{3} \sum_{\substack{i=0 \\ n \text{ even}}}^{n-2} (f_i + 4f_{i+1} + f_{i+2}) + O(h^3)$$

Recall that Simpson's Rule requires that you take segments in pairs, involving three points per step. Although the equation above is the correct formula for Simpson's Rule, a direct implementation of it would be terrible. That's because the end of one pair of segments is the beginning of the next. Implementing Equation [9.20] directly would require you to process half the points twice. This becomes obvious when I write the summation out in its long form.

$$I = \frac{h}{3}[(f_0 + 4f_1 + f_2) + (f_2 + 4f_3 + f_4) + (f_4 + 4f_5 + f_6)$$
$$+ (f_6 + 4f_7 + f_8) + (f_8 + 4f_9 + f_{10})]$$

Notice that the interior boundary points, f_2, f_4, ..., are each repeated twice. Simplifying gives

[10.29]
$$I = \frac{h}{3}(f_0 + 4f_1 + 2f_2 + 4f_3 + 2f_4 + 4f_5$$
$$+ 2f_6 + 4f_7 + 2f_8 + 4f_9 + f_{10}).$$

This is the optimal form for implementation. If I group the terms in pairs, I can revert back to a summation notation, adding all the interior points then tweaking the two end points separately. The resulting equation is

[10.30]
$$I = \frac{h}{3}\sum_{i=0}^{\frac{N}{2}-1}(2f_{2i} + 4f_{2i+1}) - f_0 + f_N.$$

Listing 10.1 gives a general-purpose implementation of Simpson's Rule based on a verbatim translation of Equation [10.30]. It is a flexible and eminently practical addition to your toolbox. Just remember the cardinal rule (which the function will remind you of, if you forget): the number of intervals must be even.

Listing 10.1 Simpson's Rule

```
/* The function simpson provides a general-purpose
   implementation of numerical quadrature using Simpson's
   rule. This algorithm generates the integral under the
   curve:
 *
 *            y = f(x)
 *
 * Between the limits x1 and x2, dividing this range up into n
   intervals.


 *
 * The nature of Simpson's rule requires that n be even.
 */


double simpson(double (*f)(double), double x1, double x2, int n){
    if((N & 1) != 0){
        cout << "Simpson: n must be an even integer\n";
        return 0;
    }
    double sum = 0;
    double h = (x2-x1)/n;
    double x = x1;
    for(int i=0; i<n/2; i++){
        sum += f(x)+2*f(x+h);
        x += 2*h;
    }
    sum = 2*sum-f(x1)+f(x2);
    return h*sum/3;
}
```

Trajectories

The second class of problems in numerical integration present considerably more challenge. I call the problems trajectory generators, for two reasons. First, I encountered the problem while working at NASA, helping to generate the trajectories of spacecraft. Second, even if the problem doesn't involve trajectories in the traditional sense of the word it does involve a solution that depends on initial conditions, and the function to be integrated cannot be drawn out in advance.

In problems of quadrature, you're given some function $f(x)$, which is forever fixed. Whatever the function, it depends only on x; therefore, you can draw it on a graph, look at it, analyze it, and decide how best to integrate it. The integral is, as I've said so many times, simply the area under the curve of $f(x)$.

Trajectories are a different animal altogether. In problems of this type, you are not given the function to integrate but, rather, a differential equation that defines it. In this second case, you begin with an ordinary differential equation of the form

[10.31] $\qquad \dfrac{dy}{dx} = f(x, y)$.

Note that the derivative function now depends on y as well as x, which presents a dilemma. You can't simply plot the curve because you don't know the value of the next y until you know $f(x, y)$, which in turn depends on y again. The best you can do is begin with some initial condition $y = y_0$ and inch your way forward one small step at a time until the entire table of y_i is constructed.

In a very real sense, the two kinds of numerical integration are completely analogous to the two kinds of integrals you can obtain analytically. Quadrature is equivalent to the definite integral

[10.32] $\qquad y = \displaystyle\int_{x=0}^{x=X} f(x)dx$,

which produces a simple, scalar number: the area under the curve of $f(x)$. The second type of integration is equivalent to the indefinite integral

[10.33] $\qquad y(x) = \displaystyle\int f(x, y)dx$.

The result of this kind of integral is not a single number, but a new function: the history of y as a function of x. Because f is a function of y as well as

x, there is no single solution but, rather, a whole family of solutions: one "trajectory" for each starting value of y.

A problem like this is called a single-point boundary value problem. You generate a solution that gives y as a function of time, beginning with some initial condition $y = y_0$. There are no restrictions on what the independent variable x should be, but in many cases x is time, t. In this special case, Equation [10.31] would be called an equation of motion of a dynamic system, and the resulting computations would be a simulation of that motion. As simple as Equation [10.31] looks, it is an extremely powerful and useful concept. I've made no restrictions on the nature of $f(x, y)$, so any set of differential equations, no matter how nonlinear or otherwise nasty, can be cast into this form. You can predict the motion of virtually any dynamic system if you have a general scheme for solving its equations of motion.

Defining the Goal

The approach I use to obtain the integration formula is exactly the same as for differentiation: I write down the thing I'm looking for, use Equation 10.5 to express it in terms of differences, and expand the resulting series. The only tricky part is to be sure I know what form I'm really looking for.

First, I should not be looking for y_i directly. At any given point, I have already computed all the values of y (and f) up to and including the index, i. What I want is the next point, y_{i+1}. I can get this by adding a Δy to the current value, so I am seeking a formula for Δ.

On the opposite side, I have past values of y_i (and f_i) available, but no values beyond y_i. Hence my formula must be restricted to backward differences (no fair peeking into the future). Negative powers of z also work.

Finally, I'm operating on $f(x, y)$, not y. My formula should include one and only one power of D. In other words, an acceptable form is

[10.34] $\Delta = P(\nabla)hD$,

where P is a polynomial in ∇. Using this form to operate on y_i gives

$$\Delta y_i = P(\nabla)hDy_i,$$

from which I can now write

[10.35] $y_{i+1} = y_i + hP(\nabla)f_i$.

The only remaining task is to determine the polynomial P.

Begin with the expression for Δ. From Equation [9.52], I can write

[10.36] $\Delta = \nabla z.$

Because I also need a D on the right-hand side, I'll multiply and divide by D:

$$\Delta = \nabla z \left(\frac{D}{D} \right).$$

Substituting Equation [10.9] for the D in the denominator, I get

[10.37] $\Delta = \left(\frac{\nabla z}{\ln z} \right) hD.$

I'm almost there, except the expression is mixed, involving both z and ∇. Substituting for z in terms of ∇ gives

[10.38] $\Delta = \left(\dfrac{-\nabla}{(1 - \nabla)\ln(1 - \nabla)} \right) hD.$

Now I have the formula in terms of the parameters I need. I still need to turn the expression in parentheses into a power series in ∇. Because the math gets quite tedious, I'll merely outline the process here. Recall that the series for the logarithm gives

$$\ln(1 - \nabla) = -\left(\nabla + \frac{\nabla^2}{2} + \frac{\nabla^3}{3} + \frac{\nabla^4}{4} + \dots \right).$$

Multiplying this by $1 - \nabla$ gives

$$(1 - \nabla)\ln(1 - \nabla) = -\left(\nabla + \frac{\nabla^2}{2} + \frac{\nabla^3}{3} + \frac{\nabla^4}{4} + \dots \right)$$
$$+ \left(\nabla^2 + \frac{\nabla^3}{2} + \frac{\nabla^4}{3} + \frac{\nabla^5}{4} + \dots \right).$$

Collecting terms through Order 4, I get

[10.39] $(1 - \nabla)\ln(1 - \nabla) = -\left(\nabla - \dfrac{\nabla^2}{2} - \dfrac{\nabla^3}{6} - \dfrac{\nabla^4}{12} - \dots \right).$

To get the final form, I must now divide this into $-\nabla$ using synthetic division. The complete process is outlined in Figure 10.6. From this, I get the first few terms of the integration formula:

[10.40] $\Delta = \left(1 + \dfrac{\nabla}{2} + \dfrac{5}{12}\nabla^2 + \dfrac{3}{8}\nabla^3 + \dots \right) hD.$

Using this to operate on y_i, and recalling that $Dy_i = f_i$, I finally get

[10.41] $\qquad y_{i+1} = y_i + \left(1 + \dfrac{\nabla}{2} + \dfrac{5}{12}\nabla^2 + \dfrac{3}{8}\nabla^3 + \ldots\right)hf_i.$

This is the famous Adams predictor formula.

Figure 10.6 Synthetic division.

$$\nabla - \dfrac{\nabla^2}{2} - \dfrac{\nabla^3}{6} - \dfrac{\nabla^4}{12} - \cdots \overline{)\dfrac{\displaystyle 1 + \dfrac{\nabla}{2} + \dfrac{5}{12}\nabla^2 + \dfrac{3}{8}\nabla^3 - \cdots}{\nabla}}$$

$$\begin{array}{l}
\nabla - \dfrac{\nabla^2}{2} - \dfrac{\nabla^3}{6} - \dfrac{\nabla^4}{12} - \cdots \\[6pt] \hline
\dfrac{\nabla^2}{2} + \dfrac{\nabla^3}{6} + \dfrac{\nabla^4}{12} + \cdots \\[6pt]
\dfrac{\nabla^2}{2} - \dfrac{\nabla^3}{4} - \dfrac{\nabla^4}{12} - \cdots \\[6pt] \hline
\dfrac{5}{12}\nabla^3 + \dfrac{\nabla^4}{6} - \cdots \\[6pt]
\dfrac{5}{12}\nabla^3 - \dfrac{5}{24}\nabla^4 - \cdots \\[6pt] \hline
\dfrac{3}{8}\nabla^4
\end{array}$$

Because of the synthetic division, things get messy rather fast computing the higher order terms. Fortunately, a symbolic algebra program like Maple, Mathematica, or Mathcad can come to the rescue. Using these, I've computed coefficients through Order 25, which is far more than anyone should ever need. The formula through Order 7 is shown in Figure 10.7. This should be more than ample for most of your needs.

Figure 10.7 Integration formulas.

Adams predictor
$\Delta = \left(1 + \dfrac{\nabla}{2} + \dfrac{5}{12}\nabla^2 + \dfrac{3}{8}\nabla^3 + \dfrac{251}{720}\nabla^4 + \dfrac{95}{288}\nabla^5\right.$ $\left. + \dfrac{19,087}{60480}\nabla^6 + \dfrac{5,257}{17280}\nabla^7 + \ldots\right)hD$

Moulton corrector
$$\nabla = \left(1 - \frac{\nabla}{2} - \frac{\nabla^2}{12} - \frac{\nabla^3}{24} - \frac{19}{720}\nabla^4 - \frac{3}{160}\nabla^5 \right.$$ $$\left. - \frac{863}{60480}\nabla^6 - \frac{275}{24192}\nabla^7 - \dots\right)hD$$

Predictors and Correctors

The Adams formula is called a predictor because it extrapolates y to the point y_{i+1} using only the value of y_i plus f_i and its backward differences. That's as it should be: Until you know y_{i+1}, you have no way to compute more values of $f(x, y)$.

But suppose you *did* have f_{i+1}. Then you could use it, too, in the computation. Intuition suggests that you might get a better value for the new y. To see how this can help, write analogs of Equations [10.34] and [10.35] using ∇ instead of Δ.

$$\nabla = Q(\nabla)hD$$

Operating on y_{i+1} instead of y_i gives

$$\nabla y_{i+1} = Q(\nabla)hDy_{i+1},$$

which gives

[10.42] $y_{i+1} = y_i + Q(\nabla)hf_{i+1}.$

In other words, you can compute a value for the next y based on f_{i+1} instead of f_i. But what good does this do if you can't compute f_{i+1}? The trick is to compute it using the value of y_{i+1} generated by Equation [10.41]; that is, use the Adams predictor formula to predict a "trial" value for y_{i+1}. Using this trial value, you then compute f_{i+1} and advance the differences one step. Now you can compute a refined value for y_{i+1}. This approach is called the Adams–Moulton predictor-corrector method. You get the corrector formula in exactly the same way as the predictor. The corrector equivalent to Equation [10.41] is

[10.43] $y_{i+1} = y_i + \left(1 - \frac{\nabla}{2} - \frac{\nabla^2}{12} - \frac{\nabla^3}{24} - \dots\right)hf_{i+1}.$

Notice that the coefficients decrease at a much faster rate in this formula, which suggests a higher accuracy for the same order. The formula for the corrector through Order 7 is also shown in Figure 10.7.

As in the case of the derivatives, you can also cast the difference equations into forms involving powers of z. These formulas are shown for various orders in Table 10.7. The equations were developed by a direct substitution for ∇ in terms of z.

Figure 10.8 Predictor–correctors based on z.

First order
$\Delta = hD$
$\nabla = hD$

Second order
$\Delta = \frac{1}{2}(3 - z^{-1})hD$
$\nabla = \frac{1}{2}(1 + z^{-1})hD$

Third order
$\Delta = \frac{1}{12}(23 - 16z^{-1} + 5z^{-2})hD$
$\nabla = \frac{1}{12}(5 + 8z^{-1} - z^{-2})hD$

Fourth order
$\Delta = \frac{1}{24}(55 - 59z^{-1} + 37z^{-2} - 9z^{-3})hD$
$\nabla = \frac{1}{24}(9 + 19z^{-1} - 5z^{-2} + z^{-3})hD$

Fifth order
$\Delta = \frac{1}{720}(1901 - 2774z^{-1} + 2616z^{-2} - 1274z^{-3} + 251z^{-4})hD$
$\nabla = \frac{1}{720}(251 + 646z^{-1} - 264z^{-2} + 106z^{-3} - 19z^{-4})hD$

Sixth order	
$\Delta = \dfrac{1}{1440}(4277 - 7923z^{-1} + 9982z^{-2} - 7298z^{-3}$ $+ 2877z^{-4} - 475z^{-5})hD$	
$\nabla = \dfrac{1}{1440}(475 + 1427z^{-1} - 798z^{-2} + 482z^{-3}$ $-173z^{-4} + 27z^{-5})hD$	

The Adams and Adams–Moulton methods are called multistep methods because the backward differences cover several previous steps. However, note that you need only one evaluation of the derivative function $f(x, y)$ per step (two if you use the corrector). This is true regardless of the order of integration used. That's important because for most nontrivial dynamic systems, the time required to evaluate the derivative function tends to dominate the computer time. Thus, the fewer function evaluations, the better. The Adams–Moulton method is widely known as one of the fastest and most accurate ways to solve differential equations. Now you know how to do it to any arbitrary order.

Error Control

With numerical methods, control of errors is always an issue. In everything I've done here, I've assumed that data was available in tabular form with a uniform step size, h. But what is a reasonable value for h? And how do I know that I'm really computing meaningful results and not just random numbers?

The options are not many and not very appealing. A common method is to do everything at least twice: perform the computation with one step size, then halve the step size and try it again. If the results are nearly equal, you say, "that's close enough." If they differ, you must halve again and continue until things settle down.

Fortunately, in addition to its other virtues, the Adams–Moulton method gives a virtually free estimate of the truncation error. Both the predictor and corrector formulas would be exact and would give the same result if you took enough terms in each series. Because you truncate the series to a finite number of terms, both formulas contain some error. For the predictor, write

$$\Delta y_i = (\Delta y)_P + e_P ,$$

where e_P is the truncation error for the predictor. Similarly, the corrector is

$$\nabla y_{i+1} = (\Delta y)_C + e_C.$$

But remember that

$$\Delta y_i = \nabla y_{i+1},$$

so the two left-hand sides must be identical. Equating the right-hand sides gives

$$(\Delta y)_P + e_P = (\Delta y)_C + e_C$$

or

[10.44] $\quad (\Delta y)_P - (\Delta y)_C = e_C - e_P.$

The left-hand side of this equation is measurable — it's simply the difference between the integrals produced by the predictor and the corrector. You have no idea what the errors are on the right-hand side, but it's reasonable to suppose that the dominant part of both errors is the first neglected term in the series. In other words,

[10.45] $\quad e_P = P_n \nabla^n h f_i$

and

[10.46] $\quad e_C = C_n \nabla^n h f_{i+1},$

where P_n and C_n are the coefficients of the first term neglected in each case.

Now the error you're really interested in is e_C because that's the error of the last value computed and the one that you'll end up using. If you make the (reasonable) assumption that $f(x, y)$ doesn't change appreciably from step i to step $i + 1$, you can divide the two equations to get

$$\frac{e_P}{e_C} = \frac{P_n}{C_n}$$

or

[10.47] $\quad e_P = \frac{P_n}{C_n} e_C.$

Substituting this into Equation [10.44] gives

$$e_C - \left(\frac{P_n}{C_n}\right) e_C = (\Delta y)_P - (\Delta y)_C.$$

Solving for e_C gives

[10.48] $e_C = \left(\dfrac{C_n}{C_n - P_n}\right)[(\Delta y)_P - (\Delta y)_C]$.

This formula states that by subtracting the two estimates of Δy, you can compute an estimate of the truncation error left by the corrector. It's worth noting that, having computed this estimate, you can also "correct the corrector" by adding in this estimate. Thus, you not only gain a good estimate of the error, but you can actually improve the accuracy as well.

Recall that the coefficients P_n and C_n are the coefficients of the powers of the first neglected terms. In Table 10.8, these are tabulated for various orders, and the ratio

$$\frac{C_n}{C_n - P_n}$$

is given. To get the estimated error, simply take the difference of the two Δy's and multiply them by the factor corresponding to the first ignored order.

Table 10.8   Error coefficients.

Order	P_n	C_n	$C_n/(C_n - P_n)$
1	1/2	–1/2	1/2
2	5/12	–1/12	1/6
3	3/8	–1/24	1/10
4	251/720	–19/720	19/270
5	95/288	–3/160	27/502
6	19087/60480	–863/60480	863/19950
7	5257/17280	–275/24192	1375/38174
8	1070017/3628800	–33953/3628800	33953/1103970

Problems in Paradise

So far, the Adams–Moulton predictor–corrector method seems ideal. It's extensible to any order, gives fast and accurate results, and gives an almost free estimate of the error incurred at each step. Who could ask for more?

Well, for starters, you could ask for ... starters. Remember that all these high-order methods require the availability of several backward differences or, equivalently, past values of f_i. But what do you do at the beginning of a solution when there are no past values? Aye, there's the rub and the fundamental problem with multistep methods: they're not self-starting.

In a related area, you have this wonderful way of computing the estimated errors, but no way to react if you find the error growing too large. Remember, the step size is fixed. The only solution is to stop the integration and restart it again with a smaller step size. But alas, you can't restart, because you can't start in the first place!

Several solutions come to mind. Without a doubt, the worst solution (but one often used in practice) is to assume all the differences are zero at the beginning of the computation. This is a terrible idea because, in effect, it is a first-order integration for the first step, second-order for the second step, and so on. If the step size is chosen to give a reasonable error for a higher order integration, it will be far too large for that first step. You might think that a little error in just the first few steps won't matter, but you'd be dead wrong. Most dynamic systems are not self-restoring. Any error at the beginning of the trajectory will propagate forever, and the computed solution will continue to diverge from the actual one.

So here's a word of advice: never start with an empty difference table. The one time you can get away with it is in a real-time system when it can be guaranteed not to have any motion until the integration is well under way.

Most other practical implementations use some kind of single-step method like Runge–Kutta (more about this method in Chapter 11), as a "starter" for the Adams–Moulton method. A single-step method needs no past values and can use any step size. For example, for a seventh-order Adams–Moulton result, you could first take six Runge–Kutta steps, saving the values and generating differences. At this point you will have enough information to switch to the Adams–Moulton method. The Runge–Kutta helper also allows you to change step size by allowing you to restart with a new value.

Unfortunately, even then you're not completely out of the woods. Few single-step methods can match the accuracy of a high-order multistep method. Runge–Kutta integration is typically fourth order, so in a way, you have the same problem as before: The lower accuracy of the starting steps degrades the accuracy of the overall solution. If you use a step size that is adequate for the starter routine, it will be too small for the multistep method, and vice versa.

Are you beginning to get the idea that there's more to simulation than merely having a good integration formula? You're absolutely right, which is why some people have devoted whole lifetimes to developing good integration routines. Clearly, I can't go into enough detail here to solve all the problems, but at least I can identify them for you.

Some years ago I tried to invent a self-starting Adams–Moulton integrator, and the results were very encouraging. The general idea is to try to build up the set of differences that I would have gotten had I passed through the initial point from a previous time, instead of merely starting there. The method also permits step size control. I can give you the general idea here, but first, take a look at the next major application of these methods:

Interpolation

Interpolation is the act of estimating the value of a function at some non-mesh point, given a table of values of that function. You've probably done this at one time or another in school, almost certainly using linear interpolation. In the more general case, you'd like to have higher order methods that allow you to interpolate to any degree of accuracy.

Of all the methods I've looked at so far, this one is the trickiest, because I'm dealing with values of x that aren't in my mesh; that is, they aren't separated by the step size h. Nevertheless, you'll find that my methods using operator math will still serve.

For a start, go all the way back to Equation [9.57]. This time, instead of letting $x = x_i + h$, I'll introduce a "partial step" given by αh, where I assume (but don't require) that $\alpha < 1$. Going through the same process as before,

[10.49]
$$f(x_i + \alpha h) = e^{\alpha h D} f(x_i)$$
$$= z^{\alpha} f(x).$$

The new difference operators are

[10.50]
$$\Delta' f_i = f(x_i + \alpha h) - f(x_i)$$
$$\nabla' f_i = f(x_i) - f(x_i - \alpha h).$$

Now I can write Equation [10.49] in the form

$$\Delta' f_i = (z^{\alpha} - 1) f_i,$$

or in operator parlance,

[10.51]
$$\Delta' = z^{\alpha} - 1.$$

Unfortunately, I don't know how to raise an operator to a noninteger power. But one more step gets me where I'd like to be. From Figure 9.7,

$$z = 1 + \Delta,$$

so I can write

[10.52] $\Delta' = (1 + \Delta)^\alpha - 1.$

This is something I *can* use, because the term in parentheses can be expanded in a binomial series to

$$(1 + \Delta)^\alpha = 1 + \alpha\Delta + \frac{\alpha(\alpha - 1)}{2!}\Delta^2 + \frac{\alpha(\alpha - 1)(\alpha - 1)}{3!}\Delta^3 + \dots,$$

so I get

$$\Delta' = \alpha\Delta + \frac{\alpha(\alpha - 1)}{2!}\Delta^2 + \frac{\alpha(\alpha - 1)(\alpha - 2)}{3!}\Delta^3$$

[10.53]
$$+ \frac{\alpha(\alpha - 1)(\alpha - 2)(\alpha - 3)}{4!}\Delta^4 + \dots$$

If Equation [10.5] is the Rosetta Stone, this equation has to be a close second. Applying it to f_i gives the interpolated value at $x_i + \alpha h$ as desired. You can get that interpolation to any desired degree of accuracy by choosing where to truncate the series. But the formula also has more far-reaching implications, as you'll see in a minute.

Of course, as in the case of finding the derivative, you have a one-sided solution, depending only on forward differences. I just as easily could have expressed z in terms of backward differences to obtain

$$\Delta' = \alpha\nabla + \frac{\alpha(\alpha + 1)}{2!}\nabla^2 + \frac{\alpha(\alpha + 1)(\alpha + 2)}{3!}\nabla^3$$

[10.54]
$$+ \frac{\alpha(\alpha + 1)(\alpha + 2)(\alpha + 3)}{4!}\nabla^4 + \dots$$

Equation [10.53] or [10.54] can be truncated or extended to get a formula of any desired order.

As usual, you can alternatively express these interpolation formulas in terms of powers of z by substituting for Δ or ∇ the equivalent form given in terms of z (or, better yet, z^{-1}). Simply expand each term and collect the results in powers of z^{-1}. Remember, however, that whenever you're using a z-based implementation, you must decide in advance what order to use.

With forward or backward differences, the formula is given in terms of series that can be truncated at any point. However, because the formulas in z depend on expanding and collecting terms, they will be different for each order used. Equations for the first few orders are given in Figure 10.9.

Figure 10.9   Interpolation formulas.

Order 1

$$\Delta' = \alpha(1 - z^{-1})$$

Order 2

$$\Delta' = \frac{\alpha}{2}[(\alpha + 3) - 2(\alpha + 2)z^{-1} + (\alpha + 1)z^{-2}]$$

Order 3

$$\Delta' = \frac{\alpha}{3!}[2(\alpha + 3)^2 - 3(\alpha + 3)(\alpha + 2)z^{-1} + 3(\alpha + 3)(\alpha + 1)z^{-2} - (\alpha + 2)(\alpha + 1)z^{-3}]$$

Order 4

$$\Delta' = \frac{\alpha}{4!}[(\alpha + 5)(4 + (\alpha + 3)(\alpha + 2)) - 4(\alpha + 4)(\alpha + 3)(\alpha + 2)z^{-1} +$$

$$6(\alpha + 4)(\alpha + 3)(\alpha + 1)z^{-2} - 4(\alpha + 4)(\alpha + 2)(\alpha + 1)z^{-3} + (\alpha + 3)(\alpha + 2)(\alpha + 1)z^{-4}]$$

Paradise Regained

I said earlier that there were cures to the problems associated with the Adams–Moulton method, these problems being the need for a starting mechanism and the difficulty of changing step size. Both problems can be fixed using a method hinted at by Equation [10.54]. If you use this equation to operate on f_i as usual, you use it as an interpolation formula. But look at the equation again. Standing alone, it expresses the difference for a new step size, αh, in terms of backward differences for the old step size. A similar formula holds for new backward differences.

$$\nabla' = \alpha\nabla - \frac{\alpha(\alpha - 1)}{2!}\nabla^2 + \frac{\alpha(\alpha - 1)(\alpha - 2)}{3!}\nabla^3$$

[10.55]
$$- \frac{\alpha(\alpha - 1)(\alpha - 2)(\alpha - 3)}{4!}\nabla^4 \quad + \dots$$

Suppose that you're integrating along, and you have a value for y_i, f_i, and all the backward differences. At this point you decide to change the step size. Using Equation [10.55], you can transform all the differences to new values appropriate for the new step size! In other words, after the transformation, the differences will be the same as if you had arrived at point x_i using the new step size all along. Once the transformation is complete, you can simply replace h by αh and proceed with the integration as though nothing special had happened. So much for step size control.

In practice, it's a good idea to modify step sizes by doubling or halving them. This works to your advantage when using the method outlined above. For $\alpha = 1/2$, the formula simplifies, and you fill the mesh points with new points midway between the original points. For $\alpha = 2$, things are even easier: you don't have to interpolate at all. Just save enough back values of f_i to skip every other one.

Finally, Equation [10.55] also provides the key to self-starting. Consider what happens when $\alpha = -1$. In this case, the differences are transformed into those you would have gotten if you integrated backwards. In other words, you can "turn around" and integrate back to the starting point.

For a self-starting algorithm:

• initialize all differences to zero,

• start the integration and proceed for n steps (n being the order of integration),

• turn around using Equation [10.55] and $\alpha = -1$,

• integrate back to the initial time (chances are that y_i does not equal y_0),

• force y to y_0,

• turn around again and proceed forward,

• repeat until the initial differences converge.

Over the years, I've experimented with this concept and had enough promising results to convince me that the method is sound. I have a number of decisions yet to make, which is one reason I've never completed a working, production-level integration package based on the concepts. One question I have not answered is: is it better to start with the full order every time or is it better to alter the order as I build the table?

Let me explain. I can use, say, a seventh-order method and begin with zeros for all the higher order differences. After one cycle forwards and backwards, I can expect to have a pretty good value for ∇f_0. On the other hand, it's reasonable to suppose that the higher order differences will be far off their actual values — perhaps wildly off — and further iterations will be necessary to trim them up.

The other alternative is to begin with a first-order formula for the first step, a second-order for the second step, and so on. Indeed, many people start their production multistep methods in this manner, although for reasons I've already given, it's a terrible idea. However, in this case I expect to be iterating on the difference table, so it's OK to use approximations the first time through.

In a related question, maybe it would be better to use a single-step method for the entire N steps (where N is the order of the integration) for one cycle of integrate/reverse. Although this wouldn't give me a good enough table in a single cycle to start the integration in earnest, it may be the most stable way to get a first guess of the starting table. On the next pass I could use a second-order formula, the next pass a third-order, and so on, until the entire table is built.

From the tests I've run so far, it appears that altering the order as I start up is a better approach than starting with the full order. It makes early determination of the higher differences a little more stable; therefore, it converges faster. On the other hand, it also complicates the algorithm considerably.

The final decision I should make is, how do I know when to stop iterating? This is one question I think I already know the answer to: it's safe to start integrating in earnest when the estimated error is within bounds. But that answer leads to yet another question: what if the estimated error is *never* within bounds? In other words, what if I start with too large a step size in the first place? A key duty of any numerical integration algorithm is to make sure the initial step size is small enough to get good accuracy on that first step. But if I'm busy iterating on a starting table and using the error as a measure of when to start moving away from the origin, I can't also use it to decide the step size.

To summarize, the idea of a self-starting multistep method is still a work in progress. Most available multistep algorithms are strictly fixed-step-size algorithms and are not self-starting. They depend on some external helper, usually based on a single-step (also known as a Runge–Kutta) method to get going and to restart after a step size change. From the work I've done so far, I'm convinced that a self-starting, step-size changing, multistep algorithm is eminently feasible. However, I'm still tinkering with the details of the algorithm and the most stable way to deal with the initial building of the difference table (or, equivalently, the negative time back values, in a z-based implementation). The final algorithm is not yet ready for prime time.

So what have I been using instead? The answer makes an interesting story. When I originally started out to write a numerical integration package, my intention was to write the world's best multistep integrator using a Runge–Kutta starter. After doing some of the analysis you've seen here on the Adams–Moulton method, I began a library search to find a really good Runge–Kutta integrator to use as that starter. I found one — a variable-step method — and implemented it. The darned thing worked so well, my need for a multistep method suddenly didn't seem so nearly so important after all. I have been using the Runge-Kutta method ever since, and have never gotten around to writing that killer multistep method.

A Parting Gift

As I close this odyssey through the topic of numerical methods, I'd like to leave you with a very simple integration algorithm that's served me well in real-time simulations for more than 20 years.

Normally, in real-time systems you're limited to using only the predictor equations, because data comes in as a stream of numbers and you have no way of knowing what the future values will be. Recall that corrector algorithms depend on being able to estimate the value of the derivative function at time $t + h$. However, in a real-time system that derivative function almost surely depends on measurements that haven't been made yet. Hence, corrector formulas don't usually help much in real-time systems.

Fortunately, there's an important exception to this rule, which depends on the fact that most dynamic systems have equations of motion of second order.

[10.56] $$\frac{d^2x}{dt^2} = f(t, x, \frac{dx}{dt})$$

You can reduce this to two equivalent first-order equations by introducing a new variable v.

$$\frac{dx}{dt} = v$$

[10.57] $$\frac{dv}{dt} = f(t, x, v)$$

In a real-time system, you can't evaluate $f(t, x, v)$ at the new point t_{i+1} until you have v_{i+1} and x_{i+1}. This means that you can only use the predictor formula on the integral of *dv/dt*, to give v_{i+1}. On the other hand, once you've integrated the first equation using a predictor, you have the new value v_{i+1}, so you can use the corrector formula for integrating *x*. The results for a second-order integral are

[10.58] $$v_{i+1} = v_i + \frac{h}{2}(3f_i - f_{i-1})$$

and

[10.59] $$x_{i+1} = x_i + \frac{h}{2}(v_{i+1} + v_i).$$

I first derived these equations around 1970 while working for a NASA contractor, and therein lies a tale. At the time, I was asked to modify one of NASA's real-time simulations. Looking inside, I saw that they were using the form of Equation [10.59] for both position and velocity. This is just plain wrong; because they didn't yet have a value for f_{i+1}, they had no business using the formula for a corrector. Stated another way, they were generating data that was one step off of where it should have been.

The funny thing about errors like this is that they are easy to miss, because to all intents and purposes, the program seems to be working. If you make *h* small enough, even an incorrect formula is going to work just fine. The only effective way to really test such things is to plot the error versus the step size and examine its slope on a log–log scale, as I did for the errors in Figure 9.3. The programmer for NASA hadn't done that.

This aspect — that things can appear to be right even when they aren't — led to my next difficulty: convincing the programmer that he'd goofed. After all, not only did the simulation appear to be working correctly, NASA had been using it for years. He didn't want to tell his bosses that the data it had generated was suspect.

Add to that a common programmer's failing, a reluctance to accept criticism from someone else, and you have a conflict. This one got pretty heated. I was trying to show the programmer the math in Equations [10.58] and [10.59], but he wasn't as interested in seeing a math derivation as he was in seeing me disappear. In fact, he suggested rather convincingly that I might get a bloody nose if I didn't remove it from his business. In the end, I had to go over his head to coerce him into making the change. The change worked, and worked well, but I felt I'd made an enemy for life.

The funny part is, 10 years later I met the same programmer again. As he rushed over to me, I wasn't quite sure whether I should greet him warmly or run like blazes. Fortunately, all seemed forgiven, because he greeted me warmly. He asked if I remembered the numerical integration equations and said, "Boy, do they work *great!* We've changed all our other simulators to use them, as well." Apparently he became a believer when he saw how much larger a step size he could use and still get good accuracy.

The bottom line is that Equations [10.58] and [10.59] represent the very best one can do using second-order equations for a dynamic system in real time. I was convinced of this a long time ago. NASA eventually was convinced, also, and as far as I know, they are still using the equations. Now you can have them too. Use them wisely.

Listing 10.2 shows just how simple the equations are to use. I'm assuming a call to a user-provided function that returns the value of $f(t,x,v)$. As you can see, the end result is a world-class integrator in about seven lines of code. Such a deal.

Listing 10.2 Integrator for real-time systems.

```
/* Second/Third order integrator for real-time problems
 *
 * This is the "NASA" integrator, specifically for systems
 * with second-order dynamics.
 *
 */
void solve(double (*f)(double, double, double), double &x, double &v,
double tf, double h){
    double t = 0;
    static double f1 = f(t, x, v);
    static double v1;
    double temp;
    while(t<tf){
        cout << "t = " << t << "; x = " << x << "; v = " << v << endl;
        temp = f(t, x, v);
        v1 = v;
        v += h*(3*temp-f1)/2;
        f1 = temp;
        x += h*(v+v1)/2;
        t += h;
    }
}
```

References

Crenshaw, Jack. 1993. "More Calculus by the Numbers," *Embedded Systems Programming*, 6(2): 42–60.

Pierce, B.O. 1929. *A Short Table of Integrals* (third edition). Ginn and Company.

Chapter 11

The Runge–Kutta Method

Golf and Cannon Balls

In the last chapter, I talked about two kinds of numerical integration: quadrature and a general-purpose numerical integration I called trajectory generation. The difference between the two types is profound, so it deserves emphasis here. In quadrature, you have a function to integrate of the form

$$[11.1] \qquad \frac{dy}{dx} = f(x).$$

Because this function does not depend on y, you can plot a graph of $f(x)$, and integration finds the area under the curve. More appropriate to numerical calculus, you can generate a table of values of the function and find the integral by operating on those tabular values. For this purpose, Simpson's Rule is more than adequate for most applications.

Think of Equation [11.1] as a special case of the more general and more challenging problem in which the function depends on y as well as x.

$$[11.2] \qquad \frac{dy}{dx} = f(x, y)$$

In this case, because the derivative of y depends on y itself, you can't draw the function $f(x)$ in advance. In fact, there's an infinite number of solutions of such equations. To narrow the solution down to one, you must define some *boundary condition*, which is almost always an *initial condition* — the value y_0 at some initial value of x. Other boundary conditions are sometimes required, but those problems are even more difficult to solve.

Setting the initial condition allows you to generate a solution function, $g(x)$, which is the integrated value of $f(x, y)$, given the initial condition y_0.

This class of problems is so important because it reflects the way in which the universe operates. Objects in the real world obey the laws of physics, and those laws almost always can be cast into the form of differential equations like Equation [11.2]. What's more, the independent variable is almost always time. To reflect this reality, I'll switch notation and rewrite Equation [11.2] in the form

[11.3] $$\frac{dx}{dt} = f(\mathbf{x}, t),$$

where $\mathbf{x}$ is some (scalar or vector) variable, and t is time. (I haven't talked about vectors here, but that doesn't matter; you will not do vector algebra here. For now, just think of a vector as a set of variables.)

Because the equations of motion depend on the value of $\mathbf{x}$ at any instant, you need to start with some set of initial conditions

[11.4] $$\mathbf{x}(0) = \mathbf{x}_0.$$

Multistep Methods

In Chapter 10, I gave one method for solving equations of this type: the Adams–Moulton (A-M) predictor–corrector method. This method is based on building a table of differences of previous and current values of $f(\mathbf{x}, t)$ and summing them to get a predicted value of $\mathbf{x}(t)$, thereby inching forward in time using small steps.

Recall in Chapter 10 that I derived the A-M formulas using an elegant derivation based on z-transforms. The power of z-transforms is in allowing you to derive equations that are easily extensible to any desired order, which often equates to any desired degree of accuracy.

The A-M method is called a multistep method because it depends on the results of previous steps to keep the difference table going. This gives the A-M method both its greatest strength and its greatest weakness. By retaining data from past steps in the difference tables, you need only one new evaluation of f to take the next step forward — two evaluations if you apply the corrector half of the formula. This is important because, for most real-world problems, Equation [11.3] can be very complex, so the time required to do the differences and sums is trivial compared to that required to evaluate f. The fewer evaluations per step, the better.

The weakness lies in the table of differences. Because at $t = 0$ you don't have any past values of f, you also have no table. Thus you are in the embarrassing position of having an elegant and powerful method that you can't use because you can't take the first step. Over the years, various methods have been devised to help kick-start multistep methods, but they all require extra code and lots of care to make sure the accuracy of the whole computation isn't ruined by the first few steps. In Chapter 10, I outlined a scheme for making the A-M method self-starting, and other analysts have devised similar schemes. Even so, getting such a method going is a bother and a potential source of devastating errors.

Also, because the difference table — indeed, the whole idea of z-transforms on which the method is based — depends on the assumption that all integration steps are the same size, it's practically impossible to change the step size during a run. To do so is tantamount to starting all over again with the new step. Because real-world problems aren't always cooperative, the error from step to step is often different for different values of time. So you have a serious dilemma: because you can't easily change step size, you typically must run at the step size needed for the worst case part of the run. In practice, the theoretical speed advantage of multistep methods is wasted because changing the step size is so difficult. Again, in Chapter 10 I gave a technique for changing the step size in A-M methods, but to use it requires yet another layer of transformation algorithms.

Single-Step Methods

I wouldn't be telling you all this if I didn't have a solution: a whole separate class of methods that fall under the general umbrella of single-step methods, often described as Runge–Kutta (R-K) methods. The classical way of kick-starting an A-M algorithm is to use an R-K starter. I first became involved with R-K methods almost 30 years ago, for just that reason: I intended to build a routine as a starter for a planned A-M program. As it turned out, I found an R-K method that was so stable, accurate, and easy to use that it completely met all my needs, and I've never found a compelling reason to go farther with that A-M program.

As the name implies, single-step methods are self-contained, in the sense that each step is completely independent of the others. That means you can start the process just like any other process, then change the step size at will. For reasons that I'll discuss later, the R-K method is more stable than multistep methods, which means you can use it with more confidence on problems with ill-behaved differential equations. Finally, the algorithm itself is simple, which is why most analysts reach for R-K when they need a good integration routine in a hurry. The fourth-order Runge–Kutta solution is the bubble sort of simulation algorithms: it's everywhere, easy to program, and easy to use.

What's the Catch?

If the Runge–Kutta method is so good, stable, and easy to apply, why bother to explain multistep methods? Unfortunately, R-K has its own set of problems. You may recall that a predictor–corrector method gives a virtually free estimate of the error incurred in a single step, telling you when to change step size to maintain accuracy. Classical R-K gives us no such estimate. Note the ultimate irony: the A-M method provides the means to know what step size you need, but you can't change it. R-K lets you change the step size at will, but gives no information as to when you need to. Sometimes life just isn't fair.

The advantages of single-step computation come at a price: more function evaluations per step. Whereas the A-M method needs only two evaluations per step regardless of the order, the R-K method requires more evaluations per step, and the number goes up dramatically as the order increases beyond four. Classical formulas require six evaluations for fifth order, and eight for sixth order. And the number tends to at least double if you want an error estimate to permit step size control. You might even say that the term "single-step" is cheating, because you really end up taking ministeps within each R-K step. Intermediate storage is also needed, taking the place of the multistep's difference table. There's no such thing as a free lunch.

Finally, where multistep methods can be extended to any arbitrary order, deriving the equations for higher orders of the R-K method is tedious in the extreme. In this case, there is no elegant z-transform method to fall back on because the "equal step" rule doesn't apply. Simply put, where multistep methods can be derived from total derivatives using a direct application of the Taylor series, the derivation of R-K methods involves partial derivatives, and the substitution of one Taylor series as the argument of other series. The resulting math is absolutely overwhelming, and it expands geometrically as you try to increase the order: so much so that it wasn't until the mid-1960s that Erwin Fehlberg was able to generate formulas for orders beyond six (he went all the way to the 12th order). To do it, Fehlberg had to rely on computerized symbolic algebra. Using the computer, Fehlberg was able to find solutions that not only needed fewer function evaluations than other R-K methods, but also found solutions that gave an error estimate. Even so, the formulas are so complex that one rarely goes beyond his fifth-order method in practice.

In the rest of this chapter, I'll describe how Runge–Kutta methods work. Because of the outrageously complicated math, I won't burden you with the derivation of even the fourth-order method, but the equations for lower orders are much simpler, and I can illustrate the ideas behind the method without overwhelming you with algebra.

Basis of the Method

Recall that single-step methods all come, ultimately, from Equation [9.56], which amounts to a definition of the z-transform. Runge–Kutta methods are based on an entirely different approach. To see the general idea, take another look at Equation [11.3], assuming that x is not a vector, but a simple scalar (in this particular case, extension to vectors is trivial). For simplicity of notation, physicists like to denote the derivative with respect to time as x with a dot over it: $\dot{x}$.

Equation [11.3] then becomes

[11.5] $\dot{x} = f(x, t).$

Because the function has two dimensions, I can plot $f(x, t)$ on two axes, with the value of f as a third variable. But remember that f is the same as $\dot{x}$, which is the slope of the curve of the trajectory passing through any point. Thus, I can depict the function as a two-dimensional field with little lines at each point, as in Figure 11.1, to denote the slope. It looks like a fluid flow field. As you might guess, the desired solution for a given initial value x_0 is the "streamline" that is everywhere parallel to those little lines. The Runge–Kutta method operates by probing the "flow field" at carefully selected points and using the information gained to come up with the best guess for the next point along the trajectory. Because f depends on both x and t, I have to deal with partial derivatives instead of total derivatives, as for previous derivations. The partial derivatives are denoted by $\partial f/\partial x$ and $\partial f/\partial t$. Again, I'll strive for the simplest possible notation by letting

$$\frac{\partial f}{\partial x} = f_x \qquad \text{and} \qquad \frac{\partial f}{\partial t} = f_t.$$

This notation may seem confusing at first; be careful not to confuse f_t with the total derivative f'. Partial derivatives are the derivatives taken with respect to one variable, all others held constant. In Figure 11.1, for example, f_t is the change in the slope of the lines as t varies, moving horizontally along a line of constant x, whereas f_x is the change in the slope as x varies, moving vertically along a line of constant t. The total derivative $\dot{x}$ is the slope of one of those little lines at that point.

Figure 11.1 Flow field.

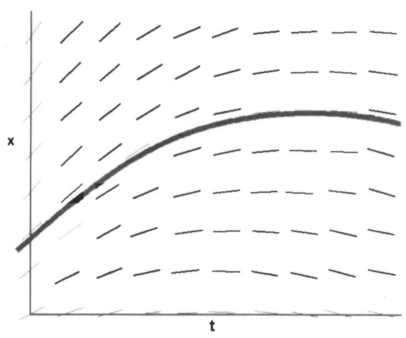

Now I ask myself how to compute the derivative of f, which I'll need for the higher order formulas. The chain rule from calculus gives the answer.

$$\frac{df}{dt} = \frac{df}{dt} + \frac{df}{dx}\frac{dx}{dt}$$

Note that the last derivative is the total derivative $\dot{x}$, which Equation [11.5] tells me is the same as f. In a more economical notation, I can write

[11.6] $\dot{f} = f_t + f f_x .$

This neat little formula is the closest I'll find to the Rosetta Stone z-transform that I used for the A-M derivation.

First Order

As in all other integration formulas, the first-order R-K is trivial. In fact, for first order, all formulas are identical. I'll still go through the process of deriving it, because doing so will give me a chance to introduce the notations and concepts that characterize all higher orders.

The Taylor series, on which all integration formulas are based, provides all the guidance needed to compute the first-order solution. That series can be written

[11.7] $x(t + h) = x + h\dot{x} + \dfrac{h^2 \ddot{x}}{2} + \dfrac{h^3 \dddot{x}}{3!} + \dots ,$

where h is the integration step size, and x and all its derivatives are evaluated at the current value of t. Because $\dot{x} = f(x, t)$, you can write

[11.8] $f(t + h) = x + hf + \dfrac{h^2 \dot{f}}{2} + \dfrac{h^3 \ddot{f}}{3!} + \dots .$

For the first-order solution, you need only the first two terms. In the peculiar notation common to R-K methods, write

[11.9] $k = hf(x, t),$

so

[11.10] $x(t + h) = x + k.$

In the more familiar notation,

[11.11] $x(t + h) = x(t) + hf(x, t),$

this formula is the same as all other first-order formulas. At higher orders, however, you'll find that the R-K method begins to show its true colors.

Second Order

The Taylor series for second order is

[11.12] $x(t + h) = x + hf + \dfrac{h^2}{2}f$.

This time, use Equation [11.6] to get f, and write

[11.13] $x(t + h) = x + hf + \dfrac{h^2}{2}(f_t + ff_x)$.

This equation gives a formula for the second order, all right, but it leaves the problem of how to find the partial derivatives. That is where the probing comes in and where the nature of the R-K approach finally becomes visible.

For an order higher than Order 1, you can be sure that you'll have to evaluate f at more than one point in the x–t space. But what should the second point be? Because the slope of the trajectory at the starting point x is know after the first evaluation of f, it makes sense to use that slope for the next probe. Indeed, it's difficult to think of any other way of finding it. But (and here's the important feature of the R-K method), you don't necessarily have to extend the slope to the time point $t + h$ — any time will do. I'll extend the slope as defined by f to a new time $t + \alpha h$, which gives the following two steps:

[11.14]
$$k_1 = hf(x, t)$$
$$k_2 = hf(x + \alpha k_1, t + \alpha h).$$

Note that, whatever scale factor α you use for t, you must also use for k_1. It's pointless to try different scale factors for the two, because you want to progress along the slope line. Finally, assume (although it's not guaranteed) that the "true" value of x at $t + h$ is given by a linear combination of the two k's.

[11.15] $x(t + h) = x + a_1 k_1 + a_2 k_2$

Now the problem is reduced to that of finding three coefficients, a_1, a_2, and α, and that's where all the rest of the work in deriving R-K equations comes.

The next key step is to recognize that k_2 can be expanded into its own Taylor series — this time based on partial instead of total derivatives. You may not have seen the Taylor expansion of a function of two variables before, but it looks very much like the series in one variable, except with partial derivatives with respect to both independent variables. Each term looks like a sort of binomial expansion.

[11.16]
$$f(x+a, t+b) = f + af_x + bf_t + \frac{1}{2!}(a^2 f_{xx} + 2ab f_{xt} + b^2 f_{tt})$$
$$+ \frac{1}{3!}(a^3 f_{xxx} + 3a^2 b f_{xxt} + 3ab^2 f_{xtt} + b^3 f_{ttt}) + \dots$$

Now you can begin to see why the derivation is so complex. Only keep the first three terms to first order, which gives

[11.17] $k_2 = h[f + \alpha(k_1 f_x + h f_t)]$.

Substituting this into Equation [11.15] gives

[11.18]
$$x(t+h) = x + a_1 hf + a_2 h[f + \alpha(k_1 f_x + h f_t)]$$
$$= x + (a_1 + a_2)hf + a_2 \alpha h(ff_x + f_t).$$

At this point, two separate equations have been derived for $x(t + h)$: Equation [11.13] and [11.18]. If the conjecture that the form of Equation [11.15] is to hold, the two equations must agree. Setting them equal, and equating coefficients of like terms, gives two conditions which the unknown constants must satisfy:

[11.19]
$$a_1 + a_2 = 1$$
$$a_2 \alpha = \frac{1}{2}.$$

These conditions constitute two equations in the three unknowns: α, a_1, and a_2. Two equations are clearly not enough to completely specify all three parameters. You are free to choose any one of them — say α — and solve Equation [11.19] for the other two.

What guidance can you get in the choice of α? Well, it's certain that a value of zero won't work. If α were zero, k2 would then be identical to k1, and the solution could not be second-order. It also makes little sense to try a value for α greater than 1, because that would be equivalent to extrapolating outside the range of the integration step. But you are free to choose any value between $\alpha = 0$ and $\alpha = 1$.

Because there is one degree of freedom to play with, you can choose some other criterion to pin down the value of α. Typically, try to pick a value that will simplify and perhaps speed up the computation by making one or more of the coefficients unity or a power of two. The three most popular values, in order, are $\alpha = 1$ (Euler-Cauchy), $^1/_2$ ("midpoint"), and $^2/_3$ (Huen). The complete formulas for each of these cases are given in Table 11.1.

A few years ago, as I was playing around with these methods, I decided to adopt a different goal in choosing the coefficients based on the value of the ignored third-order term. Because I'm working with a second-order method, there is no way I can force the third-order term to zero, in fact I don't even know its value. As it turns out, however, I *can* control the magnitude of the coefficient multiplying that term. In fact, the choice I made forces the term to zero if there exists no cross-product term, f_{xt}, a case that often occurs in practice. This algorithm is also shown in Table 11.1. Theoretically, it should produce third-order performance in many practical cases and improved performance in others. Without error comparisons I can't guarantee the performance, but it is a fact that some R-K formulas do seem to work better than others of the same order (you'll see another example later), so there is a definite precedent for such possibilities.

Table 11.1   Second-order formulas.

$\alpha = 1$	Euler–Cauchy

$$k_1 = hf(x, t)$$
$$k_2 = hf(x + k_1, t + h)$$
$$x(t + h) = x + \frac{1}{2}(k_1 + k_2)$$

$\alpha = \dfrac{1}{2}$	Midpoint

$$k_1 = hf(x, t)$$
$$k_2 = hf\left(x + \frac{1}{2}k_1, t + \frac{1}{2}h\right)$$
$$x(t + h) = x + k_2$$

$\alpha = \frac{2}{3}$	Huen

$$k_1 = hf(x, t)$$

$$k_2 = hf\left(x + \frac{2}{3}k_1, t + \frac{2}{3}h\right)$$

$$x(t + h) = x + \frac{1}{4}(k_1 + 3k_2)$$

$\alpha = \frac{1}{3}$	Crenshaw

$$k_1 = hf(x, t)$$

$$k_2 = hf\left(x + \frac{1}{3}k_1, t + \frac{1}{3}h\right)$$

$$x(t + h) = x + \frac{1}{2}(-k_1 + 3k_2)$$

As a matter of fact, this choice of coefficients is one of the more frustrating aspects of R-K methods: you have no proven way to show that any one choice is better than another, so you always have this tantalizing feeling that the truly best algorithm is yet to be found. For the second-order methods I've been discussing, it would be reasonably straightforward to try all possible values of α, but as I go to higher and higher orders, the number of degrees of freedom also increases dramatically. Instead of one free variable, I might have five or 10. An exhaustive search for such cases is out of the question.

A Graphical Interpretation

At this point, it will be helpful to try to visualize how the method works. You can do that using a "flow field" graph like Figure 11.2. First, because k_1 is always given by the slope at the starting point, extrapolation always begins along the tangent line for a distance αh. In the case of the Euler–Cauchy algorithm, you go all the way to the next grid point $t + h$. Here k_2 is defined by the new slope. At this point, you can go back and erect a second extrapolation using the slope implied by k_2. Because the two k's are weighted equally, the final estimate is given by a simple average of the two slopes.

Figure 11.2 Euler-Cauchy method.

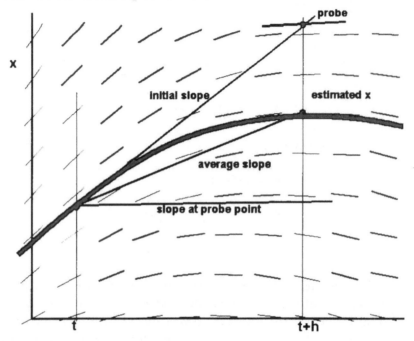

For the midpoint algorithm shown in Figure 11.3, the method is differ-ent. This time, extrapolate only to $t + h/2$, and compute k_2 proportional to the slope there. Then erect another slope line parallel to this slope and extend it all the way to $t + h$. Note that in this formula, k_1 doesn't appear in the formula for $x(t + h)$. You only use k_1 (and the slope that generated it) to find the midpoint, which in turn gives k_2.

Figure 11.3 Midpoint method.

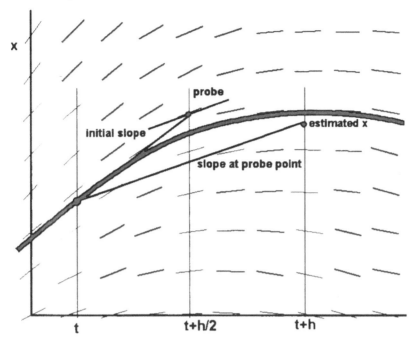

You can draw similar graphs for the other cases that can help you see what's going on, but in reality, once you've seen one such interpretation, you've seen them all. The details of the graphical interpretation don't really matter, because you only need to blindly implement the algorithm once the coefficients are known. Still, this graphical interpretation does serve an important function: It lets you see the "probing" process in action. Two important facts emerge. First, note that, unlike the case with multistep methods, you are probing at points that are not actually on the final trajectory. This gives you the idea that you are somehow learning more about the nature of the "flow field," which certainly ought to be a Good Thing. Similarly, except when $\alpha = 1$, you are also probing at nongrid spacings. Both of these facts add to the stability of the method and explain why it outperforms A-M methods in that department. This stability issue may also explain why off-grid values of α, such as $1/3$ or $2/3$, seem to work a little better than others.

Implementing the Algorithm

Deriving the formulas for R-K methods may be hard, but implementing them certainly isn't. All you need to do is write a subroutine that will evaluate f for any given values of x and t, then call the subroutine as many times as necessary to compute the k's. Finally, sum the k's to get $x(t + h)$. Extending the method to any number of differential equations is also trivial: simply make **x** a state vector instead of a scalar variable and make the subroutine return a vector **f**. Then apply the R-K formulas to each element of **x**, as though it were a scalar.

Of course, in practice you need to take care of such things as initializing the state vector; setting the step size; controlling the step size, print intervals, and so on; and deciding when to quit. Writing practical numerical integration algorithms is an art and a nontrivial exercise, but the code to take a single R-K step can be written in just about as many lines as there are equations in the algorithm. That's why the method is so popular. It's also why there are so many poor implementations out there: the method seems so simple that people tend to write their own rather than use someone else's proven algorithm.

A Higher Power

From my analysis of the second-order algorithm, I think you can see how the R-K method can be extended to higher orders. In the general case, compute the following.

[11.20]
$$k_1 = hf(x, t)$$
$$k_2 = hf(x + c_{21}k_1, t + d_2h)$$
$$k_3 = hf(x + c_{31}k_1 + c_{32}k_2, t + d_3h)$$
$$\vdots$$
$$k_n = hf(x + c_{n1}k_1 + c_{n2}k_2 + \dots + c_{n-1}k_{n-1}, t + d_3h)$$
$$x(t + h) = x + a_1k_1 + a_2k_2 + \dots + a_nk_n$$

Note that each new k_i is computed from already-existing values, and the final solution is a linear combination of all the k's. Except for one small detail, you could theoretically extend this concept to any desired order, just as with the multistep methods. That small catch is that you have to come up with formulas like Equation [11.19] to define the coefficients. To do that, you have to perform the equivalent of comparing terms in Equations [11.13] and [11.18].

You've already seen the general approach. First, expand each k as a Taylor series based on partial derivatives of f, which gives the equivalent of Equation [11.13]. Next, expand x using total derivatives, and the chain rule for derivatives as given by Equation [11.6]. Finally, sum the k terms to get the equivalent of Equation [11.18] and compare terms to get the conditions that the coefficients must meet. Conceptually, it's quite straightforward. In practice, it's nearly impossible. You've already had a hint as to how things explode by your view of the bivariate Taylor series. Remember that each time you try to stretch the order, you need to use more terms in that series. But that's a trivial problem compared to the others that you face.

In deriving Equation [11.12], you got off easy because you already had a relation for f. For higher orders, you need higher derivatives of f, which means differentiating f again and again, applying the chain rule at each step. But even this problem is not the worst, by any means. For the second-order case, k_2 depended only on k_1, which came directly from f. But, in general, every k depends on all k's that preceded it, and each of those can only be computed by substituting all the expanded Taylor series, recursively, as arguments for the next Taylor series. It's no wonder that few analysts dare to go beyond fourth order. My derivation of the third-order equations took four pages of very small writing and led to the conditions

$$a_1 + a_2 + a_3 = 1$$

$$a_2 c_{21} + a_3(c_{31} + c_{32}) = \frac{1}{2}$$

[11.21]
$$a_2 c_{21}{}^2 + a_3(c_{31} + c_{32}) = \frac{1}{3}$$

$$a_3 c_{31} c_{32} = \frac{1}{6}$$

Here are four equations in six unknowns, giving two degrees of freedom instead of one. In practice, you almost never see examples of the third-order R-K algorithm written down or used — not because there's anything wrong with using third order but because fourth order requires very little more effort. For completeness, though, I've included two possible algorithms for third order in Table 11.2.

Table 11.2   Third-order formulas.

Algorithm 1

$$k_1 = hf(x, t)$$

$$k_2 = hf\left(x + \frac{1}{3}k_1, t + \frac{1}{3}h\right)$$

$$k_3 = hf(x - k_1 + 2k_2, t + h)$$

$$x(t + h) = x + \frac{1}{4}(3k_2 + k_3)$$

Algorithm 2

$$k_1 = hf(x, t)$$

$$k_2 = hf\left(x + \frac{1}{2}k_1, t + \frac{1}{2}h\right)$$

$$k_3 = hf(x - k_1 + 2k_2, t + h)$$

$$x(t + h) = x + \frac{1}{6}(k_1 + 4k_2 + k_3)$$

Fourth Order

In the improbable case that you haven't gotten the message yet, deriving the equations for higher order R-K methods involves mathematical complexity that is staggering, which explains why Fehlberg needed a computer to deal with them. Fortunately, you don't have to derive the formulas, only use them. If there ever was a case where you should take the mathematicians' word for it and just use the algorithms like black boxes, this is it. I'm sure you already grasp the general approach used in R-K methods, so without further ado, I'll give you the two most oft-used fourth-order algorithms. They are shown in Table 11.3. The Runge formula is almost always the one used, because Runge chose the coefficients to make the formulas as simple and fast to compute as possible. However, because of the $^1/_3$ and $^2/_3$ mesh points used in the Kutta algorithm, I suspect it might be a bit more accurate and stable. Of course, other choices of coefficients are possible; in fact, an endless number of them. The Shanks formula is one sometimes quoted, but it's rarely used anymore because its only real virtue is that it requires the storage of one less k vector. In the days when memory was tight, Shanks' method was very popular, but today it has little to recommend it.

Table 11.3 **Fourth-order formulas.**

Runge

$$k_1 = hf(x, t)$$

$$k_2 = hf\left(x + \frac{1}{2}k_1, t + \frac{1}{2}h\right)$$

$$k_3 = hf\left(x + \frac{1}{2}k_2, t + \frac{1}{2}h\right)$$

$$k_4 = hf(x + k_3, t + h)$$

$$x(t + h) = \frac{1}{6}(k_1 + 2k_2 + 2k_3 + k_4)$$

Kutta

$$k_1 = hf(x, t)$$

$$k_2 = hf\left(x + \frac{1}{3}k_1, t + \frac{1}{3}h\right)$$

$$k_3 = hf\left(x - \frac{1}{3}k_1 + k_2, t + \frac{2}{3}h\right)$$

$$k_4 = hf(x + k_1 - k_2 + k_3, t + h)$$

$$x(t + h) = \frac{1}{6}(k_1 + 3k_2 + 3k_3 + k_4)$$

Error Control

Now I get to the big bugaboo of R-K methods: how to control truncation errors. Recall that the A-M predictor–corrector method gave a virtually free estimate of the error. Because the step size is so easy to change with R-K methods, it would be delightful if you could get a similar kind of estimate from them. Without an error estimate, your only option is to use a fixed step size, which is simply not acceptable.

With a fixed step size, the only way to tell whether or not you have a good solution would be to run it again with a different step size and compare the results. If they're essentially equal, you're okay. One of the first applications of the R-K method to provide step size control used a similar approach. The idea is to take a single R-K step then repeat the integration over the same interval, but using two steps with half the step size. Comparing the results tells whether the larger step is good enough. This method, called the *half-step* method, will certainly get the job done — in fact, it's probably the most reliable method in terms of giving a true error estimate — but it's also terribly expensive. Fourth order needs 11 function evaluations instead of four — almost three times the computations.

A second alternative is to use two different methods, such as the Runge and Kutta formulas, of the same order. If the two match, you have a reasonable assurance that the error is within bounds. Unfortunately, this approach is also very expensive, needing seven function evaluations. Worse yet, you really don't know that you're getting good results, because you have no way of estimating the fifth-order term that has been truncated. For all you know, *both* solutions could be way off.

What is really needed is a method that gives a reliable error estimate without burning so many function evaluations. One way to do this is to compare the results using two different orders. With a little luck, you can find two sets of formulas that have some k values in common, thereby reducing the number of function evaluations to perform.

Comparing Orders

As an example, consider the first-order method,

[11.22]
$$k = hf(x, t)$$
$$x(t + h) = x + k,$$

and the Euler-Cauchy second-order method,

[11.23]
$$k_1 = hf(x, t)$$
$$k_2 = hf(x + k_1, t + h)$$
$$x(t + h) = x + \frac{1}{2}(k_1 + k_2).$$

For both methods, k_1 is clearly the same (this will always be true). Subtracting the two expressions for $x(t + h)$ gives a measure of the second-order term.

[11.24]
$$e = x(t + h)_{\text{second}} - x(t + h)_{\text{first}}$$
$$= \left[x + \frac{1}{2}(k_1 + k_2)\right] - (x + k_1)$$
$$= \frac{1}{2}(k_2 - k_1)$$

This estimate should be accurate. Unfortunately, it's not exactly what you want. To control the error in a second-order method, you really need an estimate of the first term that's *ignored*, the third-order term, which you have no way of getting. Still, using error control based on the last term *included*, which is what you get from Equation [11.24], is not a bad approach. Because the magnitude of the terms almost always decreases rapidly with order (because of the increasing powers of h and the factorial in the Taylor series), if you keep the value of the last included term down so that its contribution is small, you can be fairly sure that the next term will be negligible. With a little practice, you can learn what range of values of e is safe.

You can extend this approach to higher orders. Again, the trick is to look for formulas that have as many terms as possible in common in order to reduce the number of function evaluations. The methods for second, third, and fourth order using this approach are shown in Table 11.4. Note that the second- and third-order algorithms don't require extra function evaluations, whereas the fourth-order algorithm requires one extra evaluation. In other words, you have what is almost the ideal integration formula: all the advantages of the fourth-order R-K method, plus a nearly free error estimate to control step size, costing only one extra function evaluation. The weak spot in the algorithm is that the estimate is of the included fourth-order term instead of the first neglected term, which is the true error. The effect of this weakness can be seen in the formula for $x(t + h)$, which does not use the information gained from k_4.

Merson's Method

Thirty years ago, when I was first browsing R-K methods for the A-M starter, I came across an obscure paper in a British journal describing Merson's method. It looks like the fourth-order method from Table 11.4 and requires the same number of function evaluations, but Merson claimed that it gave a true estimate of the neglected fifth-order term. The algorithm is given in Table 11.5, and the claim seems to be borne out by the appearance of all the k's in the formulas for both $x(t + h)$ and e.

The history of the method is interesting. Shortly after its publication, a fellow named Shampine published a rebuttal, claiming that Merson's method doesn't really estimate all of the fifth-order term. He pointed out that if the estimate were truly of the neglected fifth-order term, one could add it into the computation for x and get a fifth-order R-K formula with only five function evaluations — a possibility that has been proven to be

Chapter 11: The Runge–Kutta Method

false. As an alternative, Shampine published his own algorithm. It was more complicated, requiring seven function evaluations and using coefficients like 25,360/2187, but he claimed it gave a true fifth-order error estimate.

Table 11.4 Methods with error estimates.

Second order

$$k_1 = hf(x, t)$$
$$k_2 = hf(x + k_1, t + h)$$
$$x(t + h) = x + \frac{1}{2}(k_1 + k_2)$$
$$e = \frac{1}{2}(k_2 - k_1)$$

Third order

$$k_1 = hf(x, t)$$
$$k_2 = hf\left(x + \frac{1}{3}k_1, t + \frac{1}{3}h\right)$$
$$k_3 = hf(x - k_1 + 2k_2, t + h)$$
$$x(t + h) = x + \frac{1}{4}(3k_2 + k_3)$$
$$e = \frac{1}{4}(-2k_1 + 3k_2 - k_3)$$

Fourth order

$$k_1 = hf(x, t)$$
$$k_2 = hf\left(x + \frac{1}{2}k_1, t + \frac{1}{2}h\right)$$
$$k_3 = hf\left(x + \frac{1}{2}k_2, t + \frac{1}{2}h\right)$$
$$k_4 = hf(x - k_1 + 2k_2, t + h)$$
$$k_5 = hf(x + k_3, t + h)$$
$$x(t + h) = \frac{1}{6}(k_1 + 2k_2 + 2k_3 + k_5)$$
$$e = \frac{1}{6}(-2k_2 + 2k_3 - k_4 + k_5)$$

What is the truth? I suspect that Shampine is right. Similar to my algorithm for second order, Merson's method only gives a true estimate of the fifth-order term when the equations don't have certain cross-product partial derivatives. But when I implemented both methods and tested them on real-world problems, I found that Merson's method was about 10 times more accurate than the much more complex method of Shampine. Not only that, it was also about 10 times as accurate as the ordinary Runge–Kutta method! What's more, the error estimate seemed to be reliable. I was so impressed that I've been using it ever since, Shampine's warning notwithstanding. As a matter of fact, I accepted Shampine's challenge and added the error estimate to the computation for x. This should have made Merson's method behave with fifth-order characteristics, and indeed it did, for the test cases I used. My guess is that the estimate takes care of only part of the fifth-order term, but for most practical problems, it's the dominant part. Even though it's not truly fifth order, call it a very good fourth-order method or perhaps a four-and-a-half order method; whatever it is, it works just fine, and it's the method I recommend to you.

Table 11.5   Merson's method.

$$k_1 = hf(x, t)$$

$$k_2 = hf\left(x + \frac{1}{3}k_1, t + \frac{1}{3}h\right)$$

$$k_3 = hf\left(x + \frac{1}{6}k_1 + \frac{1}{6}k_2, t + \frac{1}{3}h\right)$$

$$k_4 = hf\left(x + \frac{1}{8}k_1 + \frac{3}{8}k_3, t + \frac{1}{2}h\right)$$

$$k_5 = hf\left(x + \frac{1}{2}k_1 - \frac{3}{2}k_3 + 2k_4, t + h\right)$$

$$x(t + h) = \frac{1}{6}(k_1 + 4k_4 + k_5)$$

$$e = \frac{1}{30}(2k_1 - 9k_3 + 8k_4 - k_5)$$

Higher and Higher

As mentioned earlier, Erwin Fehlberg, a space scientist at NASA's Marshall Space Flight Center, derived higher order equations in the mid-1960s. He was really trying to extend the R-K concept to very high orders: as far as 12th order. As you might imagine, the higher orders require lots and lots of function evaluations and complex arithmetic, so they are rarely used. His formulas for orders five through eight *do* get used, however; as a matter of fact, his fifth-order formula is quite popular these days. Like Shampine's method, it requires seven function evaluations and has some complex coefficients, but it is indeed a true fifth-order method. Fehlberg's formulas are shown in Table 11.6.

Table 11.6   Fehlberg's method.

$$k_1 = hf(x, t)$$

$$k_2 = hf\left(x + \frac{1}{4}k_1, t + \frac{1}{4}h\right)$$

$$k_3 = hf\left(x + \frac{3}{32}k_1 + \frac{9}{32}k_2, t + \frac{3}{8}h\right)$$

$$k_4 = hf(x + \frac{1932}{2197}k_1 - \frac{7200}{2197}k_2 + \frac{7296}{2197}k_3, t + \frac{12}{13}h)$$

$$k_5 = hf(x + \frac{439}{216}k_1 - 8k_2 + \frac{3680}{513}k_3 - \frac{845}{4104}k_4, t + h)$$

$$k_6 = hf\left(x - \frac{8}{27}k_1 + 2k_2 - \frac{3544}{2565}k_3 + \frac{1859}{4104}k_4 - \frac{11}{40}k_5, t + h\right)$$

$$\Delta x_4 = \frac{25}{216}k_1 + \frac{1408}{2565}k_3 + \frac{2197}{4104}k_4 - \frac{1}{5}k_5$$

$$\Delta x_5 = \frac{16}{135}k_1 + \frac{6656}{12825}k_3 + \frac{28561}{56430}k_4 - \frac{9}{50}k_5 + \frac{2}{55}k_6$$

$$e = \Delta x_5 - \Delta x_4$$

$$x(t + h) = x + \Delta x_4$$

Chapter 12

Dynamic Simulation

The Concept

In the last couple of chapters, I've discussed a number of methods for numerical integration. I've given you, in considerable detail, formulas for both multistep (Adams–Moulton) and single-step (Runge–Kutta) methods. What I haven't done is fully explain why these formulas are so important and why I'm so interested in them. You also haven't seen many practical applications of them — things you could apply to your own problems. In this final chapter of the book, I hope to make these points clear and to provide you with practical, general-purpose software for use in solving problems in dynamic simulation.

You've heard me mention this problem in passing in previous chapters, but until now, I haven't formalized the concept. In dynamic simulation, you are given a set of *equations of motion*, which can always be cast into the form

[12.1] $$\frac{dx}{dt} = f(x, t),$$

or in the "dot" notation introduced in the last chapter,

[12.2] $\dot{\mathbf{x}} = \mathbf{f}(\mathbf{x}, t)$,

where $\mathbf{x}$ is some scalar or vector (array) variable, often called the state vector, and t represents time. Because the derivative of $\mathbf{x}$ depends on $\mathbf{x}$ as well as t, you cannot solve this kind of problem using quadrature; you can only begin at the initial state,

[12.3] $\mathbf{x}(0) = \mathbf{x}_0$,

and propagate the solution forward in time. In Chapter 11, you saw that this solution is equivalent to solving for one specific streamline in a (possibly many dimensional) fluid flow (see Figure 11.1).

Why is this class of problems so important? Because this is the way the universe operates. Objects in the real world obey the laws of physics, and those laws almost always can be cast into the form of differential equations like Equation [12.2].

I should mention here that most problems in physics are given as second-order, rather than first-order, differential equations. For example, Newton's second law of motion can be written,

[12.4] $\ddot{\mathbf{x}} = \dfrac{\mathbf{F}}{m}$,

where $\mathbf{F}$ is the force exerted on a body (implicitly dependent on both its position and velocity), and m is its mass.

Because all problems in dynamics ultimately reduce to Newton's second law, you can expect all other equations of motion to have similar form. In practice, however, the presence of second-order equations causes no problems because you can always reduce them to first order by introducing new variables. For example, let

[12.5] $\mathbf{v} = \dot{\mathbf{x}}$.

Then Equation [12.4] can be rewritten

[12.6]
$$\dot{\mathbf{v}} = \dfrac{\mathbf{F}}{m}$$
$$\dot{\mathbf{x}} = \mathbf{v}.$$

Given a set of equations of motion for a dynamic system and a set of initial conditions, solving them predicts the future state of the system for all future times. In short, you can *simulate* the system. A dynamic system changes in a predictable way with time. It could be anything from a golf ball to a Mars-bound rocket; from a racing car to a decaying hunk of uranium.

If the "system" is something that flies, the equations define the path that it takes through the air or space (the *trajectory*). Although you don't normally think of the path of a tank or the motion of a washing machine as its "trajectory," the principle is the same, in that everything about the system's motion can be described by a time history of x; that is, x as a function of time, or $x(t)$.

Simulations are useful because they let you know what to expect of a system before you actually build and activate it. For example, it helps to know that the Space Shuttle will fly *before* it is launched or that a pilot taking off in a Boeing 767 for the first time will be able to get it down again.

The first person to compute a simulation was the first person who *could*: the King's own personal computer, Isaac Newton, who first applied the calculus he had invented to compute the motion of the planets (long before his twenty-fifth birthday). Considering first the case of two-body motion — that is, the sun plus one planet — Newton got a closed-form solution, which proved that planets follow elliptical orbits and obey the empirical laws known as Kepler's laws.

When he attempted to apply the same technique to more complex systems (say, the Sun, Earth, and Moon, the three-body problem), Newton discovered what most people quickly discover, much to their dismay: the problem cannot, in general, be solved in closed form. Only a very few cases such as the two-body problem and the spinning top can be solved that way, and even then, the solution is often extremely tricky. Although closed-form solutions are delightful things to have (because you then have the equations for all trajectories, not just one), they are generally out of reach. Therefore, to predict the motion of even slightly more complex dynamic systems, you have no choice but to do so via numerical integration. This is why the formulas given in Chapters 10 and 11 take on such great importance.

Newton went on to compute trajectories of cannonballs for the British Army, thereby becoming one of history's first defense contractors (but by no means the first — Archimedes has that honor).

It is not difficult to see why the trajectories of cannonballs and cannon shells are important to any society sufficiently advanced to own cannons. It's a matter of life and death to be able to know where the cannon shell is going to land. It's rather important, for example, that it lands on the enemy's troops and not your own. Artillery is targeted, at least partially, with reference to firing tables, which give the elevation needed for targets at given distances. All other things being equal, the army with the most accurate firing tables is going to win.

Another early "defense contractor," funded by Great Britain to compute firing tables, was Charles Babbage, whose dream of letting a machine do the work led to his invention of the Difference Engine, which led the way to computers, space flight, ICBMs, Intel, Microsoft Flight Simulator, and Windows NT.

The generation of firing tables for artillery is a science that goes on to this very day. In World War II, the famed Norden bombsight, one of the most carefully guarded secret weapons of that war, was a mechanical analog computer, predicting the motion of a falling bomb by solving its equations of motion in real time. This bombsight may well have represented history's first real-time embedded computer system.

For the next 200 years after Newton, astronomers applied his laws of gravitation and motion to predict the motions of the Moon, planets, comets, and asteroids. In those days, of course, there was no such thing as real-time simulations. The only existing "computers" were living and breathing flesh, and they worked a whole lot slower than real time, in most cases. Astronomers like Encke spent lifetimes computing the trajectories of comets like, well, comet Encke. Simulation methods were used with respect to other dynamic systems when they were necessary, but the whole process was so tedious that no one would have dreamed of applying simulation techniques to the more mundane systems of everyday life, such as airplane motions or auto suspension behavior. Certainly no one could have envisioned simulation as entertainment, such as the many race car and flight simulation programs and space war games available today.

If you see that the solution of equations of motion has attracted some important people and has changed the course of history, you're starting to get the message.

Reducing Theory to Practice

In the last two chapters, you've seen a lot of mathematics, a lot of theory, and a lot of derivations. This chapter is where I put the theory to work, to get numerical integration software useful for dynamic simulations. The first steps will be easy, the latter ones less so, but the whole process becomes both simpler and more effective if you make full use of what has gone before.

In my time, I've seen hundreds of implementations of real-time and non-real-time simulations, control systems, and analysis tools that required numerical integration. Very few of them were good implementations, and many (far more than half) were just plain wrong. I'd like to show you how to get it right.

Dynamic simulation, and the numerical integration that makes it work, is dear to my heart because it's the basis of the tools that were developed to figure out how to get to the Moon. My career began with trajectory analysis for Moon missions, including the Apollo program. I can tell you first-hand that without the analysis tools used to predict the motion of the Apollo spacecraft and the computers to run them on, we could never have gotten to the Moon. It's widely believed that the reasons the Russians lost the race to the moon was not because they couldn't build the rockets, but because they lacked the computers to figure out how to steer them. You'd be amazed how many new techniques, math algorithms, and so on had to be invented for this purpose. Some of the most important and powerful methods of modern control theory, including the practical use of Kalman filters, came out of the effort to meet that 1970 deadline to reach the moon set by President John F. Kennedy. Since then, dynamic simulation has become a fundamental tool in every engineer's toolbox.

Perhaps you may never have to go to the Moon, but if you ever need to predict the motion of a dynamic system, be it a drag racing car, a falling raindrop, or a production line, you'll need these tools. My guess is that once you have the tools to perform numerical integration in your toolbox, you'll find plenty of uses for them.

The Basics

The fundamental differential equation one seeks to solve in numerical integration is shown in Equation [12.2]. The variable t is the independent variable because the equation doesn't describe the variations in t, but only in x. For similar reasons, x is called the dependent variable because its value depends directly on t and indirectly on itself. The dependent variable can represent any quantity that you'd like to measure. The independent variable can also be any quantity, but it usually represents time, and that's what I'll assume in the rest of this chapter for simplicity. In the special case where t does represent time, I will use the shorthand, "dot" notation shown in Equation [12.2].

The job of a numerical integration algorithm is to integrate equations in the form of Equation [12.1]. Expect to see a solution in terms of a time history of x; that is,

[12.7] $\qquad \mathbf{x} = g(t, \mathbf{x}_0).$

(Note that the result for a given case is only a function of t; x_0 is a constant.)

A quick and dirty solution to the problem is given by the first-order method, which I described in Chapter 8 (see Equation [8.9]). To refresh your memory, note that, if Δ represents a small change in some value, then the derivative of any function can be approximated by a ratio

[12.8] $$\frac{dx}{dt} = \frac{\Delta x}{\Delta t}.$$

If you rearrange this equation a bit, you get

[12.9] $$\Delta x \approx \frac{dx}{dt}\Delta t = \dot{x}\Delta t = h\dot{x},$$

where h replaces Δt, again to simplify the notation. But don't forget that you have the value of $\dot{x}$ from Equation [12.1], so you end up with

[12.10] $$\Delta x = hf(x, t).$$

If you have some value of x, say x_n, the next value will be

[12.11] $$x_{n+1} = x_n + hf(x, t).$$

To compute the time history of x, all you need is an initial value x_0, a procedure to evaluate the derivative function $f(t, x)$, and a way to continually add the small deltas that result. I'll build a program to do that for the special case where

[12.12] $$f(x, t) = x.$$

(Why this function? First, because it's simple. Second, because it has a known analytical solution. When testing any analytical tool, it helps to test it with a problem that you already know the answer to.)

In this case, the problem is so simple, at least on the surface, that you can write the solution program in only a few lines of C++ code, as shown in Listing 12.1.

As simple as this code fragment is, it contains all the features that numerical integration programs must have: a section to set the initial conditions, a formula for computing the derivative, a loop to add successive values of Δx, and some criterion to decide when it's done. Because time is relative anyway, the initial time is usually set to zero, as I do here.

Listing 12.1 A rudimentary integrator.

```
#include <iostream.h>

void main(void){
  double t, x, h;

  x = 1.0;
  t  = 0.0;
  h = 0.01;
  while (t < 2.0){
    cout << t << ' ' << x << '\n';
    x = x + h * x;
    t = t + h;
  }
}
```

I'd like to point out two very important features of this program. First, note that I have a statement to effect the printing of the output values. After all, a program that computes the history of x isn't much good if it never shows that history. Second, note that:

Time is always updated at the same point in the code that x is updated.

Never, but *never*, violate this rule. As obvious as the point may be, getting x out of sync with t is the most common error in implementations of numerical integration algorithms, and I can't tell you how many times I've seen this mistake made. To avoid it, some go so far as to combine t with x (as a vector) and simply use the relation $\dot{t} = 1$ to let the integrator update t. This approach tends to hide the functional dependence of $\dot{x}$ on both x and t. Still, the idea does have merit, in that it absolutely guarantees that x and t can never get out of step.

How'd I Do?

If you compile and run the program of Listing 12.1, you'll see it print two columns of numbers: the first *t* and the second the corresponding value of *x* at *t*. The last two values my version of Borland Turbo C++ printed out were as follows.

t	x
1.999998	7.316017

As it happens, both of these numbers are *wrong*. I asked the integrator to run until *t* was at least equal to 2.0. As you can see, it didn't quite make it. The reason, of course, is the inherent error that creeps in when adding many floating-point numbers. Because people typically like to see their results at nice even values, like 0.1, 0.2, ..., I have a problem. The problem can't be eliminated, only lived with. But, as you can see, I must be extra careful that the program does the right thing when the time is almost, but not quite, equal to the expected final time.

For the derivative function chosen, the value of *x* can be shown analytically to be

[12.13] $x = x_0 e^t$.

At *t* = 2, the correct value is 7.3890561, so the result of my simulation is in error by about 1 percent. The difference this time is just plain error: truncation error introduced by my crude first-order approximation. Later on I'll at least reduce (you can never eliminate) the truncation error by using a higher order algorithm. Alternatively, I could reduce the value of the step size *h*, to get more accuracy at the cost of longer run time.

Before you belittle the first-order method for its poor accuracy, you should bear in mind that problems exist for which accuracy doesn't matter. A flight simulator is a good example. In such a simulator, the program is constantly getting real-time inputs (feedback) from the "pilot" in response to the picture displayed. If, at any instant, that picture is in error by 0.005 percent (the error in one integration step, at the step size I chose), the pilot will never notice. In an interactive system, pilot inputs far outweigh any errors I might generate with my integration method. This is true, in fact, for *any* system that includes a feedback loop, such as the simulation of a control system. The feedback doesn't have to come from a human pilot; any feedback control system will keep things stable, so the first-order method has a definite place in the scheme of things, though soon I will show you a much more accurate method that is not much harder to implement.

What's Wrong with this Picture?

You know those puzzles where you're supposed to find all the mistakes in a picture? Listing 12.1 is sort of like that. Yes, it works, sort of, and it gives the correct answers within the limitations of the first-order method used, but I deliberately violated just about every rule of good programming practice just to emphasize the difference between a *working* integrator and a *good* one. In the next few sections, I'll transform this program into a nice, reliable, and maintainable general-purpose numerical integrator. By beginning with the program of Listing 12.1, I hope the improvements I make later will be more obvious. For the record, almost every improvement I make will slow down the execution of the program! In a numerical integration run that may take on the order of 100,000 integration steps to complete, the time spent in performing each step is obviously important. Still, it's not the *most* important thing. Factors like ease of use, reusability, and so on are usually more important.

How many mistakes did you find? My list, below, may or may not be complete.

- The halting criterion is wrong. Only the round-off error associated with any floating-point operation saved me from missing the output point at (or near) the final time of 2.0.

- The derivative function is hard-coded into the program. To solve a different set of differential equations, I must rewrite the program.

- The integration algorithm is hard-coded into the program. To change algorithms (say, to a second or higher order) will also require recoding.

- The output statements are hard-coded into the program.

- The program as written can (or need be) run only once, because the initial data, step size, and final time are all hard-coded as literal constants. A change to any one of them requires recoding and recompiling. At the very least, I should assign variable names to these values. More to the point, they should be input values.

I'll take these problems one at a time.

Home Improvements

Changing the halt criterion is the easiest problem to correct: I'll simply replace the statement with the following.

```
while (t <= 2.0){
```

Run the modified program and see what happens.

What did happen? I still didn't get the final value. A look at the code will tell you why. Even with the revised criterion, the program arrives at the test with $t = 1.999998$, which satisfies it. In fact, it would just satisfy it whether I had changed the criterion or not. So the program takes one extra step, but does not print the output. To fix this, I must add at least one extra output statement, either at the beginning or the end of the loop. Listing 12.2 shows the preferred arrangement. Now, of course, I have two output statements, which beg even more to be pulled out of line and into a subprogram. On the other hand, having two (or even three) calls to the output routine can sometimes be a blessing rather than a curse, because you often want to do things a little differently on the first and last steps.

Listing 12.2 Fixing the halt criterion.

```
void main(void){
  double t, x, h;

  t = 0.0;
  x = 1.0;
  h = 0.01;
  cout << t << ' ' << x << '\n';
  while (t < 2.0){
    x = x + h * f(t, x);
    t = t + h;
    cout << t << ' ' << x << '\n';
  }
}
```

Unfortunately, now I have the opposite problem than before: The integrator will appear to go one step *past* the last time to $t = 2.009999$. This is a problem common to all integration routines, and there is really no cure. It's the price you must pay for using floating-point arithmetic, which is always approximate. You would not see the problem if h were not commensurate with the final time, but because people tend to want nice, even values, like 2.0 and 0.01, this problem will plague you forever. More about this later.

As I said earlier, I hope by showing you the kinds of subtle troubles you can get into, you will come to appreciate that things must be done carefully. I've seen many implementations of integration methods in which the integration package takes only one integration step, and it's left up to the user to write the code for looping and testing for the end condition. I hope you'll agree by now that this logic is much too important and tricky to leave to the user. It should be included in the integration package.

To remove the derivative function from the mainstream code, simply use the idea of the function $f(x, t)$ as it's used mathematically. You can write:

```
double f(double t, double x){
  return x;
}
```

The use of this function is also shown in Listing 12.2. In practice, it's not a good idea to have a single-character function name. There's too much chance of a name clash, so perhaps I should think of a better name. For historical reasons (my history, not yours), I always call my subroutine deriv().

Note carefully the order of the arguments in f() [or deriv()]: they're in the opposite order than in the equations. The reasons are more historical than anything else, but in general, I like to write my software with the independent variable first. If you find this a source of confusion, feel free to change the order.

A Step Function

I'll remove the integration algorithm from the mainstream code and write a procedure that takes one step of the integration process.

```
void step(double & t, double & x, const double h){
  x = x + h * deriv(t, x);
  t = t + h;
}
```

I used reference variables for *t* and *x* because they are updated by the routine. I also added a little protection to *h*, to assure that it isn't changed. Now I'll encapsulate the output functions into a procedure called out().

```
void out(const double t, const double x){
  cout << t << ' ' << x << endl;
}
```

The resulting code is shown in Listing 12.3.

Listing 12.3 Intermediate version.

```
void main(){
  double t, x, h;

  t = 0.0;
  x = 1.0;
  h = 0.01;
  out(t, x);
  while (t < 2.0){
    step(t, x, h);
    out(t, x);
  }

}
```

Cleaning Up

The final step for this exercise is to get rid of the literal constants, or at least encapsulate them. Because I have two kinds of variables, the state variables such as *t* and *x* and the control variables such as *h* and the maximum time, I've chosen to initialize these in two separate procedures. I could easily read in all four variables, but because *t* is almost always going to be initially zero, I've chosen to simply set it to zero rather than always having to input it. For this example, I've also chosen to set *h* as a constant, although this is probably a questionable practice in general.

The final form, at least for now, is shown in Listing 12.4. If you compare Listings 12.1 and 12.4, I think you'll agree that the latter is a much more readable, if slower, implementation. It's still a long way from a general-purpose numerical integrator; in fact, it's not a callable function, but a main program, and the control logic is still embedded in that main program. I'll deal with that problem in a moment. Nevertheless, the code of Listing 12.4 is clean, simple, and modular. Lest you think it's still a trivial program, please bear in mind that you can replace step() with a function that implements a single integration step to any desired order, without changing anything else. Likewise, you can implement any set of equations of motion, by making *x* in main() an array instead of a scalar.

Listing 12.4 The final version.

```
#include <iostream.h>
double deriv(const double t, const double x){
  return x;
}

void out(const double t, const double x){
  cout << t << ' ' << x << endl;
}

void get_params(double & h, double & tmax){
  h = 0.01;
  cin >> tmax;
}

void init_state(double & t, double & x){
  t = 0.0;
  cin >> x;
}

void step(double & t, double & x, const double h){
  x = x + h * deriv(t, x);
  t = t + h;
}

void main(){
  double t, x, h, tmax;

  init_state(t, x);
  get_params(h, tmax);
  out(t, x);
```

Listing 12.4 The final version. (continued)

```
while (t < tmax){
    step(t, x, h);
    out(t, x);
    }
}
```

Generalizing

At this point, I have a stand-alone computer program which, though extremely simple, has two things going for it:

- It is correct; that is, it generates no erroneous results.

- It is extensible to higher orders and different problems.

My next step will be to convert the computer program to a general-purpose, callable function that can be used as part of an analyst's toolbox. Before doing that, however, I'd like to digress and reminisce for a moment. This digression will help you to understand both the history of and the reason for such simulation tools.

I mentioned at the beginning of this chapter that the solution of equations of motion has a long history, going all the way back to Isaac Newton and Charles Babbage. Much of the uses of computers during the early days of Eniacs and Edsacs was devoted to the solution of such problems. In other words, they were used for dynamic simulation.

In those early days, it never occurred to anyone to build a general-purpose integration routine. Each simulation was built to solve a specific problem, with the equations of motion hand coded into the program, usually in assembly language. My own introduction to dynamic simulation came at NASA, ca. 1959, using an IBM 702 to generate trajectories to the Moon. At that time, I had never heard of FORTRAN; everyone used assembly language. Also in those days, computer time was very expensive, so the watchword was to save CPU time at all costs, even if it meant considerable extra work for the programmer. To gain maximum speed, programmers tended to eschew subroutines as much as possible, in favor of in-line code, hand-tailored to the current problem.

The time rapidly approached, however, when the time to produce a working product took precedence over execution time as the factor of greatest importance. We soon discovered FORTRAN, which made life easy enough to allow us the luxury of writing more elegant and reusable software.

During that period, someone a lot smarter than I took another look at Listing 12.2 and noticed that we were always solving the same kinds of differential equations every time. Why, he asked, didn't we just write one simulation, once and for all, and just change the equations to fit the problem? The idea seems obvious now, but at the time it was about as unlikely as developing draggable icons under CP/M. The technology just wasn't there. Thanks to the development of high-order languages like FORTRAN, and the development of larger and faster mainframes, the idea soon became practical. My first exposure to a general-purpose integrator was in 1961, when one of the developers of the Naval Ordnance Lab routine, NOL3, introduced it to my company. NOL3 was a FORTRAN subroutine that did all the hard work of simulation. It was the first of what you might call an executive numerical integrator; Once called, it took over the computer and remained in charge until the problem was solved. You wrote subroutines to evaluate the equations of motion [$f(t, x)$ in Listing 12.2], to write output to a printer, and so on, then you wrote a main program that simply set up the initial conditions, performed a little housekeeping, and called NOL3. The integrator did the rest. By changing the code for initialization, output, and evaluating $f(t, x)$, you could solve any problem in dynamic simulation. Thanks to the ability to call subroutines (or, more correctly, pass parameters) in FORTRAN, you really didn't give up much in the way of performance efficiencies. It was a great idea. I never used NOL3 much, but I was smitten by its concept, and later I shamelessly plagiarized it.

The Birth of QUAD1

In 1965, I was working on my Ph.D., and I needed a simulation of a spinning body to confirm my analytical results. I looked around for a good integration formula, found one in Merson's method (see page 351), and used it in my own integration routine. I cast it very much into the old NOL3 mode and unimaginatively called it QUAD1 (Hey, I only had six letters to work with, you know). I even used the same procedure names: DERIV to compute the derivatives, AUTO to manage step size control, and OUT for effecting output. The routine worked like a champ, I got my degree, and I was soon out on the market, integrator under my arm. Over the next 20 years, QUAD1 developed into a whole family of routines, all built very much alike, and which I used as the basis for dozens of dynamic simulations. Taking the concept of NOL3 one step further, I realized that even the housekeeping done by the main program didn't really change much from problem to problem. I developed routines for reading input data in a very general format (early

FORTRAN compilers were outrageously picky about things like input formats), writing tabular outputs in a compact form, drawing graphs using the line printer, and so on. My simulations began to look more and more alike, so that building a new one involved little more than redefining subroutine DERIV and supplying variable names to the input and output routines. As a result, my colleagues and I were able to develop and debug new dynamic simulations with impressive speed. Our record was a complete simulation in one 12-hour day, developed to help deal with Skylab, which began tumbling after it was launched.

Later, after discovering languages other than FORTRAN, I translated QUAD1 and its cousins into Pascal, C, and C++, refining them a bit each time. In the remainder of this chapter, you will basically see the 24-year-old QUAD1 in its latest, spiffy new clothing.

The Simulation Compiler

My next idea, I thought, was a real winner. Looking over the last few simulations I had written, I realized that about 90 percent was boilerplate code. The integration routine, by then evolved to QUAD3, never changed. For practical purposes, the rules for the control of step size didn't change and neither did the main program, except for array dimensions. The input/output routines changed only in the names of the variables and whatever conversions were needed to translate them into practical units. Only the DERIV subroutine really changed, and even it had a very specific format and structure. That's when I had my big revelation: Why not write a program, a simulation analog of a compiler-compiler, that would generate a simulation program, given the equations of motion? All I needed was a way to define the dynamics of the system under study [i.e., specify f(**x**, t)], and my sim-generator would spew out FORTRAN code that would solve the problem. (Today, I'd probably use drag-and-drop technology to define the system and C++ instead of FORTRAN as the output, but the concept is the same. I might call the result, "Simulink," except Mathworks, Inc., has already done so.) Again, there would be no run-time performance hit, because the code would be, in effect, custom-tailored to the problem under study.

I thought it was a great idea. Apparently, so did other people, notably two Redstone Arsenal analysts named Mitchell and Gauthier. They'd had the same idea before me and turned it into a whole simulation language called ACSL. (Sigh. Always the bridesmaid, never the bride.) Today, the ACSL concept has developed into a whole industry, simulation languages are widespread, and Mitchell and Gauthier are very rich. In a sense, routines like my QUAD series are anachronisms. Still, like many anachronisms, they serve a useful purpose. If you need a quick solution to a dynamic problem and you don't have the cash or the computing power to support a commercial simulation language, it's awfully nice to have a tool like QUAD3 in your pocket. That's what the current focus of this chapter is all about.

Real-Time Sims

This reminiscence wouldn't be complete without the mention of real-time simulations. When you're computing the orbits of comets, or even moon rockets, the last thing in the world you want is a real-time simulation, because you don't normally have eighty years to wait for the answer. But for airplanes, spacecraft, and the like, where a pilot is involved, only real-time simulation will do. I probably don't need to explain to you the advantages of real-time simulations. Training simulators are used today for everything from airliners to supertankers to the Space Shuttle. Pilots could not be trained to captain such sophisticated and complicated systems without them. You don't send ships captains out to run a supertanker until they've demonstrated that they can do so without running into something. The Maritime Commission's ship simulator at King's Point, Long Island, has been training future captains for decades. Today, I'm told that even Formula 1 racing drivers hone their skills with super-accurate simulations of both the racing car and the track it runs on.

Early simulations, like those of the Link Trainer, were implemented with all-analog mechanisms because it was the only option. Even well into the 1970s, plenty of simulations were still being done in analog computing labs. But everyone saw the advantages of computing digitally, and digital, real-time simulations began to be constructed almost the day after the hardware was capable of computing in real time. I recall such a system being built by NASA as early as 1960.

Not surprisingly, real-time simulations have different needs than off-line computer simulations. First and foremost is that numerical algorithms, like the Runge–Kutta method, developed for studying the motions of comets simply don't work in real time. They involve predicting ahead to future positions and values that don't yet exist; therefore, different algorithms must be devised. Fortunately, as I mentioned earlier, the demands on such an algorithm are not stringent because any human operator in the loop, or control system for that matter, will correct any errors introduced by a marginally accurate algorithm.

Next on the list, you must have some mechanism for ensuring that the simulation is, in fact, in real time. That is, you must assure that the time computed internally as part of the simulation is kept synchronized with the wall clock. This is almost always accomplished using a periodic interrupt. Needless to say, if you're going to be synchronized with an interrupt, any computations had darned well better be completed before the next interrupt comes in, so in real-time simulations, computational efficiency is paramount.

Control Systems

So far, I've talked only about simulations, but I don't want to leave you with the impression that these are the only applications for numerical integration algorithms. The other important area is control systems. Although the most modern techniques for control are based on digital (z-transform) techniques, a lot of engineers still specify control systems, filters, and so on using Laplace transforms and transfer functions. In such functions, every term involving the Laplace parameter s translates to an integral in the time domain, so control of a complex system typically involves the real-time integration of a fairly large number of internal parameters. The most famous integrating controller is the well-known PID controller for second-order systems, but more complex control algorithms are also common. Fortunately, the techniques developed for simulation of real-time systems work just as well when integrating the internal parameters of an s-domain controller, so you don't have to invent new methods to handle control systems.

Back to Work

Now that you've had a glimpse into the background and history of numerical integration modules, I hope you'll see how they apply to the real-world problems of embedded systems and why I'm dealing with them in this book. Having been thus motivated, I'll return to the problem at hand and finish developing the software. When last seen, the software was in the state given in Listing 12.4.

The routines of this listing are certainly simple enough, and they get the job done, but the architecture still leaves a lot to be desired. The main objection is that, contrary to my desired goal, I don't really have a general-purpose integrator here. All of the logic is still wrapped up in the main program, and you have already seen that the logic to decide when to stop can be tricky — mainly because of the round-off error associated with floating-point arithmetic — too tricky, in fact, to require the user to deal with it. So you should encapsulate the logic into a separate function called by main().

```
void integrate(double & t, double & x, const double tmax,
    const double h){
  out (t, x);
  while (t < tmax){
    step(t, x, h);
    out (t, x);
  }
}
```

Now the main program only needs the initial state and control parameters like tmax and the step size h before it calls integrate().

```
void main(){
  double t, x, h, tmax;

  init_state(t, x);
  get_params(h, tmax);
  integrate(t, x, tmax, h);
}
```

As simple as this program is, it's beginning to take form as the foundation for an integration mechanism that you can use for many applications. I know it doesn't look like much yet, but trust me on this: I'm closer than you think.

Two things are worth noting. First, as integrate() is written, it appears that I can always apply a higher-order formula merely by rewriting step(). As a matter of fact, I'll do that very soon. Second, note that, like its ancestor, NOL3, the integrator itself is an executive routine: Once given control, it never relinquishes it until the solution is complete. The functions deriv() and out() are called only by the integrator as it needs them and never by the main program. This turns out to be a very important point. If there was ever an argument for information hiding, this is it. Recall that single-step (Runge–Kutta) methods operate by evaluating the derivative function at times and with states that are not necessarily correct (see Figure 11.3). Likewise, even though multistep methods nominally perform only one function evaluation per step, the combined Adams–Moulton, predictor–corrector method evaluates the function twice: once for the predicted state and once for the corrected state. One of the most frustrating problems I used to have with users of QUAD*n* was their infuriating tendency to break the rules and try to write outputs from within deriv( ). That's a no-no, because only the integrator itself knows when the state data passed to deriv( ) is the true one.

An interesting issue arises with respect to the function names. To provide for more flexibility in the integration routine, the original NOL3 passed the names of the subroutines DERIV and OUT as parameters. This allowed the user to choose his own names for the routines and perhaps even call more than one version, depending on the state of the system. I copied this feature into the QUAD*n* series of integrators, but I must say that I found it to be of little use. It's too easy to simply use the names that the integrator is expecting anyway. I can think of situations where the possibility of having multiple deriv( ) routines would be nice, but in practice, the feature is used so infrequently that it hardly seems worthwhile to carry the extra baggage around. Although it's easy enough to pass function names through the calling list, for the purposes of this book, I'm going to assume that the names are fixed.

Printing the Rates

A third point is much more subtle, but also very important. Users often like to see the values of the derivatives as well as the state variables themselves. For example, for a space booster, you would probably want to see the acceleration profile as a function of time. To make the derivatives available, you must make sure that deriv( ) has been called for the *current* time step before you call the output routine. For the record, getting the derivatives out of sync with the state variables is one of the most common kinds of errors seen in integration routines, so it's important that you take care to get it right. A sure-fire way to do that would be to call deriv( ) again, from within out( ), but this wastes considerable time, because it requires two calls to deriv( ) per step instead of one and because the evaluation of the derivatives is typically one of the most costly parts of the computation. The only other option is to provide the derivative as a separate state variable, accessible to both step() and out(). A modified version of the integrator takes care of the problem:

```
void integrate(double & t, double & x, const double tmax,
    const double h){
  double v;
  v = deriv(t, x);
  out (t, x, v);
  while (t < tmax){
    step(t, x, v, h);
    out (t, x, v);
  }
}
```

Here, I've added the local variable v to hold the derivative, and this variable is passed as a reference parameter to both step() and out(). Of course, this means that these functions must be modified to deal with the extra variable.

```
void step(double & t, double & x, double & v,
    const double h){
  x = x + h * v;
  t = t + h;
  v = deriv(t, x);
}

void out(const double t, const double x, const double v){
  cout << t << ' ' << x << ' ' << v << endl;
}
```

Note carefully the convention here, because it represents a significant change: Because the integrator precalls deriv() to get an initial value for v, step() can be assured that the value of v pass on the first call is valid. To make sure it's also valid on subsequent steps, step() must call deriv() as its last action instead of its first.

Higher Orders

It may seem to you that the structure I've given is awfully busy. Including a new variable for the derivative seems a lot of bother to go through just to be able to print it, and it also seems that I'm passing a lot of parameters through calling lists.

You'd be right on all counts. The structure is rather cumbersome for the simple case I've been working with, but I think you'll see the wisdom of the approach as soon as I start getting serious and tackle nontrivial problems. To see how nicely things work, I'll go to a fourth-order Runge–Kutta integration. The formula for this algorithm is, you'll recall,

$$k_1 = hf(t, x)$$

$$k_2 = hf\left(t + \frac{1}{2}h, x + \frac{1}{2}k_1\right)$$

[12.14] $$k_3 = hf\left(t + \frac{1}{2}h, x + \frac{1}{2}k_2\right)$$

$$k_4 = hf(t + h, x + k_3)$$

$$x(t + h) = \frac{1}{6}(k_1 + 2k_2 + 2k_3 + k_4).$$

Note that I write the arguments of the function $f()$ in the opposite order than elsewhere in this chapter, so that it now matches the code. This order, in fact, is conventional for Runge–Kutta integrators. Don't let the change throw you.

From this Runge–Kutta formula, you can write the new version of step() almost by inspection.

```
void step(double & t, double & x, double & v,
    const double h){
  double k1, k2, k3, k4;
  k1 = h * v;
  k2 = h * deriv(t + h/2.0, x + k1/2.0);
  k3 = h * deriv(t + h/2.0, x + k2/2.0);
  k4 = h * deriv(t + h, x + k3);
  t = t + h;
  x = x + (k1 + 2.0*(k2 + k3) + k4)/6.0;
  v = deriv(t, x);
}
```

Again, notice the protocol required: Function step() expects to find the correct value of the current derivative in the variable v, and it leaves it there for the next cycle. So far, I've made no attempt to optimize the code in step() — I can do a few things to reduce the number of computations, but this version is clean, clear, and gets the job done. To see how well it works, compile and run the program for x = 1, tmax = 10. You'll find the resulting value to be 22,026.4657799, which is correct to nine digits. To see even more accuracy, change the step size h to 0.001 and try it again. This time, the result is 22,026.4657948049, which is exact to fourteen digits, about the limit of the precision you can expect from double-precision arithmetic. The integrator may seem simple and crude, but there's certainly nothing wrong with its accuracy!

The complete listing for the latest version of the general-purpose integrator is shown in Listing 12.5. Between step() and integrate(), you have a general-purpose numerical integrator in only 12 lines of executable code. Not bad at all. Remember, all other functions are user-provided and are really quite simple to implement, as this example shows.

Listing 12.5 A general-purpose integrator.

```
/* Uses fixed-step, fourth-order Runge-Kutta integration */

#include <iostream.h>

double deriv(const double t, const double x){
  return x;
}

void out(const double t, const double x, const double v){
  cout << t << ' ' << x << ' ' << v << endl;
}

void get_params(double & h, double & tmax){
  h = 0.01;
  cin >> tmax;
}

void init_state(double & t, double & x){
  t = 0.0;
```

Listing 12.5 A general-purpose integrator. (continued)

```
    cin >> x;
  }

  void step(double & t, double & x, double & v, const double
  h){
    double k1, k2, k3, k4;
    k1 = h * v;
    k2 = h * deriv(t + h/2.0, x + k1/2.0);
    k3 = h * deriv(t + h/2.0, x + k2/2.0);
    k4 = h * deriv(t + h, x + k3);
    t = t + h;
    x = x + (k1 + 2.0*(k2 + k3) + k4)/6.0;
    v = deriv(t, x);
  }

  void integrate(double & t, double & x, const double tmax,
      const double h){

    double v;
    v = deriv(t, x);
    out (t, x, v);
    while (t < tmax){
      step(t, x, v, h);
      out (t, x, v);
    }
  }


  void main(){
    double t, x, h, tmax;
    init_state(t, x);
    get_params(h, tmax);
    integrate(t, x, tmax, h);
  }
```

I am by no means finished. So far, I can only integrate a single scalar variable. I'll take care of that in the next section by introducing the concept of state vectors. I'll also address a whole host of pesky little problems and introduce new features that handle the kinds of things that distinguish a good routine from a great one. Finally, I'll address the important issue of step size control.

Affairs of State

For the next round of improvements to our integrator, I have four improvements in mind. Two of them are cosmetic, the other two profound.

The biggest weakness in the integrator as it stands now is that it can only integrate a single state variable. In the real world, dynamic problems almost always involve more state variables; even the simplest dynamic problem, that of Newton's motion of a single point mass, requires two (see Equation [12.6]). For such problems, the present integrator is worthless as it stands.

Fortunately, this difficulty is easily overcome. You've heard me hint before of the concept of a *state vector*. I haven't discussed vector math yet, and I will not do so until the next volume of this series. However, for the purposes here, you are not required to know much about vector math or vector analysis; the most complicated thing you need to do with vectors is to add them or multiply them by a scalar.

This is one of the profound differences I had in mind. Although the kinds of systems that obey a single, scalar equation of motion are rare, the kinds of systems that obey it when the variable is a *vector*, x, are virtually limitless. As a matter of fact, the only systems that can't be described in the more general vector form are those that require partial differential equations rather than ordinary equations. And even these are solved, in practice, by solving an equivalent set of ordinary equations that approximate the true form.

An equation of motion for a single scalar variable x would be:

$$[12.15] \qquad \dot{x} = f(x, t).$$

If the system had two such variables, the equations of motion would be

$$[12.16] \qquad \begin{aligned} \dot{x}_1 &= f(x_1, x_2, t) \\ \dot{x}_2 &= f(x_1, x_2, t), \end{aligned}$$

and in the general case of N such variables,

$$\dot{x}_1 = f_1(x_1, x_2, x_3, ..., x_N, t)$$
$$\dot{x}_2 = f_2(x_1, x_2, x_3, ..., x_N, t)$$
[12.17] $$\dot{x}_3 = f_3(x_1, x_2, x_3, ..., x_N, t)$$
$$\vdots$$
$$\dot{x}_N = f_N(x_1, x_2, x_3, ..., x_N, t).$$

Each of these equations is a simple scalar differential equation, and the functions $f_n(x)$ are independent functions. Note, however, that in general they may depend on any or all of the scalar variables x_n. The equations of motion must be solved simultaneously, because you don't know any of the future values of x_n until you've integrated them.

Without knowing anything at all about vector mathematics, you can still simplify the notation of Equations [12.17] by defining the *state vector*,

[12.18] $$\mathbf{x} = \begin{bmatrix} x_1 \\ x_2 \\ x_3 \\ \vdots \\ x_N \end{bmatrix},$$

then you can rewrite the equations of motion,

$$\dot{\mathbf{x}} = \mathbf{f}(\mathbf{x}, t),$$

which is precisely the form used in Equation [12.2]. Therefore, one form is merely a shorthand notation for the other, and no vector mathematics are implied. For our purposes, a state vector is simply a good old-fashioned FORTRAN–Pascal–C/C++ one-dimensional array.

The only difference is the way the indices are counted. You'll note that the index in Equations [12.17] and [12.18] run from 1 to N, which is the conventional numbering in mathematics. FORTRAN arrays follow the same rules. As you know, however, the authors of Pascal, C, and C++, knowing little about mathematics and caring less, chose to bless us with indices that run from 0 to $N - 1$. To mathematicians, physicists, and engineers, that's a frustration, but a minor one, and one we can live with.

You may think that the integration of simultaneous differential equations would be considerably more difficult than the integration of a single one, in sort of the same fashion that the simultaneous solution of algebraic equations is more difficult as the number of equations increases. You'd be wrong. The elements of the state vector are considered independent variables, connected only through their derivatives. Therefore, the Runge–Kutta method couldn't care less whether it's applied to one variable, or 100,000. It still works pretty much the same way. All you need do to make the method work for vectors is to change the variables from normal to boldface, to reflect their vector nature.

$$\mathbf{k}_1 = h\mathbf{f}(t, \mathbf{x})$$

$$\mathbf{k}_2 = h\mathbf{f}\left(t + \frac{1}{2}h, \mathbf{x} + \frac{1}{2}\mathbf{k}_1\right)$$

[12.19]
$$\mathbf{k}_3 = h\mathbf{f}\left(t + \frac{1}{2}h, \mathbf{x} + \frac{1}{2}\mathbf{k}_2\right)$$

$$\mathbf{k}_4 = h\mathbf{f}(t + h, \mathbf{x} + \mathbf{k}_3)$$

$$\mathbf{x}(t + h) = \frac{1}{6}(\mathbf{k}_1 + 2\mathbf{k}_2 + 2\mathbf{k}_3 + \mathbf{k}_4)$$

Now you can see the extent of the vector math you must use. The last of these equations shows it most graphically, where you must add the four vectors $\mathbf{k}_1 \ldots \mathbf{k}_4$. Even this addition is trivial: you merely add each pair of vectors, element by element.

From a programming perspective, you know that it's almost as easy to pass a vector (array) through a calling list as it is a scalar (some languages won't allow array return types), so the code changes required to accommodate state vectors are minor. About the only real complication involves adding `for` loops to do the vector additions.

Some Examples

If a picture is worth a thousand words, an example has to be worth at least a hundred. You've already seen one example: the equation of motion of a single point mass, as given by Newton's second law, in one dimension. I showed that the second-order differential equation, Equation [12.4], can be written as two first-order equations, Equation [12.6]. At the time, I didn't make much of the bold-faced symbols in these equations, used to indicate vector parameters. Equation [12.6] can be thought of as two scalar equations, if the motion is in one dimension. But Newton's law is much more general than that; it applies equally well in three dimensions.

To avoid confusion between the state vector and its components, I'll use the symbol r to denote the three-dimensional position vector and v to denote the velocity. Then I can define the state vector, now having six scalar elements, as

[12.20] $\mathbf{x} = \begin{bmatrix} \mathbf{r} \\ \mathbf{v} \end{bmatrix}$.

The equations of motion become the usual

$$\dot{x} = \mathbf{f}(\mathbf{x}, t) \, ,$$

where, in this case,

[12.21] $\mathbf{f}(\mathbf{x}, t) = \begin{bmatrix} \mathbf{v} \\ \dfrac{\mathbf{F}}{m} \end{bmatrix}$.

You can see in Equation [12.21] that the elements of the state vector may be, in fact, vectors (or matrices, or quaternions, or tensors). This becomes very important as I get into simulations of more complex systems, which might include multiple components.

The next example will look familiar to control systems engineers.

[12.22] $\ddot{x} + a\dot{x} + bx = F(x, \dot{x}, t)$

This is the equation for a simple, damped, second-order system. If it weren't for the forcing function F, you could write the solution in closed form, but this forcing function, which represents an applied external force, can be any function at all. and as I've indicated, it can be dependent on the current value of the position and its rate, as well as on time.

You can solve for the motion of such a system analytically if F is a simple function of time, such as a constant or a sine wave. But you don't have a prayer of solving the system for the general case, because any solution will be profoundly dependent upon the nature of F.

You can transform a first-order equation as before, introducing the new variable

[12.23] $v = \dot{x}$

to get

$$\dot{v} + av + bx = F(x, v, t)$$

or

[12.24] $\dot{v} = -av - bx + F(x, v, t)$.

Now you have the state vector

[12.25] $\mathbf{x} = \begin{bmatrix} x \\ v \end{bmatrix}$

and a new vector differential equation

[12.26] $\dot{\mathbf{x}} = \begin{bmatrix} v \\ -av - bx - F(\mathbf{x}, t) \end{bmatrix} = f(\mathbf{x}, t)$.

By now you should see that the definition of the elements of the state vector (and its dimension) and the function $f(\mathbf{x}, t)$ change from problem to problem. The method of solving the equations numerically, however, does not change. This is what makes a general-purpose numerical integrator so universally useful.

As noted in Equation [12.20], each "component" of the state vector $\mathbf{x}$ is a three-dimensional vector, so the state vector itself has six scalar components. Only three of these components are "independent," however. Remember that the velocity, v, was introduced to convert a second-order differential equation to first order.

The number of independent components of the state vector is called the *degrees of freedom* of the body. If the system is constrained (to move in a plane, for example), the number of degrees of freedom, and therefore the size of the state vector, will be correspondingly smaller.

Because all purely dynamic systems (i.e., systems governed by Newton's laws of motion) are described by second-order differential equations, you can expect the size of the state vector to be about twice as large as the degrees of freedom. I say "about" here, because there are important cases, especially in rotational dynamics, with representations that use even more elements. More precisely, then, the size of the state vector will be at *least* as large as twice the degrees of freedom.

In the analysis of a new dynamic system, a large part of the process is determining the degrees of freedom and choosing a state vector that is the best (usually meaning the most simple) representation of the motion. In practice, there are two steps, each simple in concept.

- Choose a state vector
- Determine the first derivative of this vector

This ensures the right form for numerical integration. You don't have to actually solve the differential equations (a good thing, too, because they are almost always not solvable). All you need to do is write the equations down. The integrator does the rest.

Vector Integration

To make all of this work in software, you need to modify the algorithm to deal with a state vector array rather than a single scalar variable. Fortunately, that's an easy change.

I'm going to take this transformation in two stages. First, I'll look at it in the more traditional form that I (and many others before me, and after) have used, in which the state vector is expressed as an ordinary array of scalars. This is the only approach available when using a language like FORTRAN, C, or non-Object Pascal, which probably includes most of you reading this book. Later, I'll look at a more modern, object-oriented approach, but because C++ or any other object-oriented language is rarely available for real-time applications, it's important to be able to get a good implementation the old-fashioned way.

The Test Case

Because I'll be dealing with more than one dimension, the previous test case won't do, so I'll use the following case.

$$[12.27] \quad \begin{aligned} \dot{x} &= -2\pi y \\ \dot{y} &= 2\pi x \end{aligned}$$

This case has a simple known solution. For the analytical solution, I will go back the other way for a change, from first- to second-order equations. Differentiating the first equation again gives

$$\ddot{x} = -2\pi y = -4\pi^2 x.$$

Perhaps you'll recognize this equation more easily in its normal form:

$$[12.28] \quad \ddot{x} + 4\pi^2 x = 0.$$

This is the equation of an undamped second-order system, also known as the harmonic oscillator. The solution is a sine wave. The exact solution depends on the initial conditions. For the initial conditions

[12.29] $x_0 = 1$ $y_0 = 0$,

the motion is given by

[12.30]
$$x = \cos 2\pi t$$
$$y = \sin 2\pi t.$$

Thus, the motion describes a circle of radius one in the x–y plane, with a period of one time unit.

The Software

Now I'll transform the integrator so it can handle state vectors. I'll start with a C-style typedef.

```
typedef double state_vector[N];
```

Using a typedef, of course, doesn't give the strong type checking that I'd like and expect from Pascal or C++, but remember, I'm trying to do things the old-fashioned way here, for the benefit of those who must. Because the derivative function is no longer a scalar, I can't return it as the result of a function call. Instead, I must redefine the derivative function to be void and add the derivative vector as an argument:

```
void deriv(const double t,  state_vector x,
    state_vector x_dot){
const double two_pi =2.0*3.14159265358979323846;
  x_dot[0] =   two_pi * x[1];
  x_dot[1] = - two_pi * x[0];
}
```

You may wonder, as you read this code, why I chose to encode the value two_pi in-line, instead of using the carefully developed constant.cpp file. The answer is simply that I took the easy route. This problem is so small that it easily fits in a single file. Doing it this way, I don't have to bother linking other object files. In a production system, I'd delete the line defining two_pi and link constant.cpp instead.

As you can see, the derivative routine is still essentially a verbatim translation from the equations of motion. I need only remember which variables (that is, x or y) I've assigned to which elements of the state vector.

Function out() will also have to be changed to make it print the results. Because my version is really crude ("Damn it, Jim, I'm a doctor, not a prettyprinter"), I won't even sully these pages with it. Suffice it to say that it depends heavily on stream I/O.

The changes to the integrator proper are minimal: only those needed to accommodate the state vector type and the new usage for deriv().

```
void integrate(double & t, state_vector x,
    const double tmax,const double h){
  state_vector x_dot;
  deriv(t, x, x_dot);
  out (t, x, x_dot);
  while (t < tmax){
    step(t, x, x_dot, h, n);
    out (t, x, x_dot);
  }
}
```

Although I mentioned this point earlier, it's worth repeating here: the derivative vector, here called x_dot, is encapsulated in the integration routine. Nobody has any business accessing it except as it is made available by the integrator, to the derivative and output routines. This is a very important rule violated only by the young and the reckless.

The big changes are in step(), which must be modified to deal with vectors. This routine is shown in Listing 12.6.

Listing 12.6 Vector version of step.

```
void step(double & t, state_vector x, state_vector x_dot,
    const double h){
  state_vector k1, k2, k3, k4, temp;
// evaluate k1
  for(int i = 0; i< 2; i++){
    k1[i] = h * x_dot[i];
    temp[i] = x[i] + k1[i]/2.0;
  }
```

Listing 12.6 Vector version of step. (continued)

```
   // evaluate k2
     deriv(t + h/2.0, temp, x_dot);
     for(i = 0; i< 2; i++){
       k2[i] = h * x_dot[i];
       temp[i] = x[i] + k2[i]/2.0;
     }
   // evaluate k3
     deriv(t + h/2.0, temp, x_dot);
     for(i = 0; i< 2; i++){
       k3[i] = h * x_dot[i];
       temp[i] = x[i] + k3[i];
     }
   // evaluate k4
     deriv(t + h, temp, x_dot);
     for(i = 0; i< 2; i++){
       k4[i] = h * x_dot[i];
     }
   // update state
     t = t + h;
     for(i = 0; i< 2; i++)
       x[i] += (k1[i] + 2.0*(k2[i] + k3[i]) + k4[i])/6.0;
     deriv(t, x, x_dot);
   }
```

Some points about this routine are worth noting. First, because you can't perform arithmetic in the calling lists as you could with scalars, I've had to introduce a new temporary variable to hold the trial values of x. As in the case of the derivatives, these values are strictly that: trial values. Remember that the Runge–Kutta method sort of pokes around in the vicinity of the current point, measuring the slopes in various directions, but these probings have nothing whatever to do with the final value. They are used only by the integrator as it implements the algorithm, and in general, they will not lie on the final trajectory. That's why I keep emphasizing that no one should ever access these intermediate values. It's also why you must make sure that the derivative vector is left in a clean state when you leave the subroutine; hence, the last call to deriv(), which finally gives a value of the derivative that corresponds to the updated state.

The only change to the main program is in the declaration of the state vector.

```
void main(){
    double t, h, tmax;
    state_vector x;
    init_state(t, x);
    get_params(h, tmax);
    integrate(t, x, tmax, h);
}
```

Does it work? You be the judge. Figure 12.1 shows the results for a crude approximation with only four steps around the circle. As you can see, the path doesn't follow the circle very well, displaying an obvious divergence that would get worse and worse each cycle. What's remarkable, though, is that the curve looks anything at all like a circle, with so few steps. That's the power of a high-order integration algorithm. A lower order method would have diverged much more dramatically.

Figure 12.1 Four steps per cycle.

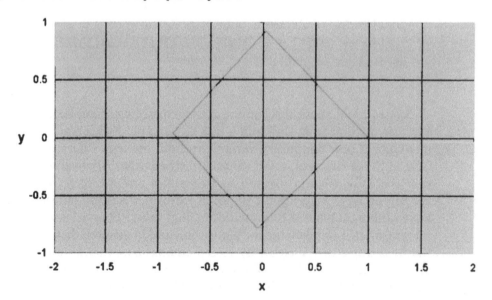

With even as few as eight integration steps, I get results that, at least to the eye, seem to stick to the circle quite well, as you can see from Figure

12.2. And the 64-step solution of Figure 12.3 is essentially perfect, diverging by only about one part per million. For single-precision solutions, I generally use about 100 to 200 steps per cycle, which is more than adequate.

Figure 12.2 Eight steps per cycle.

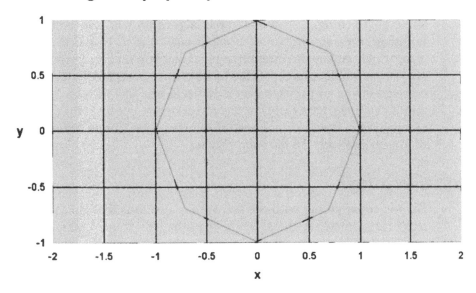

Figure 12.3 64 steps per cycle.

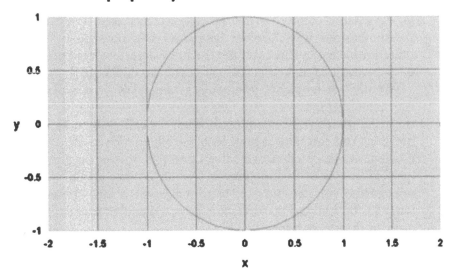

Incidentally, here's a quick and easy way to tell if anything's wrong with the integration algorithm. A fourth-order algorithm should give Order(h^4) performance. That means that every time you halve the step size, you should improve the accuracy by a factor of 16. If you don't see this kind of improvement, something's wrong — usually an erroneous term somewhere in the algorithm. Remember that most any algorithm will work if the step size is small enough. At worst, a bad algorithm will just reduce to rectangular integration, so sometimes a bug is hard to spot. Unless the error is one of sign, which makes the integrator unstable, you'll still see small errors if the step size is small enough, so the bug may not be immediately apparent. Successive halving of the step size is the acid test. It's an easy test to perform, and if the integrator displays the correct error profile, you can be sure that the algorithm is working. I tried this method for this test case and got ratios of almost precisely 16 for each halving.

What's Wrong?

Do you see anything with the software as it stands that isn't very nice? Aside from the somewhat messy code that the for loops create, my approach requires you to recompile the integrator for each new problem — hardly the kind of behavior you'd expect from a general-purpose utility. The reason has to do with those explicit for loops in function step that perform the arithmetic for the algorithm. You could avoid the explicit loop ranges by passing the dimension of the vectors as a parameter — that's the way it was done in the good ol' FORTRAN days, but FORTRAN was particularly forgiving when it came to array dimensions. Basically, it just didn't care what the dimensions were. Newer languages are more picky, and that causes more problems. In this case, the use of declarations depending on the typedef is enough to make the software problem dependent.

Wouldn't it be nice if you could arrange a way to avoid recompilation? As a matter of fact, you can. One obvious way is to simply pass a pointer to the array (that's what's coming through the calling lists, anyway), and index off of it. You can take care of local variables such as the k's in step() with dynamic memory allocation. I've done that in both my production Pascal and C versions of this integrator. I won't go into all that here, though, for one reason: you can do a lot better using the object classes of C++. That's what I'll be looking at soon. For now, I'd rather spend the time on some more utilitarian topics.

Crunching Bytes

Speaking of memory allocation: I'm sure it hasn't escaped your notice that I'm using a fair amount of intermediate storage for the four k's, plus a temporary vector. In these days of multimegabyte memory, a few wasted bytes aren't going to matter because state vectors rarely get much larger than 20 or 30 elements, anyway. There was a time, though, when I would have killed for an extra 100 bytes, and those memories still stick with me. I can remember going over my old FORTRAN integrators with a fine-toothed comb, doing my own graph coloring optimization algorithms by hand to find a few spare bytes. It's probably frivolous to bother these days, but it's also rather easy to collapse the storage somewhat using unions to overlay vectors that aren't needed at the same time. A first glance at the Runge–Kutta algorithm of Equation [12.19] doesn't seem to leave much room for compression. All four k's are needed to compute the updated position. But look again. Except for this one computation, k_1 is only used to compute k_2, k_2 is only used to compute k_3, and so on. You can remove the need to retain these vectors by keeping a running sum for the new state update, as written in pseudocode form below.

```
k1 = h * x_dot;
sum = k1;
temp = x + k1/2.0;
k2 = h * deriv(t + h/2.0, temp);
sum += 2 * k2;
temp = x + k2/2.0;
k3 = h * deriv(t + h/2.0, temp);
sum += 2 * k3;
temp = x + k3;
k4 = deriv(t + h, temp);
t = t + h;
x += (sum + k4)/6.0;
x_dot = deriv(t, x);
```

With a careful examination of this code, you can convince yourself that, except for the running sum, all the intermediate vectors collapse into the same storage location. That is, x_dot is used only to compute k1, k1 only to compute temp, temp only to compute k2, and so on. By using unions, you can use the same storage locations for all of them. This saves $5*N$ words of storage, which might be important to you if N, the size of the state vector, is especially large. If not, don't bother.

A similar approach is possible for all versions of the integrator that you'll see here. If you plan to use it for industrial-grade applications, you might find it worth your while to take the trouble to do this bit of optimization. Don't expect the compiler to do it for you. It won't.

A Few Frills

If I really want to call this a general-purpose integrator, I need more flexible control over things like stop times and print intervals. The integrator might be using step sizes like 0.00314159, but people like to see data printed at nice, simple values like 0.5, 1.0, 1.5, and so on.

As it turns out, control over the time steps is one of the more bothersome aspects of writing an integrator, and that's why so many people just do what I've done so far — take the coward's way out and cut the integrator some slack at edge conditions.

When you give the integrator a terminal time, tmax, it's reasonable to assume that it will actually stop at t = tmax. But if the integrator is always taking a full, fixed step, as I've allowed it to do so far, the best you can hope for is that it will at least stop somewhere within a step size, h, past the desired final time.

If you used a multistep method like Adams–Moulton, you would have to learn to live with such behavior. Multistep methods depend on difference tables, so the step size can't be altered easily. But Runge–Kutta is a single-step method, which means that the step size for a given step can be anything you like, within reason, of course. In this case, the "stop any time after tmax" sort of behavior should be considered downright unfriendly, and it deserves to be dealt with harshly. Personally, I find that the way the integrator stops is a good measure of its quality, and I don't have much patience with those that don't know when to stop. Similar comments apply if you specify even print intervals, as I will shortly.

Fixing the problem, however, is more difficult than you might think. Remember, you're dealing with floating-point numbers here, so after many steps, nice even values like 25 are more likely to be 24.99999. If you specify a print at t = 25, and you're not careful, you can end up getting *two* prints, one at 24.99999 and another at 25. Similar things can happen with respect to the halt condition. In short, dealing with edge conditions is important and separates the good integrators from mediocre ones.

The code in Listing 12.7 deals with the halt condition and is much more civilized.

Listing 12.7 Fixing the terminal condition.

```
void integrate(double & t, state_vector x,
    const double tmax, double h){
  state_vector x_dot;
  double eps = h * 1.0e-4;
  deriv(t, x, x_dot);
  out (t, x, x_dot);
  if(t > tmax) return;
  while (t+h+eps < tmax){
    step(t, x, x_dot, h);
    out (t, x, x_dot);
  }
  h = tmax - t;
  step(t, x, x_dot, h);
  t = tmax;
  out(t, x, x_dot);
}
```

As you can see by the listing, I've modified the loop to look ahead one step to determine if the next step will be in the vicinity of t = tmax. If so, it exits the loop and takes one more step with a step size that goes right to tmax. For safety's sake, I've also added a test at the beginning to kick you out of the loop in the off chance that you started with t > tmax. This avoids the possibility of trying to take some (potentially huge) negative time step. The presence of eps deserves some explanation. It's there to take care of the round-off problem. Without it, you would have to take a tiny final step to get back to t = tmax.

Introducing the "fudge factor" eps allows you to take a step that is marginally larger than the specified step — only 0.01 percent larger in this case. This approach neatly avoids the double-print problem, as well as the tiny-step problem, with a minimum of hassle. The whole loop is designed to eliminate strange behavior at cusp conditions. Instead of testing the end condition as the step is taken, the integrator looks ahead and calculates that only one more step is needed. Then it exits the loop, takes that one and only one step, and quits.

The assignment t = tmax in the terminal code is a controversial one, and you may prefer to leave it out. The general rule should be that nobody, but nobody, messes with the time, or the state vector for that matter, except step(). You have every right to believe that if you set

```
h = tmax - t;
```

and

```
t = t + h;
```

you should arrive right at tmax. Unfortunately, expectations are not always met when using floating-point arithmetic. To be on the safe side, and at the risk of offending purists, I just go ahead and stuff the final value of t with the value I know it should be anyway. For the record, the extra line made no difference in the few tests I made using Borland C++ and a math coprocessor. Your mileage may vary.

I have one last comment with respect to Listing 12.7. At first glance, it might put you off to see two calls to step() and three to out(), in what is basically a single-loop program. You could reduce the number of calls by using various artifices, such as a boolean "loop complete" flag often used in Pascal programs. But after trying such approaches, I discarded them. All other approaches saved duplicating the calls, but at the expense of extra lines of code to test flags. In the long run, the most straightforward and transparent solution, which faithfully represents what the code is actually doing, won out.

As a matter of fact, these duplicated calls can be a blessing in disguise. In the real world, you almost always want to do something special on the first call to the output routine, such as print a set of column headers, for example, and often you'd like to perform some terminal calculations as well. In my old production FORTRAN routine, I did just that, calling out with a control integer to distinguish between first, last, and all other calls (including error conditions). The structure of Listing 12.7 makes such an approach ridiculously simple.

Printing Prettier

If you can force the integrator to give a nice, round number the final time, why not for the print intervals as well? So far, I've simply called out after every step. For some problems, this could result in a lot of printout and waste either trees or electrons, depending on your output device. It's far better to give the user the option of controlling print interval and print frequency. For the print interval, you can use the same approach as for the terminal time. The subroutine shown below is virtually a copy of integrate().

```
void print_step(double & t, state_vector x,
    state_vector x_dot, double h, double hprint){
  double eps = h * 1.0e-4;
  double tprint = t + hprint;
  while (t+h+eps < tprint)
    step(t, x, x_dot, h);
  h = tprint - t;
  step(t, x, x_dot, h);
  t = tprint;
}
```

This routine takes the place of step(), and calls that function as needed to progress from the current time to a target time, tprint. This new routine is called, instead of step(), by the integrator. Note that it does not effect any output itself. That's still taken care of by the integrator. In effect, the integrator sees a (potentially) larger step size than that defined and prints only after one of these larger steps has been taken.

Print Frequency

While I'm dealing with print control, there's one last feature to add, which requires some explanation. Later on, I'll be adding the capability for variable step size. This permits the integrator to choose its own step size, in order to keep errors within bounds. For problems where the dynamics vary strongly, depending on the current value of the state, it's not at all uncommon to see the step size vary quite a bit during a run — sometimes by several orders of magnitude. For such cases, a fixed print interval is not good; it can cause you to miss all the interesting, rapid changes in state. On the other hand, printing every time might use up a lot of paper or CRT display scroll distance. The perfect solution is to give the user the option of printing every n steps, where n is some number the user chooses. Fortunately, you can add this feature with only a very minor change to print_step( ). The following code shows how.

```
void print_step(double & t, state_vector x,
    state_vector x_dot, double h, double hprint,
    int nprint){
  double eps = h * 1.0e-4;
  double tprint = t + hprint;
  int count = nprint-1;
  while (t+h+eps < tprint){
    step(t, x, x_dot, h);
    if(!count--){
      out(t, x, x_dot);
      count = nprint-1;
    }
  }
  h = tprint - t;
  step(t, x, x_dot, h);
  t = tprint;
}
```

Here, I've supplied print_step( ) with an extra argument: the print frequency. If the function takes that many integration steps to get to the next print interval, it will execute its own call to out( ). This approach has the disadvantage of spreading the calls to out( ) over two function, whereas before it was called only by integrate( ). Still, this is the cleanest approach.

You can use the integrator with any combination of print frequency and print interval. If you only want one of the options, simply make the other value large. Unfortunately, this usage creates a new problem. If the print interval is made larger than the value of tmax, you won't exit print_step( ) until you've passed the terminal time. To avoid that undesirable result, you must add a bit of insurance in the integrator to force a smaller hprint, if necessary. The fix involves a one-line change.

```
hprint = min(hprint, tmax - t);
```

Summarizing

The complete listing of the integrator, the test case, and the ancillary routines developed thus far is shown in Listing 12.8. In addition to giving excellent accuracy, it handles the terminal condition, as well as user-specified print intervals and print frequency in a civilized manner. You won't find a much better fixed-step algorithm anywhere. To make it a truly production-quality routine, you only need fix the dependence on the type state_ vector and pass the length of the vector through the calling list.

Listing 12.8 The complete package.

```cpp
#include <iostream.h>
#include <stdlib.h>

typedef double state_vector[N];

void deriv(const double t,  state_vector x,
    state_vector x_dot){
const double two_pi =2.0*3.14159265358979323846;
  x_dot[0] =   two_pi * x[1];
  x_dot[1] = - two_pi * x[0];
}

void out(const double t,  state_vector x,
    state_vector x_dot){
  cout << t << ' ' << x[0] << ' ' << x[1] << ' '
      << x_dot[0] << ' ' << x_dot[1] << '\n';
}

void get_params(double & h, double & hprint, int & nprint,
    double & tmax){
  cout << "Enter step size, print interval, print frequency,
      final time: ";
  cin >> h >> hprint >> nprint >> tmax;
}

void init_state(double & t,  state_vector x){
  t = 0.0;
  cout << "Enter initial state vector: ";
```

Listing 12.8 The complete package. (continued)

```
      cin >> x[0] >> x[1];
    }

    void step(double & t, state_vector x, state_vector x_dot,
        const double h){
      state_vector k1, k2, k3, k4, temp;
// evaluate k1
      for(int i = 0; i< 2; i++){
        k1[i] = h * x_dot[i];
        temp[i] = x[i] + k1[i]/2.0;
      }
// evaluate k2
      deriv(t + h/2.0, temp, x_dot);
      for(i = 0; i< 2; i++){
        k2[i] = h * x_dot[i];
        temp[i] = x[i] + k2[i]/2.0;
      }
// evaluate k3
      deriv(t + h/2.0, temp, x_dot);
      for(i = 0; i< 2; i++){
        k3[i] = h * x_dot[i];
        temp[i] = x[i] + k3[i];
      }
// evaluate k4
      deriv(t + h, temp, x_dot);
      for(i = 0; i< 2; i++){
        k4[i] = h * x_dot[i];
      }
// update state
      t = t + h;
      for(i = 0; i< 2; i++)
        x[i] += (k1[i] + 2.0*(k2[i] + k3[i]) + k4[i])/6.0;
      deriv(t, x, x_dot);
    }
```

Listing 12.8 The complete package. (continued)

```
void print_step(double & t, state_vector x,
    state_vector x_dot, double h,
    double hprint, int nprint){

  double eps = h * 1.0e-4;
  double tprint = t + hprint;
  int count = nprint-1;
  while (t+h+eps < tprint){
    step(t, x, x_dot, h);
    if(!count--){
      out(t, x, x_dot);
      count = nprint-1;
    }
  }
  h = tprint - t;
  step(t, x, x_dot, h);
  t = tprint;
}


void integrate(double & t, state_vector x, const double tmax,
    double h, double hprint, double nprint){

  state_vector x_dot;
  double eps = h * 1.0e-4;
  deriv(t, x, x_dot);
  out (t, x, x_dot);
  if(t > tmax) return;
  while (t+hprint+eps < tmax){
    print_step(t, x, x_dot, h, hprint, nprint);
    out (t, x, x_dot);
  }
  hprint = tmax - t;
  print_step(t, x, x_dot, h, hprint, nprint);
  t = tmax;
  out(t, x, x_dot);
}
```

Listing 12.8 The complete package. (continued)

```
void main(){
  double t, h, tmax, hprint;
  int nprint;
  state_vector x;
  init_state(t, x);
  get_params(h, hprint, nprint, tmax);
  integrate(t, x, tmax, h, hprint, nprint);
}
```

A Matter of Dimensions

Before proceeding with the development of the integrator, I need to talk about an important impediment to progress.

To an old FORTRAN programmer, one of the more frustrating, infuriating, and puzzling aspects of modern programming languages, and a scandal of at least minor proportions, is their lack of support for arrays with varying dimensions, so-called conformant arrays. The lack is significantly more puzzling because FORTRAN has had them from the get-go. This fact is one of the more compelling arguments math analysts and engineers give for hanging onto FORTRAN.

How could FORTRAN II support conformant arrays, but Pascal and C++ cannot? Two of the reasons are "features" of these modern languages (particularly Pascal), which do have much to recommend them, but can sometimes get in the way of progress: strong variable typing and range checking. Pascal compilers traditionally check to make sure that every variable is used in the manner that is appropriate for its type, as declared. They also (some optionally) make sure that index values in arrays cannot go out of bounds. Without some tricky coding, it's well nigh impossible to defeat these attempts to save us from ourselves.

FORTRAN has no such problems because at the time it was invented (a) the compilers weren't sophisticated enough to check such things and (b) the prevailing attitude then, unlike now, was not to try to save a programmer from himself. The compiler translated code from source to object form. If you happened to write stupid source code, you got stupid object code. Although the compiler made every attempt to generate tight code (the original FORTRAN compiler was among the most efficient ever written), optimization was never thought to include turning bad source code into good object code.

When you write library routines, you try to do them in such a way that they can be used in any situation without change. Thus, vector operator packages should work on vectors of any size and matrix multipliers or inverters should work with matrices of arbitrary size (within reasonable upper limits, of course). When FORTRAN became available, it didn't take long to figure out how to trick the compiler into allowing such uses, even though it wasn't explicitly written to support them. The trick had to do with the fact that in FORTRAN, all variables in a subroutine's calling list were/are passed by reference, rather than by value. Modern languages do exactly the opposite: by default, they pass by value, and C can do nothing else.

If a variable is passed by reference, it can be changed, and that can be a dangerous practice. It led to a very famous and insidious bug in FORTRAN programs, in which a literal constant, such as 1, was passed to a subroutine, which changed it. From then on, every use of a 1 in the program used the new value instead, thereby leading to some very confusing results. Perhaps that's the reason the designers of C chose to pass by value, although I suspect that the real reason was more mundane: Unlike the old IBM mainframes for which the FORTRAN compilers were developed, the DEC computer, on which C was developed, had a stack.

In any case, passing by reference is just what is needed for vector and matrix arithmetic, because no local storage is required for the arrays; the storage is allocated by whatever program calls the subroutines, and the subroutine modifies only the referenced array.

When you program in FORTRAN, its syntax requires that you declare that an array is an array by using a dimension statement.

```
DIMENSION ARRAY(20)
```

However, it didn't take long to discover that the compiler really didn't care what dimensions an array had; it only needed to know that it was an array. No range checking was done, either, so the compiler really didn't care if array range bounds were exceeded or not. To a FORTRAN compiler, the code

```
DIMENSION  X(1)
...
X(300) = 4.5
```

is perfectly legal and acceptable code. Although it may look dangerous, it was just what the doctor ordered for general-purpose vector and matrix operators. It also worked just fine as long as the programmer made sure the index never got out of the bounds of the storage allocated for the array. Here's a short example giving a typical use.

```
      SUBROUTINE VADD(A, B, C, N)
      DIMENSION A(1), B(1), C(1)
      DO 10 I = 1, N
10        C(I) = A(I) + B(I)
      RETURN
      END
```

There was only one problem. This trick only works for one-dimensional arrays (vectors). It doesn't work for matrices, because the compiler has to compute the equivalent index by calculating an offset from the beginning of the array (vector or matrix. They're all just a set of numbers to the compiler). For an array dimensioned M by N, with indices I (row) and J (column), the offset is

[12.31] $index = I + (J - 1)M.$

FORTRAN stores matrices column-wise, as any good math-oriented compiler should, because mathematically, a vector is a column matrix. Pascal, C, and C++ do it backwards, which is another source of frustration. For these languages, and also because indices count from zero instead of one, the equivalent of Equation [12.31] would read

[12.32] $index = j + i \cdot n.$

Notice that N is not used in Equation [12.31] and m is not used in Equation [12.32]. Just as for vector arrays, the compiler doesn't really care how many columns (FORTRAN) or rows (everything else) there are.

Those who deal with matrices a lot aren't deterred by a compiler that doesn't know how to handle them. You know that the compiler should be using Equation [12.31], and you can apply it manually almost as easily as the compiler could. The resulting code may look cryptic, but it works efficiently for matrices of any size.

FORTRAN programmers who used such tricks understood full well that they were taking advantage of a loophole in the compiler by using a characteristic that was never formalized and couldn't be counted on to work in future releases. Fortunately, the practice became so universal that any compiler that didn't work that way would not survive in the marketplace. Because of its widespread use, the programming trick became the de facto standard.

When FORTRAN IV came out, the standard became formalized by a mechanism called *variable dimensions*. You no longer had to resort to tricks: the compiler writers built Equation [12.31] into the language. FORTRAN IV allowed the compiler to use the dimension data passed to it; thus, you could write the following, for example.

```
SUBROUTINE MMULT(A, B, C, L, M, N)
DIMENSION A(L,M), B(M, N), C(L,N)

...

   SUM = A(I,K)*B(K,J)

...
```

This feature removed the need for programming tricks. Although the code ran only marginally faster (because the compiler could optimize the indexing better), it certainly was a lot easier to read.

Sad to say, none of these tricks work in either Pascal or C. I think my greatest disappointment with Pascal was to learn that this magical, beautiful, and modern language couldn't handle the simplest matrix math unless the sizes of the matrices were declared in advance. If anyone had told me in 1960, when I was using the matrix-as-vector trick in FORTRAN and implementing Equation [12.31], that I'd still be doing it in 1999, I'd have laughed right in their face. But, for C and C++ programs today, that's exactly what I'm doing. C and C++ will not even allow me to treat a matrix as a vector. The two declarations

```
double x[9];
```

and

```
double A[3][3];
```

each allocate exactly the same storage: nine double words in sequential locations. However, if you pass either array through a function's calling list and try to use one array as though it is the other inside the function, the compiler will smack your hands with a ruler.

I know I can get around the problem by using a typedef to declare a union, but that in itself is a kludge and is also clumsy to work with. Sometimes a compiler can be too smart for its own darned good and for yours, as well.

Back to State Vectors

What does this have to do with numerical integration? Simply that you'd like to make the integrator work like my old vector routines; that is, a general-purpose numerical integrator should be able to integrate N simultaneous equations, where N is a (potentially large) number, not known at compile time. To require the integrator to be recompiled for each new problem, as done here, is truly unacceptable for a production-level integrator.

You almost get your wish using C or C++ by passing the state vector as an array of indeterminate size.

```
void integrate(double & t, double x[], const double tmax,
    double h, double hprint, double nprint){
```

But look again. In addition to x, the integrator and its called functions are littered with local vectors of equal size: k1, k2, k3, k4, temp, and x_dot (two copies). A production-level integrator should allow you to allocate these vectors at startup. My production versions of the quadN series, as written in C and Pascal, use dynamic allocation to do just that (they also have the number of local vectors needed, optimized down to the absolute minimum).

Fortunately, most practical problems have only one state vector, which has the same size throughout the entire program, and require only one instance of the integrator. For the purposes here, the trick of defining the size of the vectors in the typedef statement is sufficient.

Incidentally, the genius who wrote that old NOL3 integrator, which I so shamelessly plagiarized, had what I considered a slick way of dealing with variable-length, local storage: he didn't require it. Instead, he passed the integrator a matrix of scratch-pad storage, equal to x in its row dimension. I took that same idea a step further and simply required the user to allocate x to be a vector of dimension $6N$, where N was the "real" size of the state vector. The user was supposed to access only the first N elements of x;

indeed, the rest of the vector didn't even exist as far as other functions, such as deriv( ), out( ), were concerned.

I found that approach to be eminently workable, and it relieves the integrator of any need to allocate storage dynamically (something FORTRAN would have been unhappy to do for me, anyhow). My only problem came from nosy users, who were forever poking around in that scratch pad and trying to use it for other things. Users can be a pain in the neck, sometimes.

While I'm discussing state vectors, there's another source of frustration that you should be aware of. You saw in Equation [12.20] that practical state vectors often include components that are themselves vectors. Alternatively, you might choose to declare the elements of the state vector to be individually named scalars, as in the following example taken from a six degrees of freedom (6DOF) simulation.

$$[12.33] \qquad x = \begin{bmatrix} x \\ y \\ z \\ vx \\ vy \\ vz \\ q1 \\ q2 \\ q3 \\ q4 \\ p \\ q \\ r \end{bmatrix}$$

The problem is that to the integrator, the state vector is no different than any other vector. But to the analyst who writes the equations of motion implemented by deriv( ), it's a set of 13 scalars, each of which has a name. The programmer of deriv( ) would like to refer to these variables by name, instead of looking them up in a data dictionary in order to remember that variable p is really $x[10]$. The end result is code inside deriv( ) that is very difficult to read. You can't tell the variables without a scorecard.

I must tell you, I have never really found a good solution to this problem. FORTRAN does not allow you to reference individual elements of a vector by name, although it does have equivalence statements. However, they are not allowed in the context of passed parameters.

With Pascal and C, you can use nice, readable data structures to access individual elements of the state vector by name. But now you're back to recompiling the integrator for each application. You also must somehow persuade the compiler to allow the integrator to treat the structure as simply an array of doubles, while deriv() treats it is as a structure. Because of Pascal's strong type checking, I'm not really sure it's possible to get around this problem. With C, I expect you can, given enough combinations of structs and unions, but it won't be easy, and you'll end up having to do some type-casting and dereferencing to get it all to play.

On the Importance of Being Objective

At this point, we can see the light at the end of the tunnel, and it's time to finish the integrator and this book. I have two major improvements: one structural and the other functional. Structurally, I'll modify the code to make better use of the object-oriented character of the C++ programming language. Functionally, I'll greatly increase the usefulness and flexibility, as well as the performance, of the integrator by the inclusion of the all-important automatic step size control. Both changes are significant, and the final result should be a general-purpose routine that will look good in any toolbox.

Although I'm an old FORTRAN programmer at heart, I'm also very excited about C++ and the advantages it affords scientific programmers. This language can simplify rather than complicate the development of scientific programs, primarily because of its operator overloading features. So far, I haven't used these features; the code of Listing 12.8 is very much traditional C despite being programmed in C++. This wasn't an oversight. I wanted you to see how the integrator should be structured in a traditional language. For that reason, the structure so far has been very, very traditional. That's now going to change drastically.

The traditional view has also had traditional problems. In Listing 12.8, you can see how I've had to add for loops at every step to perform the arithmetic for each element of the state vector. With operator overloading, I can make those loops magically disappear, at least explicitly. To do so, I'll define the class State_Vector as a true C++ class, complete with constructors, destructors, and operators. Using the definition of this class, I can greatly simplify the integrator as well as the user-supplied subroutines. In general, you can expect to be able to avoid passing so many parameters through calling lists. For example, C++ allows you to return objects as function values. With the definition of State_Vector as a bona fide class, you can return the value of the derivatives from a deriv(), as I did in the earlier, scalar version of the integrator. What's more, by defining deriv() as a friend function, you can have direct access to the members of the state vector.

What does this mean? It means, mercifully, that the problem of getting to the elements of the state vector by name goes away for good; you can refer to the parts of the state vector by name, even if those parts happen to be arrays.

For the moment, I'll just assume that I've defined such a class, with all the parts that it needs to operate. The only part you need be concerned with is the definition of its member data items.

```
class State_Vector{
  double x,y;
public:
...
};
```

Because I've given the elements of the state vector names, I can refer to them by name in the software I must write, such as in deriv().

```
State_Vector deriv(const double t, const State_Vector &X){
const double two_pi =2.0*3.14159265358979323846;
  State_Vector temp;
  temp.x =   two_pi * X.y;
  temp.y = - two_pi * X.x;
  return temp;
}
```

You must admit, this is a lot neater than the traditional version, where x is a simple array. Notice that the judicious use of const and &, as well as the class definitions, provide security and efficiency. While I was developing this code, the const keyword saved me more than once from the generation of unnecessary temporaries. I got messages, warning me of the problem that I wouldn't have seen otherwise.

If I define the stream output operator << for the class, I can also create a somewhat more civilized version of the output routine.

```
void out(const double t,  const State_Vector x,
    const State_Vector x_dot){
  cout << "t =     " << t << endl;
  cout << "x =     " << x << endl;
  cout << "x dot = " << x_dot << endl;
  cout << endl;
}
```

Most importantly, using the state vector class and operator overloading, the function that performs the fourth-order Runge–Kutta (R-K4) integration becomes a direct expression of the R-K4 algorithm, just as it was when I was dealing with scalars.

```
void step(double &t, State_Vector &x, State_Vector &x_dot,
    const double h){
  static State_Vector k1, k2, k3, k4;
  k1 = h * x_dot;
  k2 = h * deriv(t + h/2.0, x + k1/2.0);
  k3 = h * deriv(t + h/2.0, x + k2/2.0);
  k4 = h * deriv(t + h,     x + k3);
  t += h;
  x += (k1 + 2.0*k2 + 2.0*k3 + k4)/6.0;
  x_dot = deriv(t, x);
}
```

Here I've declared the local variables k1 to k4 as static. This avoids the overhead of thrashing the memory management support and of creating and deleting them as temporaries.

One moment to compare the above version of step() with the that in Listing 12.8 should convince you of the value of operator overloading. Not only have I eliminated the for loops, but also the temporary variable and the code to compute it. Because of operator overloading, you can write algebraic expressions involving state vectors and the temporary values are hidden from you. This simplicity will pay off in spades later when I add automatic step size control.

The advantages of object orientation are by no means limited to the state vector. The integrator itself has a number of internal parameters other than the state vector, such as the rate, step size, print control parameters, and so on that the outside world has no business dinking with, except to set to their initial values. To realize how many parameters this is, you need only look at the calling list of integrate(). What's more, the internal procedures of the integrator, such as step() and print_step(), should never be called by anyone except the integrator. Data and functions such as these beg to be encapsulated inside an object. This approach helps control one of the classic objections to C data integrity.

Unlike C, Pascal permits nested procedures, and it's possible to define data that are local to a main procedure but globally accessible to subprocedures. This allows you to encapsulate things like x, x_dot, and the k's inside the integrator, where they're safe, but still allows them to be available to procedures without having to pass them through calling lists. Classical C is more like FORTRAN in that data are either global or local, and procedure nesting is not allowed; hence, you see long calling lists. I've mentioned in previously one of the dangers of the FORTRAN approach: no matter how much they're warned, users can't seem to resist the temptation to go in and change things, like the state vector, that they have no business changing.

By using the object-oriented features of C++, you can have your cake and eat it too, making internal data readily available to those functions that use it, but safe from meddling hands. You may worry that adding a layer of indirection, as is usually the case when something is "objectified," might also make the algorithm more inefficient. This is a serious concern with something as time-critical as a dynamic simulation. Worry not. In the process of objectifying, you'll find that you remove a horrible inefficiency hidden in step(): the creation and destruction of the intermediate state vectors embodied in the k's. I could have (and should have) removed this inefficiency in other ways, but putting them into the object format takes care of the problem much more elegantly.

My best guess at the form for the integrator object is shown below.

```
class Integrator{
  double t;
  State_Vector x, x_dot;
  State_Vector k1, k2, k3, k4;
  double h, hprint, nprint;
  void out();
```

```
      void step(double h);
      void print_step(double hp);
  public:
    Integrator(){};
    Integrator(State_Vector &a,
      const double _t = 0.0,
      const double _h = 0.001,
      const double hp = 1.0,
      const double np = 1000);
    ~Integrator(){};
    void integrate(double tm);
  };
```

Notice again how the parameter lists for the procedures are reduced to almost nothing. In some cases, I've kept a parameter because the routine is not always called with the same value as for the internal parameter; for example, with the step size. Note also the default values in the constructor, which allow you to set only those parameters you really care about.

How does this approach simplify main()? See it below in its entirety.

```
void main(){
  Integrator I(State_Vector(1.0, 0.0));
  I.integrate(3);
}
```

Maybe I cheated a bit by hiding default values of the parameters in the constructor and using hard-coded values in main(). Still, you have to admit that this object approach has a certain elegant appeal. Indeed, most users don't really care what things like the print frequency are. They're content to let the integrator choose the default values. By giving every parameter a default in the constructor, you give the user the option of being as picky or as lenient as he chooses. You can tinker around with the structure and the member functions of class Integrator, but the first cut shown here is already pretty nice.

The code for the class State_Vector is shown in Listing 12.9. In practice, I'd probably derive the class from a general vector class, but the one shown here will suffice. The code for the rest of the package is shown in Listing 12.10. (Discussion continues on page 415.)

Listing 12.9 State vector class.

```
/* Definition of State Vector class for the numerical
   integrator

   The structure of this class should not change from
   problem to problem.  However, the member data items,
   and the functions that operate on them, are very much
   problem-dependent.  Use this class as a template for
   your own.
*/

class State_Vector{
  double x,y;
public:
  State_Vector(){};
  State_Vector(double _x, double _y){x=_x; y=_y;}
  ~State_Vector(){};
  State_Vector(const State_Vector &a){x=a.x; y=a.y;};
  State_Vector & operator = (State_Vector &);
  void operator += (const State_Vector &);
  void operator -= (const State_Vector &);
  void operator *= (const double);
  void operator /= (const double);
  State_Vector operator + (const State_Vector &);
  State_Vector operator - (const State_Vector &);
  State_Vector operator * (double);
  State_Vector operator / (double);
  friend State_Vector operator * (double, const State_Vector
&);
  friend ostream& operator << (ostream &os, State_Vector &);
  };
```

Listing 12.9 State vector class. (continued)

```
State_Vector & State_Vector::operator = (State_Vector & a){
  x = a.x;
  y = a.y;
  return *this;
}


void State_Vector::operator += (const State_Vector & a){
  x += a.x;
  y += a.y;
}


void State_Vector::operator -= (const State_Vector & a){
  x -= a.x;
  y -= a.y;
}


void State_Vector::operator *= (const double a){
  x *= a;
  y *= a;
}


void State_Vector::operator /= (const double a){
  x /= a;
  y /= a;
}


State_Vector State_Vector::operator + (const State_Vector & a){
  State_Vector temp(*this);
  temp += a;
  return temp;
}
```

Listing 12.9 State vector class. (continued)

```
State_Vector State_Vector::operator - (const State_Vector & a){
  State_Vector temp(*this);
  temp -= a;
  return temp;
}


State_Vector State_Vector::operator * (double a){
  State_Vector temp(*this);
  temp *= a;
  return temp;
}


State_Vector State_Vector::operator / (double a){
  State_Vector temp(*this);
  temp /= a;
  return temp;
}


State_Vector operator * (double s, const State_Vector & a){
  return a * s;
}


ostream& operator << (ostream &os, State_Vector &a){
  os << a.x << ' ' << a.y;
  return os;
}
```

Listing 12.10 Objectified integrator.

```
        Integrator::Integrator(State_Vector &a, const double _t,
          const double _h, const double hp, const double np){
        x = a;
        t = _t;
        x_dot = deriv(t, x);
        h = _h;
        hprint = hp;
        nprint = np;
        }

        void Integrator::out(){
          cout << "t =            " << t << endl;
          cout << "x =           " << x << endl;
          cout << "x dot = " << x_dot << endl;
          cout << endl;
        }

        void Integrator::step(double h){
          k1 = h * x_dot;
          k2 = h * deriv(t + h/2.0, x + k1/2.0);
          k3 = h *  deriv(t + h/2.0, x + k2/2.0);
          k4 = h * deriv(t + h,      x + k3);
          t += h;
          x += (k1 + 2.0*k2 + 2.0*k3 + k4)/6.0;
          x_dot = deriv(t, x);
        }

        void Integrator::print_step(double hp){
          double eps = h * 1.0e-4;
          double tprint = t + hp;
          int count = nprint-1;
          while (t+h+eps < tprint){
            step(h);
            if(!count--){
              out();
              count = nprint-1;
```

Listing 12.10 Objectified integrator. (continued)

```
          }
        }
        h = tprint - t;
        step(h);
        t = tprint;
      }

    void Integrator::integrate(double tmax){
      double eps = h * 1.0e-4;
      out();
      if(t > tmax) return;
      while (t+hprint+eps < tmax){
        print_step(hprint);
        out();
      }
      hprint = tmax - t;
      print_step(hprint);
      t = tmax;
      out();
    }
```

Step Size Control

The R-K4 integration scheme I've been using is accurate and often adequate for well-behaved models, but it does suffer from one important limitation. Although the single-step method allows you to change the integration step size easily from step to step, you have no way of knowing when you need to do so to maintain accuracy because R-K4 gives no estimate of the integration error. That's why everything I've done so far is still limited to maintaining a constant step size throughout the duration of the integration. In practical terms, this means that you must either know enough about the problem in advance to know what a reasonable choice for the (fixed) step size should be over the entire range of the solution, or you must find out by trial and error for each new problem. This limitation makes the appellation "general-purpose" somewhat debatable. A truly general-purpose integrator should solve problems with a wide variety of step sizes. More to the point, it should work efficiently with problems in which the optimal step size is drastically different over the solution range.

To develop a variable-step integrator, you need an algorithm that offers an estimate of the single-step truncation error. Although other algorithms exist, I'm partial to Merson's method.

$$\mathbf{k}_1 = h\mathbf{f}(\mathbf{x}, t)$$

$$\mathbf{k}_2 = h\mathbf{f}\left(\mathbf{x} + \frac{1}{3}\mathbf{k}_1, t + \frac{1}{3}h\right)$$

$$\mathbf{k}_3 = h\mathbf{f}\left(\mathbf{x} + \frac{1}{6}\mathbf{k}_1 + \frac{1}{6}\mathbf{k}_2, t + \frac{1}{3}h\right)$$

[12.34]
$$\mathbf{k}_4 = h\mathbf{f}\left(\mathbf{x} + \frac{1}{8}\mathbf{k}_1 + \frac{3}{8}\mathbf{k}_3, t + \frac{1}{2}h\right)$$

$$\mathbf{k}_5 = h\mathbf{f}\left(\mathbf{x} + \frac{1}{2}\mathbf{k}_1 - \frac{3}{2}\mathbf{k}_3, t + h\right)$$

$$\mathbf{x}(t + h) = \frac{1}{6}(\mathbf{k}_1 + 4\mathbf{k}_4 + \mathbf{k}_5)$$

$$\mathbf{e} = \frac{1}{30}(2\mathbf{k}_1 - 9\mathbf{k}_3 + 8\mathbf{k}_4 - \mathbf{k}_5)$$

As before, thanks to operator overloading, you can code this algorithm by inspection.

```
void Integrator::step(double h){
  k1 = h * x_dot;
  k2 = h * deriv(t + h/3.0, x + k1/3.0);
  k3 = h * deriv(t + h/3.0, x + k1/6.0 + k2/6.0);
  k4 = h * deriv(t + h/2.0, x + k1/8.0 + (3.0/8.0)*k3);
  k5 = h * deriv(t + h, x + k1/2.0 - 1.5 * k3 + 2.0 * k4);
  t += h;
  e = (2.0*k1 - 9.0*k3 + 8.0*k4 - k5)/30.0;
  x += (k1 + 4.0*k4 + k5)/6.0;
  x_dot = deriv(t, x);
}
```

See how easy this is? If you doubt the value of using operator overloading, try coding this same routine using for loops and intermediate variables.

The problem that you're faced with now, though, is how do deal with the error estimate. The first thing you must recognize is that the format of step() is wrong. In general, you cannot assume that you'll want to accept the step, because having computed it, you may find that the error was too large. No matter how tempting it is, you must never take a step that would lead to an error out of range, so you can't let this function actually update the state until you're satisfied that the step was legit.

Fortunately, this problem is easily fixed. Simply let step() compute the increment in x, and decide externally whether the step was suitable or not.

```
void Integrator::step(double h){
  k1 = h * x_dot;
  k2 = h * deriv(t + h/3.0, x + k1/3.0);
  k3 = h *  deriv(t + h/3.0, x + k1/6.0 + k2/6.0);
  k4 = h * deriv(t + h/2.0, x + k1/8.0 + (3.0/8.0)*k3);
  k5 = h * deriv(t + h, x + k1/2.0 - 1.5 * k3 + 2.0 * k4);
  e = (2.0*k1 - 9.0*k3 + 8.0*k4 - k5)/30.0;
  dx = (k1 + 4.0*k4 + k5)/6.0;
}
```

The variables e and dx are new state vector variables declared in the body of the class. Function step() is, in turn, called from a new function called good_step(). The idea is that good_step() will never accept a step that is not accurate, within some definition of that term. To keep from breaking the integrator, you can define a temporary, stub version of this new function. This stub version always accepts the step.

```
void Integrator::good_step(double h){
  step(h);
  t += h;
  x += dx;
  x_dot = deriv(t, x);
}
```

The error, e, is a vector estimate of the error incurred in the state vector in a single step.

One Good Step...

The obvious question is, just how should the real good_step() work? Clearly, it must adjust the step size, but how and by what logic? A typical approach is to double and halve the original step size — halving when the error is found to be too large and doubling when it's "acceptable." This approach is popular because multiplying and dividing by powers of two tends to be not only faster, but also less prone to floating-point round-off errors. But that answer only leads to the next question. How do you decide when a vector of errors in the individual components of the state vector is acceptable or not?

The easiest way to do this is to convert the error vector into a single scalar, usually by taking a vector norm or some equivalent operation. The exact nature of the appropriate error vector can be dependent on the problem. For example, should the error be considered as absolute or relative? For flexibility, you should encapsulate the computation of the error criterion in yet another user-supplied function.

Over the years, I've tinkered with several definitions of a user-defined function called auto() (yet another idea, complete with function name, shamelessly borrowed from NOL3). In my earliest implementations, auto() included the switching logic and altered the step size. Later, I had it return a control integer, either 0, 1, or 2 (nowadays, I'd use an enumerated type), that instructed the integrator what to do. Both approaches, though, turned out to be unsatisfactory because they put too much burden on the user to get the logic right. More recently, I've had better results by having auto() return a simple scalar that is a positive number representing a measure of the composite error, which the integrator proper then uses to make its decisions.

The vector e returns values that represent an estimated error in each element of the state vector. Conceptually, you can think of the elements of e as defining an error volume in N-space (a rectangle, for a 2-D problem). The idea behind auto is to return a single scalar that represents the worst case value of the error.

But what scalar should you return? If all the parameters of the state vector have the same magnitude, you might consider returning something like a diameter.

[12.35] $$\sqrt{e_1^2 + e_2^2 + \ldots + e_N^2}$$

However, recall that the elements of e are (hopefully) very small. For that reason, I've had trouble in the past with floating-point underflow. Perhaps more importantly, the square root function is going to burn some unnecessary clock cycles. For this reason, I prefer to simply use the largest element of e as the returned value. In effect, think of the elements of e as defining a hyper-rectangle, rather than a hypersphere.

For most practical problems, the elements of the state vector have units, which are almost certainly not the same for all. Some may be quite a bit larger, in numerical magnitude, than others. So you should also normalize the error value by the magnitude of the variable associated with it. If, for example, one of the elements of the state vector is itself a vector — perhaps position — a reasonable return value might be

$$[12.36] \qquad \text{error} = \frac{|\mathbf{e_x}|}{|\mathbf{x}|} \, .$$

Again, you're tempted to use the usual vector definition of the magnitude, as the square root of the sum of the squares. But then you are back to square roots again. Also, the same solution presents itself again. Use the largest element in magnitude.

You now have a potentially serious problem. What if $|\mathbf{x}| = 0$? Things are then going to get nasty. But there's an easy way around this problem, also: Limit $|\mathbf{x}|$ to be at least one. In effect, use relative error when $|\mathbf{x}|$ is large, absolute error when it is small. Using a little pseudomath, you can write

$$[12.37] \qquad \text{error} = \frac{\max(|e_i|)}{\max(\max(|x_i|, 1))} \, .$$

Mathematically, I don't think this algorithm has much of an argument to defend it. On a practical basis, however, I've found it to be more robust than any other scheme I could find. Remember, however, that auto( ) is a user-defined function, so feel free to experiment on other schemes. Just remember that the value returned by auto( ) should be a measure of the worst case error reported by vector e.

For the test case, it's much simpler. The state vector has a magnitude constrained to unity, so no computation of relative error is needed. The following is an appropriate auto( ) function (renamed in deference to C's keyword auto).

```
double error_size(const State_Vector &x, const State_Vector
&e){
   return max(abs(e.x), abs(e.y));
}
```

To allow this function to access the elements of the state vector directly, it must be declared a friend function to class State_Vector.

Given the availability of a scalar measure of the error, the logic is as follows.

- If the error is larger than some epsilon, reject the step, halve the step size, and repeat the step.
- If the error is less than epsilon, accept the step.
- If the error is smaller than epsilon/64, double the step size for the *next* step.

You have to do this with a little care, because the step size is temporarily altered for the partial steps that come just before print points or the final point. You must take care that you don't end up using an extra small (and probably strange) step size in your logic.

The following implementation of good_step() takes care of all this.

```
void Integrator::good_step(double _h){
  double err;
  for(int i=0; i< 20; i++){
    step(_h);
    err = error_size(x, e);
    if(err <= max_err){
      t += _h;
      x += dx;
      x_dot = deriv(t, x);
      if((err < max_err/64.0) && (_h == h))
        h *=2.0;
      return;
    }
    h /=2.0;
    _h /= 2.0;
  }
  cout << "Step size halved twenty times\n";
  exit(1);
}
```

Notice the for loop, which bails out with an error message if the step size is halved too many times. In general, it's not a good idea to bail out of a general-purpose routine so precipitously, but in this case, it happens to make perfect sense. A loop count of 20 corresponds to reducing the step size by a factor of one million, which should be more than adequate, unless you're really off on your original estimate. In the years that I've been using this integration algorithm, I've seen this error message many times. It has never occurred except in the development process, when someone was doing something crazy in the evaluation of the derivatives. In short, it's almost certainly caused by a programming bug. Nothing whatever would have been gained by trying to soldier on, rather than aborting.

As an aside, I discovered that Borland C++ didn't give me the right results for error_size, using the code shown above because it was using the integer version of abs(). That won't happen if you use my jmath.cpp library functions. I solved the problem in this simple test case by defining my own overloaded abs().

The final version of the integrator, featuring automatic step size control, some rather sophisticated handling of print control and termination, and maximum use of object-oriented approaches, is shown in Listing 12.11. On the test cases that I've run it works very well, just as its FORTRAN ancestors did. I hope you like it. It's not perfect yet. I'm still not quite satisfied with the treatment of print times. I'm thinking that perhaps a target time, rather than a delta time, might give slightly cleaner code, but what's here works. One thing's for sure: you're not likely to find an integrator more accurate or more robust. (Discussion continues on page 427.)

Listing 12.11 Variable-step integrator.

```
class State_Vector{
  double x,y;
public:
  State_Vector(){};
  State_Vector(double _x, double _y){x=_x; y=_y;}
  ~State_Vector(){};
  State_Vector(const State_Vector &a){x=a.x; y=a.y;};
  State_Vector & operator = (State_Vector &);
  void operator += (const State_Vector &);
  void operator -= (const State_Vector &);
  void operator *= (const double);
  void operator /= (const double);
  State_Vector operator + (const State_Vector &);
  State_Vector operator - (const State_Vector &);
```

Listing 12.11 Variable-step integrator. (continued)

```
        State_Vector operator * (double);
        State_Vector operator / (double);
        friend State_Vector operator * (double, const State_Vector
&);
        friend ostream& operator << (ostream &os, const State_Vector
&);

// Members Specific to State Vectors
        friend State_Vector deriv(const double t, const State_Vector
&X);
        friend double error_size(const State_Vector &x,
          const State_Vector &e);
};

State_Vector & State_Vector::operator = (State_Vector & a){
    x = a.x;
    y = a.y;
    return *this;
}

void State_Vector::operator += (const State_Vector & a){
    x += a.x;
    y += a.y;
}

void State_Vector::operator -= (const State_Vector & a){
    x -= a.x;
    y -= a.y;
}

void State_Vector::operator *= (const double a){
    x *= a;
    y *= a;
}
```

Listing 12.11 Variable-step integrator. (continued)

```
void State_Vector::operator /= (const double a){
  x /= a;
  y /= a;
}

State_Vector State_Vector::operator + (const State_Vector & a){
  State_Vector temp(*this);
  temp += a;
  return temp;
}

State_Vector State_Vector::operator - (const State_Vector & a){
  State_Vector temp(*this);
  temp -= a;
  return temp;
}

State_Vector State_Vector::operator * (double a){
  State_Vector temp(*this);
  temp *= a;
  return temp;
}

State_Vector State_Vector::operator / (double a){
  State_Vector temp(*this);
  temp /= a;
  return temp;
}

State_Vector operator * (double s, const State_Vector & a){
  State_Vector temp(a);
  return temp * s;
}
```

Listing 12.11 Variable-step integrator. (continued)

```
ostream& operator << (ostream &os, const State_Vector &a){
  os << a.x << ' ' << a.y;
  return os;
}

State_Vector deriv(const double t, const State_Vector &X){
const double two_pi =2.0*3.14159265358979323846;
  State_Vector temp;
  temp.x =   two_pi * X.y;
  temp.y = - two_pi * X.x;
  return temp;
}

class Integrator{
  double t;
  State_Vector x, x_dot;
  State_Vector k1, k2, k3, k4, k5, dx, e;
  double h, hprint, nprint;
  double max_err;
  void out();
  void step(double h);
  double error();
  void good_step(double _h);
  void print_step(double hp);
public:
  Integrator(){};
  Integrator(State_Vector &a,
    const double _t = 0.0,
    const double _h = 0.001,
    const double hp = 1.0,
    const double np = 1000);
  ~Integrator(){};
  void integrate(double tm);
  friend error();
};
```

Listing 12.11 Variable-step integrator. (continued)

```
Integrator::Integrator(State_Vector &a,
   const double _t, const double _h,
   const double hp, const double np){
 x = a;
 t = _t;
 x_dot = deriv(t, x);
 h = _h;
 hprint = hp;
 nprint = np;
 max_err = 1.0e-10;
}

void Integrator::out(){
 cout << "t =      " << t << endl;
 cout << "x =      " << x << endl;
 cout << "x dot = " << x_dot << endl;
 cout << endl;
}

void Integrator::step(double h){
 k1 = h * x_dot;
 k2 = h * deriv(t + h/3.0, x + k1/3.0);
 k3 = h *  deriv(t + h/3.0, x + k1/6.0 + k2/6.0);
 k4 = h * deriv(t + h/2.0, x + k1/8.0 + (3.0/8.0)*k3);
 k5 = h * deriv(t + h, x + k1/2.0 - 1.5 * k3 + 2.0 * k4);
 e = (2.0*k1 - 9.0*k3 + 8.0*k4 - k5)/30.0;
 dx = (k1 + 4.0*k4 + k5)/6.0;
}

double error_size(const State_Vector &x, const State_Vector
&e){
 return max(fabs(e.x), fabs(e.y));
}
```

Listing 12.11 Variable-step integrator. (continued)

```
void Integrator::good_step(double _h){
  double err;
  for(int i=0; i< 20; i++){
    step(_h);
    err = error_size(x, e);
    if(err <= max_err){
      t += _h;
      x += dx;
      x_dot = deriv(t, x);
      if((err < max_err/64.0) && (_h == h))
        h *=2.0;
      return;
    }
    h /=2.0;
    _h /= 2.0;
  }
  cout << "Step size halved twenty times\n";
  exit(1);
}

void Integrator::print_step(double hp){
  double eps = h * 1.0e-4;
  double tprint = t + hp;
  int count = nprint-1;
  while (t+h+eps < tprint){
    good_step(h);
    if(!count--){
      out();
      count = nprint-1;
    }
  }
  good_step(tprint-t);
}
```

Listing 12.11 Variable-step integrator. (continued)

```
void Integrator::integrate(double tmax){
  double eps = h * 1.0e-4;
  out();
  if(t > tmax) return;
  while (t+hprint+eps < tmax){
    print_step(hprint);
    out();
  }
  hprint = tmax - t;
  print_step(hprint);
  t = tmax;
  out();
  cout << "Step size = " << h << endl;
}

void main(){
  Integrator I(State_Vector(1.0, 0.0));
  I.integrate(3);
}
```

What About Discontinuities?

The two integrators you've seen, the fixed-step and variable-step versions, represent two of the four integration routines in the stable I tend to call quadN.

quad0: fixed-step, fourth-order Runge–Kutta
quad1: variable-step, "4.5-order" Merson's method
quad2: quad1 with scalar termination function
quad3: quad1 with vector termination function

The last two descriptions deserve an explanation.

Recall that Runge–Kutta methods — indeed, all methods for numerical integration — ultimately boil down to fitting a derivative function with a polynomial of some degree. To perform this fit, you must assume that the function is continuous and has continuous derivatives. Proper functioning of the integrator depends on this.

Yet dynamic simulations of many practical problems have, buried within them, discontinuities of one sort or another. The simplest example might be a bouncing ball. When the ball hits a boundary, it rebounds with a component of its velocity changed.

In real-world problems, there are constraints of various kinds that might be required of the system; perhaps a wall exists in the real world too. Rockets drop stages and therefore change their mass characteristics, not to mention thrust. Spacecraft attitude control systems and other systems have "bang-bang" controllers, which fire rockets or otherwise do something drastic to urge the system to stay within some reasonable bounds of its commanded position. Although the equations of motion are well behaved on either side of such events, they change abruptly at the boundaries.

Physicists talk about impulsive changes in the state, meaning that the changes take place instantaneously as discontinuities are encountered. The simplest example I can think of is the collision of two billiard balls. Strictly speaking, the collision isn't really impulsive, meaning that it doesn't really happen in zero time. If you were to look at the collision in great detail, you'd find that the balls deform a bit, squish into each other, and then rebound. In the same way, a basketball doesn't rebound impulsively; it deforms and then recovers.

For practical reasons, you don't want to deal with the details of these deformations or their dynamics. For all practical purposes, you can say that the interaction is instantaneous and, therefore, impulsive.

I must tell you in all candor that most dynamic simulations I've seen don't handle such cases very well at all. I said earlier that of all the errors I've seen in numerical integration algorithms, getting things out of step by one time step, h, is the most common. Surely the second most common cause of errors comes from the way a simulator deals with discontinuities.

Most implementers take the easy way out. In deriv(), they simply test to see if the condition for an edge has been met or not and change the equations of motion to match. I must admit, sheepishly, that I've done this myself. If I'm using a quad1-class integrator, meaning one with automatic step size control, the results are actually not too bad. The integrator cuts the step size drastically as it approaches the boundary, then it cautiously probes across it like a kid dipping a toe in a cold lake. Eventually, the step size is so small that the error, even though still finite, is within the error criterion. The integrator then carefully steps across the boundary and accelerates away from it with doubled step sizes at each new step.

A fixed-step method has a different response: it makes an error. Remember that R-K methods work by probing the "flow field" described by the equations of motion. If a discontinuity is present, some of these probes will be on one side of it, some on the other. The end result will be a completely bogus computation for the updated state. That bogus computation may not cause a problem, but more likely, it will introduce an error that compounds itself as the run continues.

Even though you can force a good variable-step method to work through discontinuities, to do so borders on criminal negligence. Strictly speaking, you should never try to integrate across such discontinuities; rather, you should break the solution into parts that are piecewise continuous. There is only one reasonable and mathematically correct way to deal with discontinuities: stop the integrator, apply the impulsive change, and restart. That's why all my production integrators have the capability to restart at the previous state.

The process of stopping the integration is easier said than done, however. The trick is finding out just where the discontinuities occur. In the case of a rocket staging, the problem is easy, because the staging probably occurs at a specific and predefined time. For cases such as this, you can simply program the solution in two or more calls to the integrator.

Hitting walls is much more difficult, because you can't predict when in time you'll hit them. In my quad2 series, I solved this problem by defining some function of the state that goes to zero at the boundaries:

[12.38] $f(\mathbf{x}, t) = 0.$

This function, which can be anything from one (which, of course, never crosses zero and thus never triggers an action) to as complicated and nonlinear a function as you choose, is computed by yet another user-definable function to the routine called term( ). I added code to the integrator to search for places where the function passes through zero (thus changing sign). When the integrator sees a sign change, it backs up and iterates to find the precise time of the zero crossing. This scheme turned out to be extremely useful, and it's become a fixture, almost a trademark, in my FORTRAN integrators. It's worked beautifully.

Unfortunately, the implementation details are a bit much for this chapter; the math is rather complicated, and I haven't talked yet about methods for finding the zeroes of a function, for which I have a robust, second-order method. To be frank, I have not incorporated it into quad2, where I still use linear interpolation, which works well because the step size control presumably keeps things from changing too much in a single step.

Multiple Boundaries

The notion of terminating on a zero of a function is nice, but in general, you're likely to have more than one edge condition or boundary to watch for. A ball in a box, for example, has at least six such conditions and more if you count corners. So how do you deal with multiple boundaries using only a single scalar function? If you have multiple edge conditions, define multiple functions and create a single function by multiply them together. Then, $f(x)$ will go to zero when any one of the functions does. I've used this approach quite often with quad2, and it works very nicely.

There's only one problem. Once the integrator returns, having found an edge condition, you have no idea which one it found. This means that you must provide code to find out which wall you hit. Remember, the integrator has terminated, so it's up to the main program to decide what to do before it restarts the solution.

The problem is complicated by floating-point arithmetic, which means that you may not be *exactly* on the edge when the integrator stops. You might be at $t = 9.999999999$ instead of 10. Although quad2 is quite useful, I found that searching for the reason it stopped was sometimes tricky, and the complexity compounds factorially as the number of such conditions increases.

That's when it occurred to me that the one entity that best knew why quad2 stopped would be quad2 itself. I just needed it to tell me why it stopped. Unfortunately, quad2 didn't know either, because information about which of the multiple boundary functions [multiplied together inside term( )] was responsible was obscured by that multiplication.

The answer to this problem is to return a vector function f(x, t). The integrator could stop when any element of this function crossed zero, and it could tell which element it was.

Enter quad3, which does exactly that. In addition to monitoring all the elements of the vector function f(x, t), it returns a vector of Boolean flags telling me which condition(s) had been met. That solves the problem for good.

If you think the logic for finding a single edge condition inside an integrator is tricky, you should see the logic for the vector case. There is no way I can go into such detail here; it's far outside the scope of this chapter, and I have to hold something back for my paying customers!

Integrating in Real Time

At this point, you might be justified in thinking that I have nothing left to say about numerical integration. You'd be wrong.

As I've hinted previously, single-step methods such as the R-K methods are not suitable for real-time computations. Although I've touched on the reasons before, I'll now discuss these reasons in more detail and find out what can be done to fix the situation. I'll wrap up the study of integrators with a series of algorithms that will work for just about any real-time application you will ever need.

Why R-K Won't Work

Remember that the R-K method uses a "what if" approach. It gives high-order (fourth or better) solutions by performing a series of "probes" into the space described by the derivative function. A typical R-K step looks like Equation [12.39].

$$[12.39] \qquad \mathbf{k}_2 = h\mathbf{f}\left(t + \frac{1}{2}h, \mathbf{x} + \frac{1}{2}\mathbf{k}_1\right)$$

Translating into English, you are "asking" the derivative function the question, "If I have a state vector $\mathbf{x}$ of so-and-so at such-and-such a time, what would my derivative vector be?" In general, the trial state vectors used in the R-K method are fictitious values. In the final solution, the state vector will never actually take on these values.

Asking questions like this are appropriate when the derivative function is known analytically and can be evaluated for any possible input value, including times that haven't occurred yet. I think you can see, though, that such questions are completely inappropriate for a real-time system. At a given time, the state vector is what it is, and you can do nothing to change it. More importantly, you can't "ask" the system what the derivatives will be at some future time. At least some of the derivatives are likely to come from sensors like flow meters or accelerometers, and you have no idea what their readings will be until you're actually at the time point in question. The R-K method depends on knowledge of future values that you simply never can have in a real-time system. In real time, you absolutely must restrict yourself to values of the derivatives when they actually occur.

The First-Order Case

If first-order accuracy is good enough, the problem is easy, because to a first-order approximation, the change in the state vector is directly proportional to its rate

[12.40] $\Delta \mathbf{x} = h\mathbf{f}(t, \mathbf{x})$.

If you denote your current value of the state vector as $\mathbf{x}_i$, the next value is

[12.41] $\mathbf{x}_{i+1} = \mathbf{x}_i + h\mathbf{f}(\mathbf{x}, t)$.

This equation translates into code in a straightforward manner. For the sake of conciseness, I'll stick to the C++ format that I've been using. As in the past, I'll assume that the derivatives are given by a C++ function, which in this case would presumably include reading the values of sensors.

```
State_Vector deriv(const double t, const State_Vector &X){
   State_Vector temp;
   ...
   // read sensors
   // compute the derivative vector
   ...
   return temp;
}
```

Now the code for updating the state vector is a real-time integrator in one line.

```
x += h*deriv(t, x);
```

What value do you use for the step size, h, and how do you control it? In almost every case, the answer is to evaluate the derivatives at a fixed time interval; therefore, h is a constant. You must provide a periodic interrupt that causes the code execute. On a PC, most people do this by making use of the 18.2Hz real-time clock built into the PC hardware. Users can hook into this clock by providing a user-supplied interrupt handler for INT 1Ch. In a real-time system, getting such interrupts is not a problem. Almost any real-time system worthy of the name will have at least one task that executes at a specific frequency driven by a (usually nonmaskable) external clock.

In either case, just set the step size equal to the period of the fixed frequency, insert the code shown above for updating the state vector into the interrupt handler, and you've got yourself a real-time integrator. You'll find this approach or one very much like it in almost all PC-based simulators.

Notice that it's not absolutely necessary that the update rate be a fixed frequency. As an alternative, you could simply put the update code into an infinite loop and measure the time between successive passes to determine the step size. To do this, though, you must have a clock with sufficiently high resolution so that the step size can be measured accurately. Most computers aren't equipped to give this kind of information, so the fixed-step approach is almost universal. For all practical purposes, it's also the only approach that can be extended to higher orders.

For the step size, a good rule of thumb is to provide at least 100 integration steps per cycle of the highest frequency involved. In other words, if your system must respond to disturbances of 10Hz, for example, your update frequency should be at least 1,000Hz. At that rate, don't expect the kind of accuracy you're used to getting out of computers. Figure 12.4 shows a graph of the solution of the "standard" test case

$$[12.42] \qquad \begin{aligned} \dot{x} &= -2\pi y \\ \dot{y} &= 2\pi x. \end{aligned}$$

where I've taken 128 steps per cycle. At the end, the x-coordinate is in error by more than 15 percent. The error is due to truncation error in the first-order algorithm and is typical of what you can expect of this method. By contrast, you saw earlier that 64 steps of R-K4 gave essentially perfect results. You can improve the accuracy by using more steps: 8 percent error for 256 steps, 4 percent error for 512 steps, 2 percent error for 1,024 steps, 1 percent error for 2,048 steps, and so on. If you want accuracy better than 1 percent, you'd better figure on adding to the update frequency by at least another factor of 10. This crude first-order method will never give you the kind of perfection that you're used to from other computer math applications. That's just a fact of life. Consider that if you're using the 18.2Hz PC clock to set the update frequency, you'd better not be modeling frequencies in the system that are higher than 0.2Hz (five seconds per cycle). That should give you pause to think if you're implementing a fast-moving jet plane simulation.

Figure 12.4 First-order integration.

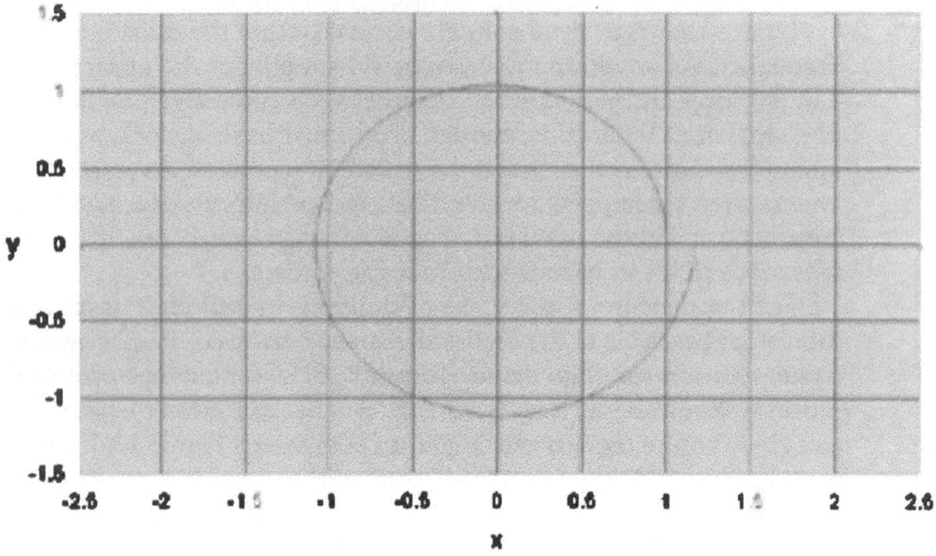

Is First-Order Good Enough?

At this point, you may wonder why I'm telling you about the first-order method if I can never get enough accuracy from it. Surprisingly, there are plenty of cases in which 15 percent error over a single cycle is perfectly acceptable. These are the cases I've discussed before, in which the system is controlled by feedback, either from a human operator or a control system. Suppose, for example, that Figure 12.4 represents the ground track of an airplane that is supposed to be flying in a circle. Seeing the radius of the circle increase, the pilot would apply just a hair more aileron, crank in the needed control, close the circle, and never even notice that the model was in error. Similarly, a control system that parks an elevator, for example, would add a bit more torque to arrive on-station. Whenever you're modeling a system with feedback, the first-order algorithm is often perfectly acceptable. For those of you whose problems fall into this category, you can stop reading now. The rest of you read on.

Higher Orders

Although the first-order method may be perfectly acceptable for systems controlled by feedback, plenty of situations where accurate solutions are imperative don't have that luxury. One example is navigation, in which case you're strictly dependent on the data received from the navigation sensors, and any error incurred is going to propagate into an error in final position. Clearly, for problems of this type, a higher order integration formula is needed. However, I've already noted that the higher order integration formulas based on the Runge–Kutta method won't work because they require knowledge of the future that simply isn't available in real time.

At first glance, it seems that there's nothing you can do to solve this problem. Remember, though, that although it's not fair to look at future values of the derivatives, nothing says you can't remember *past* values, and that notion is the key to higher order integrations in real time.

The Adams Family

In Chapter 10, I discussed the Adams and Adams–Moulton multistep methods based on difference equations, in which I gave the famous Adams formula in operator notation.

[12.43]
$$x_{i+1} = x_i + h\Big(1 + \frac{\nabla}{2} + \frac{5}{12}\nabla^2 + \frac{3}{8}\nabla^3 + \frac{251}{720}\nabla^4$$
$$+ \frac{95}{288}\nabla^5 + \frac{19,087}{60,480}\nabla^6 + \dots\Big)f_i$$

The important thing about the backward difference operator for my purposes is strongly hinted at by its name. It depends only on past values of events, not future values, which is just the kind of thing you need for a real-time system.

For some reason, most programmers don't like to deal with differences directly, but rather with past values of the derivatives. In Chapter 10, I gave the resulting formulas in terms of z-transforms. They are reproduced in Table 12.1 in somewhat different form. The first of these equations is identical to Equation [12.40]. For completeness, I've shown formulas through order six, although I'd be amazed if you ever need accuracy in real-time problems beyond that given by fourth order.

Table 12.1 Adams formulas.

First order
$x_{i+1} = x_i + hf_i$
Second order
$x_{i+1} = x_i + \dfrac{h}{2}(3f_i - f_{i-1})$
Third order
$x_{i+1} = x_i + \dfrac{h}{12}(23f_i - 16f_{i-1} + 5f_{i-2})$
Fourth order
$x_{i+1} = x_i + \dfrac{h}{24}(55f_i - 59f_{i-1} + 37f_{i-2} - 9f_{i-3})$
Fifth order
$x_{i+1} = x_i + \dfrac{h}{720}(1,901f_i - 2,774f_{i-2} + 2,616f_{i-2} - 1,274f_{i-3} + 251f_{i-4})$
Sixth order
$x_{i+1} = x_i + \dfrac{h}{1,440}(4,277f_i - 7,923f_{i-1} + 9,982f_{i-2} - 7,298f_{i-3}$ $+ 2,877f_{i-4} - 475f_{i-5})$

Doing It

To show how it's done, I'll implement the Adams algorithm for the fourth order. I'll need past values of the derivatives, so instead of a single vector **x_dot**, as I had in the R-K integrator, I'll define four values.

```
State_Vector f0,f1,f2,f3;
```

The code to update the state vector now must also rearrange the past values, shifting them by one position.

```
f3 = f2;
f2 = f1;
f1 = f0;
f0 = deriv(t, x);
x += h*(55.0*f0 - 59.0*f1 + 37.0*f2 -9.0*f3)/24.0;
```

Figure 12.5 shows the results, for a step size of 64 steps per cycle (the same as that used for R-K4). As you can see, the results are much more accurate than the first-order method, as you might expect.

Figure 12.5 Fourth-order Adams.

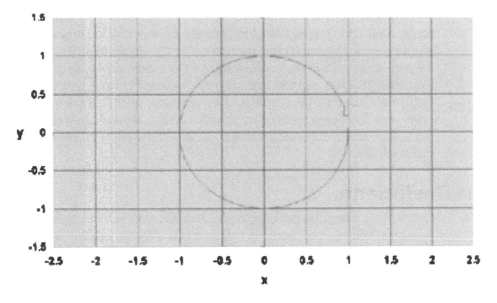

That funny little wiggle at the beginning of the graph is caused by the Achilles' heel of multistep methods: they not only use but require past values; in this case, the values of the derivative function represented by f1, f2, and f3. Unfortunately, at the beginning of the run, I have no idea what those values should be, so I just set them to zero. Note that, with these initial "past" values, the formula for the first step reduces to the following line of code.

```
x += h*(55.0*f0 )/24.0;
```

This is clearly wrong. The step is actually 230 percent too large, but as you can see, things soon sort themselves out as proper values are shifted into the slots. After three steps, the function up to full strength.

I've seen a number of attempts made to avoid the wiggle at the beginning. The most obvious approach is to change the order of the integration method: first order for the first step, second order for the second step, and so on. This does indeed avoid the problem, but it doesn't (and can't) avoid the fact that you don't have enough information to take a proper fourth-order step. Any error incurred in that first step, because only a first-order method is used, is there for good.

Fortunately, it doesn't matter in most cases how you handle the initial steps. Real-time systems tend to start up in very specific ways: airplanes are sitting on the tarmac, ships are tied up at the dock. In such cases, zero is a very good guess for the initial values of the derivatives. In any case, a few milliseconds after startup, the startup transient is long gone, so the whole thing turns out to be a nonissue in practice, all of which is a very good argument for doing as little as possible in the way of special startup practices. Just make sure to set those initial values to zero. Never leave them uninitialized.

The Coefficients

In both the formulas and the code, I've shown the coefficients as rational numbers, that is, the ratio of two integers. That's fine for illustrative purposes, but in practice you should compute the coefficients to avoid the division. The reason is not so much for speed, as for accuracy. Assuming that you're using integer arithmetic, as most real-time programmers do, you can't afford to multiply by constants like 59, because you'd be giving up six bits of precision. I know that you're every bit as capable as I am of dividing two integers and writing down the resulting coefficient, but I will show them to you (Table 12.2) to save you the trouble.

Table 12.2   Adams coefficients.

First order	1.000000000
Second order	1.500000000
	−0.500000000
Third order	1.916666667
	−1.333333333
	0.416666667
Fourth order	2.291666667
	−2.458333333
	1.541666667
	−0.375000000
Fifth order	2.640277778
	−3.852777778
	3.633333333
	−1.769444444
	0.348611111
Sixth order	2.970138889
	−5.502083333
	6.931944444
	−5.068055556
	1.997916667
	−0.329861111

If I've done my arithmetic properly, the coefficients for each order in Table 12.2 should add up to one. As you can see, the individual coefficients can be, and usually are, larger than one (almost seven in the sixth-order formula). The coefficients are trying to tell us something; the formulas get more and more unstable, and more prone to errors at startup and because of the subtraction of nearly equal terms, as we go to higher orders. These characteristics are typical of the Adams formulas because you can only use past values. In the Moulton corrector formulas (which you can't use, unfortunately, because they require forward data), the coefficients are much better behaved.

Before leaving this implementation, I want to point out that if the state vector is large, copying vectors from one place to another at each step can get a little expensive, especially for the higher orders. If time is at a premium in your implementation, you may want to look at alternatives to moving the data. The obvious approach is to put the derivative vectors in a circular buffer so that only the latest derivative is written and the past values are accessed through indices. It remains to be seen whether or not you gain anything, though, because now you have to access the vectors through an extra level of indirection. Although I've done a lot of these programs and seen a lot more written by others, I've never seen the shifting implemented by any way other than the brute-force approach.

Interestingly enough, this is one area where, again, C++ comes to the rescue. Remember, if you're developing the C++ class State_Vector, it's up to you to define the steps taken for each operator, including the assignment operator. Implementing this operator via reference counting takes care of the problem, because you end up changing only pointers, rather than the data itself. The end result is no different than what you'd get by implementing a circular buffer, but the code would be much easier to read.

Doing it by Differences

I must tell you that every real-time implementation of the Adams method that I've seen was done using the formulas of Table 12.1. I must also tell you that this is not the best way to do it because coefficients have values greater than one and alternate in sign. That's the price you pay for wanting to deal only with past values, rather than differences. The original Adams formula (Equation [12.43]) has no such limitation. All the coefficients in that formula are well behaved. What's more, you no longer need different values of the coefficients for the different orders; you only need to add one more term to increase the order. In practice, the differences get rapidly smaller with higher orders, so instead of subtracting nearly equal numbers, you end up adding very small ones, which is a much better approach.

To round out this study, I'll implement the Adams formula directly.

```
State_Vector x,f,df,d2f,d3f, temp1, temp2;

x += h*(f + df/2.0 +5.0*d2f/12.0 + 3.0*d3f/8.0);
t += h;
temp1 = f;
f = deriv(t, x);
temp2 = df;
```

```
df = f - temp1;
temp1 = d2f;
d2f = df - temp2;
d3f = d2f - temp1;
```

Now, I think you can see why most implementers prefer the past-values method rather than the difference table method: the computation of the differences is awkward and requires the use of temporary variables. But look at the resulting graph (Figure 12.6). The initial dipsy-doodle is gone! The reason is clear when you look at the formula. With all the differences (df, d2f, d3f) initialized to zero and no special effort on my part, the formula reverts to the familiar first-order formula. Similarly for the next step, the formula is equivalent to the second-order formula using past values, and so on. In other words, I get the variable-order startup I was seeking, with no extra work. Another powerful argument for the difference approach is seen if you look at the maximum values of the differences.

$$df = 0.61685$$
$$d2f = 0.0608014$$
$$d3f = 0.005959$$

As promised, these values get smaller at higher orders, so any errors made in multiplying them by their coefficients become less critical. Dealing with all the temporary values is a bother and can cost computer time, but the improvements in accuracy and startup behavior seem worth the price. Again, using reference counts for the assignment operation should minimize the cost in computer time.

Figure 12.6 Using differences.

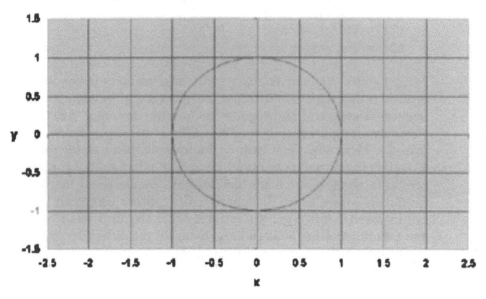

No Forward References?

Throughout this chapter you've heard me say that forward references, that is, references to points ahead of the current point, are not allowed. In one special case, though, you can actually get away with it. This is the important case in which you deal with second, rather than first, derivatives. I showed earlier (Equation [12.6]) how a second-order differential equation such as Newton's second law can be converted to two first-order equations.

Although the statement that you can't look ahead at future values to determine the new state is still true enough, in this case, you can seem to look forward because the two components, r and v, can be decoupled. Instead of integrating the entire state vector at once, first integrate only the v portion using the Adams formula in the normal manner. At this point, you have v at the new time, v_{i+1}. In effect, you have looked ahead. This means you can use the more stable and accurate Moulton corrector formula to update r.

$$x_{i+1} = x_i + h\left(1 - \frac{\nabla}{2} - \frac{\nabla^2}{12} - \frac{\nabla^3}{24} - \frac{19}{729}\nabla^4 - \frac{3}{160}\nabla^5\right.$$

[12.44]

$$\left. - \frac{863}{60,480}\nabla^6 - \dots\right)f_{i+1}$$

If your differential equations are second order, as many are (virtually all in dynamic simulations), you can find no better or more accurate method for integrating them in real time that the combination of Equation [12.43] for the velocity and Equation [12.44] for the position. Even if the equations of motion don't give the position and velocity in simple Cartesian form, you can often define "velocity-like" variables that can be expressed in terms of the derivatives of "position-like" variables, which still allows you to use the same method. Here's a useful example based on second-order formulas that I originally gave in Equations [10.58] and [10.59].

[12.45]

$$v_{i+1} = v_i + \frac{h}{2}(3f_i - f_{i-1})$$

$$x_{i+1} = x_i + \frac{h}{2}(v_{i+1} + v_i)$$

Let me reemphasize the point that if you can split the state vector into the form of Equation [12.6], Equation [12.45] is the best possible second-order algorithm you will ever find. Because it achieves better performance than any other second-order method and much better performance than any first-order method with only a modest increase in complexity, it represents perhaps the most promising compromise between speed and accuracy. Remember it, and use it wisely.

Wrapping Up

I have at last reached the end of my study of numerical integration formulas and methods. I hope you got some good out of them and learned something you will be able to use in your future projects.

Appendix A

A C++ Tools Library

The following files are intended for general purpose use in embedded or analytical software. They provide often-used constants and functions, and get rid of some of the strangeness in the standard C/C++ math libraries. See the CD-ROM for additional files.

Table A.1 Constant.h

```
/* File constant.h
 *
 * Programmed by Jack W. Crenshaw, Crenshaw Technologies
 * CIS #72325,1327
 * email jcrens@earthlink.net
 *
 * Defines commonly used constants for use by other modules
 * The constants are actually declared in file constant.cpp,
 * and are global.
 *
 */
```

```
#ifndef CONSTANT_H
        #define CONSTANT_H

extern const double pi;
extern const double halfpi;
extern double pi_over_four;
extern double pi_over_six;
extern const double twopi;
extern const double radians_per_degree;
extern const double degrees_per_radian;
extern const double root_2;
extern const double sin_45;
extern double sin_30;
extern double cos_30;
extern double BIG_FLOAT;

#endif
```

Table A.2 `Constant.cpp`

```
/* File constant.cpp
 *
 * Programmed by Jack W. Crenshaw, Crenshaw Technologies
 * CIS #72325,1327
 * email jcrens@earthlink.net
 *
 * Defines commonly used constants for use by other modules
 * The method used here requires only one copy of each constant.
 * The disadvantage is, all the constants are global, and the file
 * must be linked into the program as an ordinary .cpp file.  If you
 * don't like this idea, use it as a header file. In that case,
 * don't forget to declare the variables as static, to avoid name
 * conflicts.
 */
```

```
#include <math.h>

double pi = 4.0 * atan(1.0);
double halfpi = pi / 2.0;
double pi_over_four = pi / 4.0;
double pi_over_six = pi / 6.0;
double twopi   = 2.0 * pi;
double radians_per_degree = pi/180.0;
double degrees_per_radian = 180.0/pi;
double root_2 = sqrt(2.0);
double sin_45 = root_2 / 2.0;
double sin_30 = 0.5;
double cos_30 = sqrt(0.75);
double BIG_FLOAT = 1.0e20;
```

Table A.3   Jmath.h

```
/* File jmath.h
 *
 * Programmed by Jack W. Crenshaw, Crenshaw Technologies
 * CIS #72325,1327
 * email jcrens@earthlink.net
 *
 * Defines commonly used primitive macros and functions. Some macros
 * are redefined to make sure we get them right when porting to
 * different platforms. The functions are declared in file jmath.cpp.
 * Some of the functions are "safe" functions, intended to be used in
 * embedded systems where error halts are undesirable. Instead of
 * halting, the functions return reasonable values and keep going.
 * Function cmod fixes a problem with the library modulo functions,
 * which give incorrect values if one or both arguments are negative.
 *
 */

#ifndef JMATH_H
        #define JMATH_H
```

```
// define logicals for convenience
#ifndef FALSE
        #define FALSE 0
#endif

#ifndef TRUE
        #define TRUE 1
#endif

#undef max
#undef min
#undef abs
#undef sign

#define max(x,y)    (((x) > (y)) ? (x) : (y))
#define min(x,y)    (((x) < (y)) ? (x) : (y))
#define abs(x)      (((x) < 0  ) ? -(x): (x))
#define sign(x,y)   (((y) < 0  ) ? (-(abs(x))): (abs(x)))

double mod(double x, double y);
double radians(double x);
double degrees(double x);
double ang_360(double x);
double ang_180(double x);
double square_root(double x);
double tangent(double x);
double arcsin(double x);
double arccos(double x);
double arctan2(double y, double x);
double exponential(double x);

#endif
```

Table A.4 Jmath.cpp

```
/* File jmath.cpp
 *
 * Programmed by Jack W. Crenshaw, Crenshaw Technologies
 * CIS #72325,1327
 * email jcrens@earthlink.net
 *
 * Defines commonly used primitive macros and functions. Some macros
 * are redefined to make sure we get them right when porting to
 * different platforms.
 *
 * Some of the functions are "safe" functions, intended to be used in
 * embedded systems where error halts are undesirable. Instead of
 * halting, the functions return reasonable values and keep going.
 * Function cmod fixes a problem with the library modulo functions,
 * which give incorrect values if one or both arguments are negative.
 *
 */
#include <math.h>
#include <constant.h>

#undef max
#undef min
#undef abs
#undef sign

#define max(x,y)    (((x) > (y)) ? (x) : (y))
#define min(x,y)    (((x) < (y)) ? (x) : (y))
#define abs(x)      (((x) < 0  ) ? -(x): (x))
#define sign(x,y)   (((y) < 0  ) ? (-(abs(x))): (abs(x)))


/* Return a number modulo the second argument
 * Unlike most library routines, this function works
 * properly for all values of both arguments.
 */
```

```
double mod(double x, double y){
      f(y == 0) return x;
      int i = (int)(x/y);
      if(x*y < 0)--i;
      x = x-((double)i)*y;
      if(x==y)x -= y;
      return x;
}
```

```
// convert angle from degrees to radians
double radians(double x){
      return x * radians_per_degree;
}
```

```
// convert angle from radians to degrees
double degrees(double x){
      return x * degrees_per_radian;
}
```

```
// Reduce an Angle (in radians) to the range 0.. 360 degrees
double ang_360(double x){
      return mod(x, twopi);
}
```

```
// Reduce an angle (in radians) to the Range -180..180 degrees
double ang_180(double x){
      double retval = mod(x + pi, twopi) - pi;
      if (retval > -pi)
              return retval;
      else
              return retval + twopi;
}
```

```cpp
// Safe square root
double square_root(double x){
    if(x <= 0) return 0;
    return sqrt(x);
}


// safe tangent
double tangent(double x){
    x = ang_180(x);
    if(abs(x) == halfpi)
        return BIG_FLOAT;
    return tan(x);
}


// Safe inverse sine
double arcsin(double x){
    if(x>1)
        return halfpi;
    if(x<-1)
        return -halfpi;
    return asin(x);
}


// Safe inverse cosine
double arccos(double x){
    if(x>1)
        return 0;
    if(x<-1)
        return pi;
    return acos(x);
}
```

```
// Safe, four-quadrant arctan
/*
double arctan2(double y, double x){
      if(x == 0)
             return sign(halfpi, y);
      return atan2(y, x);
}
*/


// Safe exponential function
double exponential(double x){
      if(x > 46)
             return BIG_FLOAT;
      return exp(x);
}



/* The Functions Time Forgot
 *
 * Standard C libraries now include the functions which follow,
 * so they are not needed. Use them only if your system does not
 * support them, or use them as templates for programming the
 * functions in some other programming languaqe. Unlike the
 * safe functions in jmath.cpp, the functions below do not call
 * library functions.
 */


// tangent function
double tan(double x){
      double s = sin(x);
      double c = cos(x);
      if(c != 0)
             return s/c;
      return BIG_FLOAT * s;
}
```

```
// Inverse sine
double arcsin(double x){
        if(abs(x) >= 1.0)
                return sign(halfpi, x);
        return atan(x/sqrt(1.0 - x*x));
}

// Inverse cosine
double arccos(double x){
        return(halfpi - Arcsin(x));
}

// Four-quadrant arctan
double arctan2(double y, double x){
        double retval;
        if(x == 0)
                retval = halfpi;
        else
                retval = atan(abs(y/x));
        if(x < 0)
                retval = pi - retval;
        if(y < 0)
                retval = -retval;
        return retval;
}
*/
```

Table A.5 Arctangent with optimized coefficients.

```
/* Prototype for arctangent function using continued
 * fractions. This is Listing 6.7 of text.
 */

/* Rational approximation for atan(x)
 * This polynomial uses a minimaxed 5th-order
 * ratio of polynomials.  Input angles are restricted
```

Table A.5 Arctangent with optimized coefficients. (continued)

```
   * to values no greater than 15 degrees.  Accuracy is
   * better than 4e-8 over the full range of inputs.
   */
double arctan(double x){

  // constants for segmentation
  static const double b = pi/6;
  static const double k = tan(b);
  static const double b0 = pi/12;
  static const double k0 = tan(b0);

  // constants for rational polynomial
  static const double A = 0.999999020228907;
  static const double B = 0.257977658811405;
  static const double C = 0.59120450521312;
  double ang;
  double x2;
  int comp = FALSE;
  int hi_seg = FALSE;

  // make argument positive
  int sign = (x < 0);
  x = abs(x);

  // limit argument to 0..1
  if(x > 1){
    comp = TRUE;
    x = 1/x;
  }

  // determine segmentation
  if(x > k0){
    hi_seg = TRUE;
    x = (x - k)/(1 + k*x);
  }
```

Table A.5 Arctangent with optimized coefficients. (continued)

```
/* argument is now < tan(15 degrees)
 * approximate the function
 */
x2 = x * x;
ang = x*(A + B*x2)/(1 + C*x2);

// now restore offset if needed
if(hi_seg)
  ang += b;

// restore complement if needed
if(comp)
  ang = halfpi - ang;

// restore sign if needed
if(sign)
  return -ang;
else
  return ang;
}
```

Index

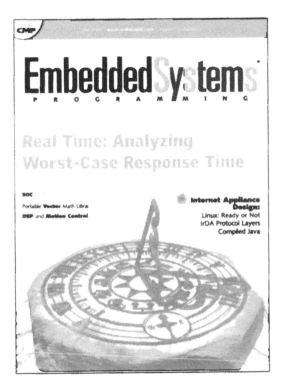

What's on the CD-ROM?

The files on the CDROM for **Math Toolkit for Real-Time Programming** fall into three main categories:

1. Utility files that you will probably find useful in general-purpose computing, for either real-time or non-real-time applications. These include the two files constant.cpp and jlib.cpp, together with their corresponding header files. Link them in with all your programs. They are small enough to take up negligible space, and should make your life easier. These files also include the safe versions of functions such as sqrt, asin, acos, and atan2.

   Note that jlib.cpp and jlib.h redefine the C/C++ library function abs() to be a macro rather than the standard subfunction. This is to allow this function to operate in a manner more nearly parallel to, say, min() and max(). Some of the other modules in this book make use of this character, so will not work properly if you fail to link jlib.cpp.

2. Code examples, not intended to be used for general-purpose applications. These include most of the examples of the fundamental functions such as square_root, sine, cosine, etc. Since any modern C or C++ compiler includes a math library that already implements these functions, usually coded in highly-optimized assembly language, there is obviously no advantage to implementing the same functions in C/C++. These functions are intended to be used mainly to illustrate the techniques for generating optimally organized functions. Use them as a guide for your own library functions for small compilers, or for assembly-language versions for microcomputers. These functions are collected in file function.cpp. Do not expect this file to compile, however, since it may contain multiple definitions of the same function.

3. Code that is neither of the above. These include the integer versions of sqrt (several versions, depending upon performance requirements), sin, cos, etc. They include the bitlog function and its inverse, iexp. While these functions are also given mainly for their tutorial value, you may also find the integer square root and log/exp functions to be useful in their own right. The file forgot.cpp gives algorithms for functions usually, but not always, included in modern math libraries. Use these functions only if you need them to round out your library.

T - #0292 - 101024 - C0 - 235/184/26 [28] - CB - 9781138412477 - Gloss Lamination